MEANING OF LIFE & THE UNIVERSE

TRANSFORMING

MAE-WAN HO
Institute of Science in Society

MEANING OF LIFE & THE UNIVERSE

TRANSFORMING

World Scientific

NEW JERSEY • LONDON • SINGAPORE • BEIJING • SHANGHAI • HONG KONG • TAIPEI • CHENNAI • TOKYO

Published by

World Scientific Publishing Co. Pte. Ltd.
5 Toh Tuck Link, Singapore 596224
USA office: 27 Warren Street, Suite 401-402, Hackensack, NJ 07601
UK office: 57 Shelton Street, Covent Garden, London WC2H 9HE

Library of Congress Cataloging-in-Publication Data
Names: Ho, Mae-Wan.
Title: Meaning of life and the universe : transforming / by Mae-Wan Ho
(Institute of Science in Society, UK).
Description: New Jersey : World Scientific, 2017. |
Includes bibliographical references and index.
Identifiers: LCCN 2016058645| ISBN 9789813108851 (hardcover : alk. paper) |
ISBN 9789813108868 (pbk. : alk. paper)
Subjects: LCSH: Biology--Philosophy. | Art and science. |
Science--Philosophy. | Intellect. | Genetics.
Classification: LCC QH331 .H578 2017 | DDC 570.1--dc23
LC record available at https://lccn.loc.gov/2016058645

British Library Cataloguing-in-Publication Data
A catalogue record for this book is available from the British Library.

Typeset by Stallion Press
Email: enquiries@stallionpress.com

This book is dedicated to

Peter, Li, Adrian, Tracy, Jasmine and Jade

Preface

This collection of essays is selected from about a thousand works written over a period of 46 years. They elaborate on what it means to live in a quantum universe as opposed to a mechanistic universe. With quantum physics and chemistry as the pervading theme and reference point, the essays range over philosophy, anthropology, psychology, evolution, genetics and epigenetics, non-equilibrium thermodynamics, consciousness, neuroscience, art and the humanities, quantum electrodynamics of water, fractal mathematics and cosmology. In the process, practically every subject is rewritten in a new synthesis that redefines and transforms life itself. It is indeed my personal quest for the meaning of life and the universe.

I was fortunate to have encountered many teachers and mentors in my quest, most of whom are mentioned in the essays. In particular, I am indebted to Fritz-Albert Popp, who taught me almost everything I know about quantum theory. Another towering figure was Kenneth Denbigh, who gave me a thorough lesson in thermodynamics, a subject I singularly failed to learn as an undergraduate. He also most generously encouraged and helped me to extend his own theory of the steady state to the far-from-equilibrium regime, which eventually became the circular thermodynamics of organisms and sustainable systems. James Clegg's work initiated me to the special state of water in organisms, leading to our discovery that it is liquid crystalline. Emilio del Giudice, honoured with a special essay in this collection, taught me quantum electrodynamics and the crucial importance of light-matter interactions. Last, though by no means the least, Mohamed El Naschie introduced me to the fascinating world of discontinuous and non-differentiable mathematics, which may well describe the fabric of the quantum universe.

I also have numerous collaborators who are listed as co-authors of papers cited in the references. Among them are those with whom I have shared key discoveries, dreams and adventures, and many exhilarating hours in and out of the laboratory. Franco Musumeci, a great friend, teacher and fellow dreamer, generously invited me to work in his Catania laboratory for many years with Agata Scordino and Antonio Triglia on biophotons (pioneered by Fritz-Albert Popp), which has not been supported by any funding agency to this day. Michael Lawrence made it possible for us to discover the liquid crystalline Rainbow Worm. Julian Haffegee and Zhou Yu Ming were my graduate students whose ingenuity and enthusiasm carried us through good times and bad. Stephen Ross volunteered in our laboratory for some years and produced the first quantitative analytical imaging software for collagenous and other liquid crystalline tissues.

Outside of scientific circles, I was most fortunate to have met pioneering ecologist Edward Goldsmith, who introduced me to indigenous knowledge, anthropology and more; and Martin Khor, Vandana Shiva, and Chee Yokeling, who turned me into a science activist, and led me to the view that science is reliable knowledge of nature that enables us to live sustainably with her. They were instrumental in helping to set up the Institute of Science in Society to reclaim science for the public good in all respects. We have been successful thanks to Julian Haffegee, one of our founding members, who has remained with us ever since, and especially to fellow scientist Joe Cummins, already a champion of the environment, who joined us in the first years, and wrote numerous submissions to the US Environment Protection Agency, Department of Agriculture and Food and Drug Administration, on our behalf. He is no doubt responsible for any good these agencies have ever done. Joe was among the first to warn of genetically modified organisms, the threat of neonicotinoid pesticides to bees and many other things that escape the mainstream media. Sadly, Joe died on 8 January 2016 as this volume was going to press. He was a great friend, a kind, gentle and generous soul, and a pillar of strength when it comes to defending the environment and the public good.

Finally, without the love and support of Peter Saunders, my husband, collaborator and co-conspirator, none of the essays would have been written, nor the scientific research carried out, nor the inspiration that gave birth to them.

Sources

With the exception of the poem "Why Say Consciousness?" published here for the first time, all other essays are based on previously published works. They have all been edited for style and to minimize repetition, updated and annotated. Some essays are combinations of several articles. For these reasons, they depart from the published works to varying extents, while remaining faithful to the original themes.

The sources are as follows, in order of their appearance in the volume.

1. Ho, M.W. (1992). "Natural Being and Coherent Society". This paper was intended to be part of *Social and Natural Complexity* (special issue of *The Journal of Social and Biological Structures*), edited by E.L. Khalil and K.E. Boulding, but was rejected by the co-editor after Kenneth Boulding's death in 1993. The article went through several versions before it was finally published under the same title in *Gaia in Action* (P. Bunyard, ed.), pp. 286–307, Floris Books, Edinburgh, 1996a.
2. Ho, M.W. "Toward an Indigenous Western Science". In *Reassessing the Metaphysical Foundations of Science* (W. Harman, ed.), pp. 179–213, Noetic Sciences Institute Publications, Sausalito, California, 1994a.
3. Ho, M.W. "Organism and Psyche in a Participatory Universe". In *The Evolutionary Outrider* (D. Loye, ed.), pp. 49–65, Adamantine Press, Twickenham, 1998.
4. Ho, M.W. "Development and Evolution Revisited". In *Handbook of Developmental Science, Behavior and Genetics* (K. Hood,

C. Halpern, G. Greenberg and R. Lerner, eds.), pp. 61–108, Wiley-Blackwell, Chichester, 2010.

5. Ho, M.W. "Nurturing Nature: How Parental Care Changes Genes". In *Genetic Explanations: Sense and Nonsense* (S. Krimsky and J. Gruber, eds.), pp. 256–269, Harvard University Press, Harvard, 2013a.
6. Ho, M.W. "Mystery of Missing Heritability Solved?" *Sci Soc* 53 (2012a): 26–27+31; Ho, M.W. No Genes for Intelligence". *Sci Soc* 53 (2012b): 28–31; Ho MW. No genes for intelligence in the fluid genome. *Adv Child Dev Behav* 45 (2013b), special issue: *Embodiment and Epigenesis: Theoretic and Methological Issues in Understanding the Role of Biology within the Relational Developmental System* (R. Lerner and J.B. Benson, eds.), 68–91.
7. Ho, M.W. "The New Genetics and Natural versus Artificial Genetic Modification". *Entropy* 15 (2013c): 4748–4781; Ho, M.W. "Why GMOs Can Never Be Safe". *Sci Soc* 59 (2013d): 14–17; Ho, M.W. "Artificial versus Natural Genetic Modification and Perils of GMO". *Sci Soc* 64 (2014a): 9–13.
8. Ho, M.W. "Non-Random Directed Mutations Confirmed". *Sci Soc* 60 (2013e): 30–31.
9. Ho, M.W. "The Biology of Free Will". *J Conscious Stud* 3 (1996b): 231–244.
10. Ho, M.W. "Quantum Coherence and Conscious Experience". *Kybernetes* 26 (1997): 263–276.
11. Ho, M.W. "Why Say Consciousness?" Unpublished poem, 1994b.
12. Ho, M.W. "In Search of the Sublime". *Metanoia* (introductory issue, Spring 1994c): 9–16; Ho M.W. "In Search of the Sublime: Significant Form in Science and Art". *Sci Soc* 39 (2008): 4–11.
13. Ho, M.W. "Why Beauty Is Truth and Truth Beauty". *Sci Soc* 50 (2011): 32–37.
14. Ho, M.W. "Sustainable Cities As Organisms". *Sci Soc* 64 (2014b): 28–31; Ho, M.W. "Sustainable Cities As Organisms: A Circular Thermodynamics Perspective". *Int J Des Nat Ecodyn* 10 (2015a): 127–139.
15. Ho, M.W. "Water Is The Means, Medium, and Message of Life". *Int J Des Nat Ecodyn* 9 (2014c): 1–12.

16. Ho, M.W. "Illuminating Water and Life". *Entropy* 16 (2014d): 4874–4891.
17. Ho, M.W. "Story of Phi Parts 1–3". *Sci Soc* 62 (2014e): 24–31+44.
18. Ho, M.W. "Story of Phi Parts 4 and 5". *Sci Soc* 62 (2014f): 32–39; also Ho, M.W., El Naschie, M. and Vitiello, G. "Is Spacetime Fractal and Quantum Coherent in the Golden Mean?" *Global Journal of Science Frontier Research: A. Physics and Space Science* 15 (2015b): 61–80.

Contents

1

Anthropology, Philosophy & Psychology

1

Natural Being and Coherent Society

The myth of the "Darwinian man" is deconstructed by re-examining the biological roots of human nature which are inextricably tied to the social. Studies from animal and plant communities to "primitive" societies show that sociality is universal to the living world. It is a direct consequence and expression of the unity and interconnectedness of all nature, which is reaffirmed in the new physics of quantum coherence for life and the universe at large. From this perspective, culture is the creation of meaning and knowledge in partnership with nature, in which every social being participates. The coherent society is the society of natural beings living in harmony with nature's creative process.

Conference of the Birds

According to ancient legend in Persia, all manner of birds gathered for a conference one fine day and were persuaded to disperse to the four corners of the world in search of the meaning of life. After many long and arduous years, they returned home, only to discover that what they were seeking had been right there all along. They were blind to it, and the journey away was necessary to open their eyes.[1] The legend is in many ways also the parable of Western science. After centuries of intellectual wanderings that increasingly led away from nature, we are irresistibly drawn back to her in the realization that there is no authentic knowledge, and hence no meaning in life, apart from nature.

Perhaps the single most decisive factor in the evolution of the knowledge system of the West is that it depends on severing our intimate, manifold connections with nature at the outset. Mind-matter dualism — the idea that mind and body are distinct — most closely associated with French mathematician and philosopher René Descartes (1596–1650) simultaneously divides mind from body and isolates the observer, as disembodied mind, from an "objective" nature observed. The result is a desolate universe of inert, indifferent matter acted on by the push and pull of extraneous forces. Green grass and trees, fins and wings, are so many illusory "secondary qualities". Human feelings, likewise, can have no dominion; relegated as they are, to the realm of poetic fancy that hangs ever like a veil over objective reality.[2]

This exile from nature is entirely self-imposed; and is neither necessary nor inevitable. We can find our way back, if not to paradise, then surely to a more fulfilling and humane future through recovering our natural being, which is also the vehicle to authentic knowledge. But first, we must deconstruct the Darwinian man.[3]

Origins of the Darwinian Man

He bought white ties, and he brought dress suits,
He crammed his feet into bright tight boots.
And to start in life on a brand new plan,
He christened himself Darwinian Man!
But it would not do.
The scheme fell through,
For the Maiden fair, whom the monkey craved,
Was a radiant Being.
With a brain far-seeing.
While Darwinian man, though well-behaved,
At best is only a monkey shaved![4]

English naturalist Charles Darwin (1809–1882) is without doubt the most influential scientist of our age, his intellectual legacy thriving long after his death. Darwin's theory says that organisms evolve on earth as the result of the natural selection of random variations. There were three immediate sources of inspiration for the theory.[5] The first was the "argument from design" due to English theologian William Paley (1743–1805), how it is that organisms so perfectly adapted to their way of life could be explained, without invoking a supernatural designer or maker. The second was artificial selection, practised by generations of plant and animal breeders who selectively bred from organisms with the desired characteristics so as to produce new improved varieties. In answer to Paley, Darwin proposed that organisms tend to vary at random; these random variations are then subject to selection. But how could selection take place in nature where no obvious "selector" exists? A chance reading of a paper by English cleric and scholar Thomas Malthus (1766–1834) provided the third element, which was just the natural mechanism of selection. Malthus noted that human beings, like all organisms, have a propensity to increase exponentially, generally outstripping the rate at which food supply can increase. Consequently, he proposed that population numbers are kept down by starvation, famine, disease and war, which now and again take their toll, weeding out the weak and the sick. In Malthus' theory, Darwin not only found the selective mechanism but also the answer to Paley's problem of how adaptation could be explained.

To recapitulate, all organisms have a tendency to vary at random and to increase their numbers exponentially, outstripping the carrying capacity of the environment. Thus, only those organisms with characteristics that favour them in the competitive struggle for existence will survive to reproduce. Heredity ensures that the offspring of those organisms will have the same favourable or adaptive characteristics. In this way, the population will become more and more adapted to the environment in subsequent generations.

As we shall see later, competition for scarce resources is hardly the norm for natural animal or human populations. In general, natural populations do not increase exponentially because many social and biological factors come into play to keep reproductive rates in balance with death rates so that the Malthusian scenario is seldom realized. Nevertheless, the Darwinian metaphor took hold of the imagination in the Western world, and became incarnated in the Darwinian man, who proceeded to remake the world in his image.

The full text of Darwin's epoch-making book of 1859 was *On the Origin of Species by Means of Natural Selection, or The Preservation of Favoured Races in the Struggle for Life*.[6] If Darwin liberated the Victorian era from the domination of religion and superstition, he also delivered it well and truly to a nature painted "red in tooth and claw". Our continuing disharmony with nature derives ultimately from this unedifying image, which Darwin clothed with the dignity of a scientific theory. At the same time, the emphasis on competition between individuals and the implied superiority of the "favoured races" in the "struggle for life" easily provided political justification — on the basis of natural law — both for exploiting the underclasses and for colonizing and oppressing the "inferior races", as French-born American historian Jacques Barzun (1907–2012) commented.[7]

However, Darwin did not invent Darwinism; he simply imbibed the dominant ideology of his age and gave it a name. The famous phrase, "survival of the fittest", was coined earlier by English philosopher and sociologist Herbert Spencer (1820–1903), who was also inspired by Malthus' theory of human populations. Spencer is the father of Social Darwinism, in which Darwinian ideas are imported *back* into human society, giving rein unfortunately to eugenic persecutions — of indigenous peoples, Jews and other politically dispossessed human beings — that have blighted the history of the 19th and 20th centuries. Thus, a metaphor borrowed from life in Victorian English society steeped in ideas of progress through unbridled competition in the free-market, imperialist conquests and expansion of trade, became enshrined as a scientific truth. It dictates how we should see reality, ultimately shaping reality in accordance to its dictate — that Darwinian man shall rule the world.

English biologist Thomas Huxley (1825–1895), often referred to as "Darwin's bulldog" for his zealous promulgation of Darwinism, invented a birthplace for the Darwinian man in the "primitive" society, where[8]

> "... the weakest and stupidest went to the wall, while the toughest and shrewdest, those who were best fitted to cope with their circumstances, but not the best in another way survived. Life was a continuous free fight, and beyond the limited and temporary relations of the family, Hobbesian war of each against all was the normal state of existence."

One has to realize that there is not a shred of evidence to support his claims. Nevertheless, this gruesome picture was echoed by the father of psychoanalysis, Austrian neurologist Sigmund Freud (1856–1939), whose theory of the savage, patricidal primal hordes is so far-fetched and ridiculous that it hardly bears repeating here. The modern Freudian man is the bulwark of the Western industrialized society. According to a summary given by a sympathetic exponent,[9]

> "... Freud's view of human nature as interpreted by his critics is that of Hobbes and Darwin, which depicts society as a mass of isolated individuals whose most natural emotion is hostility, pushing and jostling each other in the name of the survival of the fittest, but willing under certain circumstances to band together for self-protection. Their ivory towers conceal the inner stinking cave by the entrance of which they ruthlessly trade physical needs or personal relationships for private gain, returning to the innermost recesses to enjoy them without interference — and this, after all, is not surprising, since they ceased to develop emotionally at the age of five and any trait presented in later life is mere camouflage to conceal what goes on within. Outside the tower are displayed their paintings, their collections of *objets d'art*, their musical skill and wit, or their scientific curiosity, when the psychoanalyst knows perfectly well that inside they are smearing the walls with ordure or enjoying 'retention pleasure', satisfying their autoerotism, or preparing to bite and rend any source of frustration ...
>
> More objectively expressed, Freud believed in the person as a social atom requiring community only as a means to the satisfaction of his needs; in a primary hostility so strong that only sheer necessity or common hatred directed elsewhere could join

> people in love; in a certain biological inevitability of hereditary constitution, anatomy, and development, which strictly limits human possibilities; in an inner private existence which, although in part the result of early personal relationships, seems in later life to make only indirect contact with external reality ... and finally in civilization as the result of thwarted libidinous impulses which have been deflected to symbolic ends ..."

Freud's basic ideas came exclusively from the study of pathological cases, themselves symptomatic of a society already sickened by a context unfit for life, even for the relatively privileged middle classes from where most of his patients were drawn. It is an artificial context — developed to its logical conclusion from a state of alienation from nature — that nevertheless became mistaken for nature herself, and hence regarded as both inevitable and inescapable.

However you look at it, Freud presents an abysmal, genetic determinist view that dovetails neatly with the dominant biology. It continues to validate the competitive, profit-seeking consumerist society of the industrialized west that in turn reaffirms and reinforces it until no alternative is conceivable.

The Darwinian man is the constant, unchangeable parameter that must enter into every social equation. There can be no consideration other than cost and benefit, which creates at best, an uneasy truce between loveless, self-serving individuals. "Scratch an 'altruist', and watch a 'hypocrite' bleed", a staunch defender of neo-Darwinism wrote.[10] Edward O. Wilson at Harvard University reinvented Social Darwinism for our generations in the discipline of *sociobiology*, in a book of the same name, which applies neo-Darwinist principles to explain the evolution of social behaviour. In the opening pages of his book, he wrote,[11]

> "This brings us to the central theoretical problem of sociobiology: how can altruism, which by definition reduces personal fitness, possibly evolve by natural selection?"

Indeed, much of present-day evolutionary biology and sociobiology is devoted to explaining how apparently altruistic behaviour is just disguised selfishness. Why do sociobiologists find such common and commonplace human qualities so difficult to accept that they need to do their utmost to explain them away? It has been suggested that part of the reason lies buried deep in their own psychology, which reflects the warped society that has nurtured them. It is a slavish commitment to an abysmal view of human nature for which there is little empirical evidence, based on a scientific theory that has already been invalidated by scientific findings, not just within the past 20 years, but starting soon after Darwin's theory was announced.[12]

Altruistic versus Competitive Nature

In response to the claims of sociobiologists and others that competitiveness, aggression, and worse, the propensity for rape and murder in males, are universal human characteristics, American author Robert Clairborne pointed out that in reality, the overwhelming majority of human beings readily engage in activities to help or benefit others, whereas only a tiny minority have ever committed criminal acts.[13] Therefore, it may be argued that altruism, rather than aggression, is universal. Clairborne does not regard altruism to be innate, however. Rather, he saw it as a learned behaviour based on the universal human capacity for empathy, i.e., deriving pleasure from other people's pleasure and distress from their distress. And hence, "satisfying the needs of others, and thereby sharing their satisfaction, is intrinsically rewarding." This empathy, as I shall show, comes from the experience of connectedness with kin, with fellow creatures, and ultimately with all nature.

Mutual Aid versus Mutual Struggle

Peter Kropotkin (1842–1921) was a social anarchist and Russian prince who, writing in the early part of the 20th century, drew attention to the abundant evidence of cooperation in the natural world. In his book *Mutual Aid*[14] he tells us how under the influence of Darwin's *Origin of Species*, he began to study animal life in Siberia in order to find evidence of competition among animals within a species. He found none. Indeed, there was extreme severity of "struggle for existence" against inclement nature, as one would expect in Siberia. But, even among the most abundant animal life, and under the most extenuating circumstances, there was no struggle for existence among the animals *against one another*.

On the contrary, there were numerous examples of mutual aid and mutual support among animals throughout the animal kingdom, which Kropotkin went on to document,[14] from ants and termites to birds and mammals. He quoted widely from published sources as well as from his own experience.

Ants regularly regurgitate food to feed hungry comrades, while pelicans always fish together, typically forming a wide half-circle facing the shore, then narrowing it by paddling towards the shore. Significantly, cooperation does not stop within species boundaries. Species may combine together to repel attacks, as gulls and terns do, to drive away the sea-hen. Lapwings, *Vanellus cristatus*, attack birds of prey so bravely that the Greeks gave them the name "good mother". Cranes live in excellent relationships not only with their congeners but with most aquatic birds. Their sentries keep watch around a mixed flock which is feeding or resting together.

A considerable body of present-day sociobiological theory is devoted to explaining, or explaining away, cooperation in terms of the selective advantage that after all, must accrue to the cooperating individuals. The most frequent explanation is that although the altruist has less chance of passing on its own genes, it nevertheless increases the chance of survival of relatives, which carry a proportion of genes that are the same as its own. So, a sibling will have half of its genes in common with the altruist, an aunt, a quarter of its genes and so on. This is called "kin selection"[11] and has been hailed as one of the most important consolidation to the neo-Darwinian edifice. Another "explanation" is "reciprocal altruism", which is not altruism at all, but simply that the one doing a good turn now expects that it is reciprocated at some future date. That is a misreading of nature. In many instances, help is freely given to others from whom no return can ever be expected, and with whom the individual shares no genetic relatedness.

Among mammals, dolphins are well renowned for their intelligence and friendship towards humans.[14] They will actually help fishermen drive fish into their nets if, after a long day, the fishermen have netted nothing, and they call to the dolphins for help. However, I am told that if the fishermen are greedy and do it too often, the dolphins will ignore their call.[15]

Germany's most revered writer and natural scientist Johann Wolfgang Goethe (1749–1832) was once told by a friend that two little wren-fledglings that had run away, were found the next day in the nest of robin redbreasts who fed the little ones together with their own. Goethe saw in that anecdote a confirmation of his own pantheistic view,[14] the idea that God is in all of nature. It is the universal tendency of birds to look after other's young that enables the cuckoos to exploit their hosts, and not because the latter are too stupid or mesmerized to distinguish foundlings from their own offspring.[16] No wonder neo-Darwinists are often criticized for telling "just-so" stories.

In my own experience, female domestic cats will readily adopt and look after kittens that are not their own. The love of young is so strong among Indian langur monkeys that as soon as a newborn arrives, the troop's females will cluster around the mother, all reaching out gently to try to touch and lick the infant. During its first day of life, the infant will have passed through the loving arms of as many as eight female monkeys.[13]

Much more fundamental than cooperation or help is the tendency of animals to seek out and enjoy the society of others. The crane is continually active from morn till night, of which only a few hours are devoted to finding food. All the remainder of the day is given over to society life.[14]

The most impressive display of interspecific mutual aid among birds is associated with their mass migration (p. 36)[14]

> "... Birds which have lived for months in small bands scattered over a wide territory gather in thousands; they come together at a given place, for several days in succession, before they start ... Some species will indulge every afternoon in flights preparatory to the long passage. All wait for their tardy congeners, and finally

they start in a well-chosen direction ... the strongest flying at the head of the band, and relieving one another in that difficult task. They cross the seas in large bands consisting of both big and small birds. And when they return next spring, they repair to the same spot, and in most cases, each of them take possession of the very same nest which it had built or repaired the previous year."

Social mammals were highly successful and associated in large numbers before they were decimated by humans. The numbers of solitary carnivores are trifling in comparison with the social herds of wild horses, donkeys, camels and sheep that used to roam in central Asia, and elephants, rhinoceroses, monkeys, reindeer, musk oxen and polar foxes in northern Asia and Southern Africa. Kropotkin wrote (p. 39):[14]

"And how false, therefore, is the view of those who speak of the animal world as if nothing were to be seen in it but lions and hyenas plunging their bleeding teeth into the flesh of their victims! One might as well imagine that the whole of human life is nothing but a succession of war massacres."

Above all, animals, like human beings, derive pleasure and satisfaction from life in society. Society was not created by human beings as our anthropocentric view would lead us to believe, but is antecedent to our own species. Sociality — the love of society for society's sake — is at the very basis of animal life. Not only do numerous species of birds assemble together habitually to indulge in antics and dancing performances, but nearly all mammals and birds indulge frequently in more or less regular or set performances with or without sound, or composed of sound exclusively. One has only to listen to the regular-as-clockwork chorus of bird song in the mornings and evenings during the warm seasons. The habit of singing in concert is most strikingly developed with the chakar *(Chauna chavarria)*,[14] but other species are equally impressive.

In the spring of 1990, I was a founding faculty member for the International Honours Program on Ecology of Bard College (New York, United States), the theme of which was Integrating Nature and Culture. As part of the programme, I stayed with the students in a Thai village near the suburb of Chiang Mai. Like everyone else, I was sleeping on the hard wooden floor. At dawn, I was woken up by a strange sensation of being afloat on an undulating chorus of croaking sounds so well blended and coordinated that it was almost impossible to make out individual contributions. It built up to a crescendo, then gradually died away, and the cycle would begin again. It was some time before I realized that the chorus was coordinated in space. It travelled in concentric waves from where I was lying to the furthest horizon and back again. There must have been tens of thousands of bullfrogs taking part, perhaps celebrating the return of spring.

There are numerous examples of larger animals helping others in distress, but apart from ants and bees, little is known about such behaviour among other insects.

In September 1997, my technician and student Julian Haffegee and I brought some cultures of fruit flies with us to the laboratory of our collaborator, Franco Musumeci, in Catania University, Sicily. We collected batches of freshly laid eggs in order to measure the low level of light emitted from the embryos during the first few hours of development. As the temperature of the laboratory was not well controlled, water would condense on the side of the glass bottles when it got cool in the late afternoon, and the fruit flies would get caught in the water. While trying to figure out what to do, I was watching the flies intently. I saw that wherever a fly was caught on its back in a droplet of water and struggling to free itself, one or more flies would invariably gather around it, so I began to wonder what they were doing. To my astonishment, I discovered that the other flies were actually helping to free the one in distress, using their legs or pushing with their head, and succeeding quite often. In fact, the best strategy to save the flies, we decided, was simply to put the bottles in a warm, dry place, and let the flies get on with the rescue operations themselves. Can we doubt that even fruit flies can respond empathically to the distress of their fellows? And that therefore, cooperation is by far more common in the living world than competition?

Many years after Kropotkin, the American ecologist Warder Clyde Allee (1885–1955) was stimulated to re-examine Kropotkin's thesis when, by chance, he discovered that even such lowly animals as isopods aggregate most eagerly to form social clusters. From this, he was led to review abundant evidence of swarm formation in the living world, starting with the single-celled photosynthetic organism *Euglena*, through to insects, birds and mammals.[17] He too, concluded that sociality is indeed universal (p. 23):

> "The growing weight of evidence indicates that animals are rarely solitary, that they are almost necessarily members of loosely integrated racial and inter-racial communities, in part woven together by environmental factors, and in part by mutual attraction between the individual members of different communities, no one of which can be affected without changing all the rest, at least to some slight extent."

Allee was speaking of mutual support among members in ecological communities. As an example of how multispecies communities can effectively cooperate, *without intending to do so*, he referred to the grassland bison community of the Great Plains in North America. The bison herds kept the grasslands closely cropped, preventing the invasion of herbs and shrubs. This provided a rich habitat for grasshoppers, crickets, mice and prairie dogs, all of whom converted the grass into meat, on which the plain Indians, the buffalo wolves, hawks, owls and prairie chickens fed. The plants of the community, therefore, cannot be considered in isolation from the animals. This is but the age-old wisdom of the ecological interdependence of living things that all traditional indigenous cultures possessed. The dominant modernist culture of the industrialized west is unique in its persistent denial of such ecological interdependence.

Allee and his colleagues also carried out numerous experiments demonstrating that society has important effects on the behaviour and physiology of individuals within the society, not all of which can be interpreted as contributing to an increase in survival value. The ill effects of crowding are well known and clearly documented for animals such as fruit flies and laboratory mice. The animals exhibit various signs of stress which lead to aggression and other pathological behaviour as well as failure to thrive. What is not so well known is that under-crowding is also deleterious for the survival of individuals, possibly because they need society for its own sake. Goldfish and planarians (flatworms), when isolated, succumb to poisoning more readily than when grouped. Embryonic development in sea urchins is significantly accelerated when the eggs are massed together. And ciliate protozoa reproduce faster in groups than when isolated, the reproductive rate also dependent on the density of bacteria on which they feed.

Of especial interest is Allee's demonstration that goldfish learn faster in groups than as individuals, through a combination of imitation and group cohesion. From this arises the concept of "social facilitation", which may have important implications for our own species in which so many individuals are living lonely, isolated lives.

Mutual Aid in "Primitive" Societies

Having demonstrated that cooperation and sociality are the most natural state of being among animals, Kropotkin went on to cite abundant evidence of mutual aid, compassion and moral feelings among so-called primitive human societies.[14] The relative lack of competition and strife in most traditional cultures has long impressed anthropologists. The point is not that competition or rivalry never occurs. Competition, like cooperation, is a social phenomenon; it does not follow thereby that corresponding qualities of competitiveness or cooperativeness, for that matter, actually exist. There is, at bottom, a feeling of being connected with others, a desire for society that we refer to as sociality or love by another name. Kropotkin pointed out that sociality not only offers the greatest advantage in the struggle for life under any circumstances, as opposed to competition, which is only advantageous under some circumstances, it also favours the growth of intelligence, language, social feelings and a "certain sense of collective justice" akin to morality.

Sociality is the regulating and cohesive principle in both animal and human society. It exists prior to any consideration of selective advantage. In a sense, Kropotkin, and also English ethologist Patrick Bateson at Cambridge University[18] (a strong advocate of cooperation among contemporary neo-Darwinists), have inverted cause and effect in trying to explain why cooperation or mutual aid could have evolved by natural selection. Qualities such as compassion or empathy, based on the same experience of being connected with others, are also antecedent to life in organized society. Life in society may, of course, reinforce and enhance those qualities. But they would never have arisen through any externally imposed social order, were they not already heartfelt and integral to the natural being.

Origins of Love and Hate

Scottish psychologist Ian Suttie (1898–1935) wrote his famous book, *The Origins of Love and Hate*,[19] which was first published in 1935 a few days after his death. In direct opposition to Freud and his many disciples, for whom sex is the single most important human instinct on a par with survival, Suttie saw love as primary, and distinct from sex. The idea of love comes from the ministrations of the mother or caretaker during infant life. The infant soon develops a feeling of tenderness, reciprocating that which it experiences. This is then generalized until people in the entire world are considered possible companions, who are to be enjoyed and loved, and from whom approval is sought. Hate or aggression has precisely the same source: it arises when love is lost, or threatened with loss, frustrated or thwarted. Thus, only the capacity to love predisposes us to hate; the stronger the love, the deeper the hatred that comes into being should love fail. Like Kropotkin, Suttie came to his conclusions from studying social behaviour among animals as well as "primitive" societies. Sociality is congenital to human beings as much as it is to all animals (even those that are not obviously social). And sociality is in turn, the root of culture and creativity.

Just as play is universal among animal societies, it is an integral part of human development. Play gives the individual reassuring contact with fellow human beings lost when the mother's nurturing services are no longer required or offered. From play arises creativity: play therefore, and not necessity, is the mother of invention. English paediatrician and psychoanalyst Donald Winnicott (1896–1971), a contemporary of Ian Suttie, located play, and by extension, creativity and culture, in the "virtual space" between mother and infant who, through the realization of love, remain connected as they become separate.[20] *What is it to live?* Winnicott asks; the answer: it is to live creatively. Thus, cultural and creative activities do not result from the "sublimation" or suppression of the sexual instinct, as Freud supposed. Instead, they are the *raison d'etre* of human existence; the very meaning of life and a direct extension of the primal, irreducible feeling for love. Love is indeed the primary motivation for all human activities.

This is how I see the real original motive for gifts: they were not solely nor primarily exchanges, even less so a primitive form of trade, as some anthropologists seem to believe.[21] The item given is above all a sign of love. It refers to all other loves by association. It overflows with meaning, referring without bounds to all of nature, which is fully connected with, and accessible to the primitive consciousness. The meaning is irretrievably lost when gifting becomes exchange, and exchange is reduced finally to money: money which changes hands impersonally and indifferently, obliterating all reference to value, to labour, or to love, because it is itself valueless and formless.

Science of Love

Suttie began his book with some questions for his colleagues (p. 1):[19]

> "In our anxiety to avoid the intrusion of sentiment into our scientific formulations, have we not gone to the length of excluding it altogether from our field of observation? Is love a fiction, an illusion of a weak mind shrinking from reality, and if so how and why should our minds ... ever have created the 'idea' of love?"

Science, he argues, should be concerned with the whole range of our experience. In its failure to deal with sentiment and human attachments, mechanistic science is but a form of sublimated intellectual game at best. At worst, it impoverishes our human experience, destroying all that we value as human beings and substituting a dysfunctional social reality through a self-fulfilling prophecy. Suttie himself demonstrated that it is possible to have a science of feeling, but only with feeling. It requires a genuine re-integration of feeling with intellect for us to experience and understand nature fully from within.[22]

We need a knowledge system based explicitly and firmly on natural human values, a knowledge system that is already implicit in many aspects of contemporary Western science.

Nature from Within

To know nature from within is to recover the indigenous natural wisdom that sees nature as she really is: the evolving plenitude that affords the existence of things, the source and sustenance of all life, and the ultimate inspiration for the human consciousness striving to know and to create. In a culture that has lost none of this feeling of real participation in nature's creative process, science, as much as art, is a quest for greater intimacy with nature that involves our whole being. The ideal state of true knowledge and inspiration is a state of total quantum coherence with nature in which the knower and the known are mutually transparent, as described in my book, *The Rainbow and the Worm, the Physics of Organisms*.[23]

In ancient China, this entails the spiritual union of the knower with the *tao*, the creative principle that generates the multiplicity of things.[24] As the *tao* is eternal, the knower partakes of the eternal in all things through the *tao*. Similarly, in ancient Greece, true knowledge was regarded as unobscured participation in the divine mind from which all creation springs.[25] In this quantum coherent state, one's actions are guided not by a disembodied objective intellect, but on the contrary by a passionate total involvement and harmony of mind and body in nature.

The feeling for the unity and interconnectedness of nature is not just a romantic notion entertained by poets and mystics and the so-called primitive consciousness. It is a universal intuitive insight that contemporary science is driven to validate in all respects.

In biology, by far the most tenaciously held dogma for the whole of the present century is that the genes of organisms are immune from environmental exigencies and are passed on practically unchanged to the next generation. Since the mid-1970s, however, as the tools of molecular genetics have become more and more precise, people have begun to discover that the genes can change as readily as many other characteristics of the organism in response to the external environment. So much so that molecular geneticists have coined the term "fluid genome" to describe the large variety of processes that can chop and change the genes, and expand or shrink different parts of the genetic material.[26] Recent experiments also indicate that adaptive genetic mutations are non-random in that they are much more likely to occur than non-adaptive ones.[27] All the evidence indicates that organism and environment are intimately interconnected, from the sociocultural domain right down to the genes. Stable inheritance depends on this very interconnection, rather than on a mythical unchangeable genome. The process of heredity has a dynamic stability that resides in the feedback interrelationships that can propagate from the external environment through the physiological system to the genes. Organism and environment, like figure and ground, engage in ceaseless rounds of mutual definition and transformation, which is the essence of evolution.[28] Cycles of feedback between the biosphere and the physicochemical environment are the basis of stability for the global ecosystem.

The present global environmental crisis is the direct consequence of a knowledge system based on a denial of the unity of nature. And nature is responding loud and clear: she is one indivisible ecosystem; whatever insult is perpetrated in one part of the planet will have repercussions, not only locally but globally. British space scientist James Lovelock's Gaia hypothesis encapsulates the ancient ecological wisdom in a contemporary form: the collective activities of the biosphere as a whole maintain Earth's atmosphere and temperature far away from thermodynamic equilibrium in conditions that are fit for life.[29] Instead of every individual organism working for its own selfish ends as envisaged in neo-Darwinian theory, it is the extent to which they effectively cooperate in cycles of mutual feedback and interdependency that life for all is possible. This is a generalization of the principle of mutual aid among animals and plants that Kropotkin and Allee have dwelt on (see earlier). More importantly, organisms are not so much passively adapted to the environment by natural selection, as actively adapting the environment to themselves.[30] In other words, organisms actively participate in shaping their own evolution. This arises naturally through the interconnections not just between all lifeforms, but also between the biological and the physical realms, each shaping the other in successive rounds of mutual transformation and regeneration. Every species is endowed with powers that are given by all the rest. In a very real sense, each is implicated in every other by material and energy flow, and also the flow of information, as we shall see. There is an irreducible wholeness of being and becoming on Earth. This wholeness encompasses our relationship to nature at the most fundamental physical level.

The universal wholeness of being and becoming is most clearly brought home to us in quantum physics. Quantum physics is the culmination of a long series of attempts to fragment reality into the smallest particle; only when physicists got down to the infinitesimal, indivisible quantum, did they find that the whole exercise was futile, for it cannot be done at all. It turns out that in order to have a consistent representation or theory it must be supposed that observer and observed are one indivisible system, and that the very act of observation transforms reality from an indefiniteness of multiple superimposed potentialities of being to a state of definiteness, which, however, cannot be predicted in advance. Moreover, the same act of observation can simultaneously determine the state of an entity widely separated from the one observed, as though reality were indeed, an organic universal whole. Theoretical physicist David Bohm (1917–1992) and colleagues have reformulated quantum theory on the basis of universal wholeness: every particle or being is embedded in a field, or super-quantum potential, due to every other particle in the entire universe.[31] From this perspective, *wholenesss and interconnectedness are actual and primary, while fragmentation and separation are illusory.*

Nature has negated all attempts to describe her simplistically, in terms of a flat, common sensible literalness. The reductionist, atomistic science whose aim it was to do just that, when pushed to the very limit, can only reaffirm that reality has breadths and depths beyond our comprehension or description. As our knowledge of nature deepens, so too the magic and the mystery, the same magic and mystery our ancestors had access to. Nature is both wave and particle, both here and everywhere at the same time. To know her requires not only the analytic intellect of the scientist, but also the vision of the mystic, the imagination of the poet and the sensitivity of the artist. It requires our undivided being participating fully in knowledge.[2, 23, 32]

The Coherence of Being

I said that organisms may be interconnected with one another and with their physicochemical environment by information flow, as well as by material and energy flow. It is already generally accepted that physical parameters such as day length and other seasonal variables are informational in that organisms respond to them physiologically. One hypothesis is that the biological rhythms are closely attuned to the rhythms of Earth,[33] which are in turn attuned to those of the sun and the moon. Many of these natural rhythms are electromagnetic in nature. There is now a substantial literature on the sensitivity of organisms to weak electric and magnetic fields occurring either naturally, or close to power lines and other electrical appliances; although the mechanism involved in this sensitivity is not fully understood.[23, 34]

Most molecular biologists assume that the answer to biological organization will come when all the molecules in organisms are isolated and analysed. But biological organization is a dynamic, macroscopic order extending over astronomical numbers of molecules, spanning distances at least millions of times the size of individual molecules. This organization enables organisms to transform energy with the rapidity and efficiency rarely achieved elsewhere, if at all, and to be extremely sensitive to specific signals in the environment.[23, 35] Thus muscle contraction can be as efficient as 98% in converting chemical energy to mechanical work; and it is estimated that our eye is sensitive to a single quantum of light falling on the retina.

In 1960, Nobel laureate biochemist Albert Szent-Györgyi (1893–1986) pointed out that we can only begin to understand the characteristics of living systems if we take into account the collective properties of molecules akin to superconductivity and superfluidity.[36] This idea was developed at about the same time by German-born British solid-state physicist Herbert Fröhlich (1905–1991), who suggested that living systems have collective modes of activity somewhat similar to superconductors operating at physiological temperatures.[37] Metabolic energy, instead of being lost as heat, is stored in the form of collective or coherent electromechanical and electromagnetic excitations. These "coherent excitations" could be responsible for generating and maintaining long-range order. They also make possible highly efficient energy transfer and transformation of energy and the detection of very weak electromagnetic signals.[23]

Evidence for the existence of coherent excitations in living systems comes from the work of German quantum physicist and biophysicist Fritz-Albert Popp and his co-workers, who showed that practically all organisms and cells emit light (biophotons) at very weak intensities.[38] Organisms and cells also re-emit light at higher intensities as delayed luminescence after exposure to a brief pulse of light. As the result of some 15 years of experimental work, Popp became convinced that biophotons come from a coherent electrodynamical field within the living system. This field has a wide range of frequencies that are coupled together to give effectively a single degree of freedom, and that may be the basis of biological organization. Living systems are thus both emitters and receivers of electromagnetic signals originating from the physicochemical environment as well as from other organisms. We have recently demonstrated, for example, that synchronously developing fruit fly embryos can interact nonlinearly to generate coherent light emissions that are orders of magnitude higher than the emission rate of individual embryos.[39] We are left in no doubt that organisms are interconnected by information flow as well as energy and material flow.

A key aspect of quantum coherence is that it suggests a relationship between local and global (or individual and collective) that has been deemed contradictory or impossible within a mechanistic perspective. It turns out that a coherent field shows space-time correlations between different points that are precisely the products of self-correlations at those points. In other words, any number of points in a coherent field will behave statistically as though independent of one another, even though they are acting synergistically.[40] A coherent state is thus one of maximum global cohesion and also maximum local freedom! The inevitable conflict between the individual and the collective — which serves as the starting point for all social as well as biological theories of Western society — is not so inevitable after all. Perhaps it is time for social theorists to adopt a new set of premises.

Coherent Society

Can we envisage a society that is consonant with our new knowledge of the natural being and the coherence of the organism? I shall call it the coherent society to resonate with our new knowledge, in the hope that we can ultimately live and act coherently with our knowledge. It also carries the notion of a life coherent with nature, and with our own natural being.

Biological organization has long served as the metaphor of social organization for utopians and other social theorists alike; for example, Herbert Spencer in England (see earlier), Auguste Comte (1798–1857) and Saint-Simon (1760–1825) in France, to name but a few.[41] But they were to varying degrees under the influence of mechanistic biology. A mechanistic view of life visits on society a whole set of unfounded and mistaken assumptions, of which Spencer's Social Darwinism has had the most devastating consequences. Two unspoken, deeply held beliefs encapsulated in Darwinism are that man is above all an isolated individual motivated solely by self-interest, if not outright aggression, and that in the absence of an externally imposed social and moral order, chaos will reign supreme.

These Darwinian assumptions actually originated from English political philosopher Thomas Hobbs (1588–1679) two centuries earlier, in the age of powerful absolute monarchs, the beginning of the empire and mercantilism, which reached their heights in the reign of Queen Victoria. In his book, *The Post-Corporate World*,[42] David Korten, American economist, former professor at Harvard Business School and staunch critic of the existing socio-economic regime, contrasts the idea of a civil society according to ancient Greek philosopher Aristotle (384–322 BC) with that of Hobbs (p. 141):

> "For Aristotle life was a defining reality, ... humanity, as a creation of the divine, must have an inherent capacity for goodness and wisdom. Freedom and self-rule were not only possible but were essential defining principles of a civilized society ...
>
> "For Hobbes a dead and disintegrating universe was the defining reality. Life — the accidental creation of a mindless machine — could aspire to no higher purpose than survival, reproduction, and material gratification ... Thus for Hobbes, all that stood between man and the chaos of the wild state of nature was the coerced order imposed by a strong monarch."

From our present vantage point of "nature from within", all nature is a unity in which we ourselves participate in shaping. Sociality and love are primary, just as aggression and hate are the result of frustrated and failed love. The Darwinian/Freudian man is the construct and product of a patriarchal, oppressive Hobbesian society built upon the denial of love at every turn. It is neither the universal nature of human beings, nor the ineluctable human condition. A consciousness fully indigenous to nature is grounded in nature and connected to all beings. It is never isolated or alone; hence it roams freely and without fear. It is kind and loving and ever in possession of the highest moral feelings; for morality itself is derived from the experience of being connected with kin, with fellow creatures and ultimately with all of nature (see my essay on a naturalistic ethic based on the physics of organisms).[43]

What do "primitive" indigenous cultures really tell us concerning human nature? One of the most relevant studies was that conducted by American cultural anthropologist Margaret Mead (1901–1978) and her colleagues on cooperation and competition among primitive peoples.[44] They posed the question: what does the literature on primitive peoples yield on the subject of competitive and cooperative habits which throw light on the problem of culture and personality? It is a monumental work, probably the only of its kind. *Yet it is misdirected from the start.* In the book, cooperation is defined as "the act of working together to one end"; whereas competition is defined as "the act of seeking or endeavouring to gain what another is endeavouring to gain at the same time". Note how the typical way of defining "goal" or "end" in the West is *in terms of something gained*. What can it possibly mean in cultures — of which there are many — that do not value material gain *per se*?

It is significant that in summarizing the studies, Mead admitted that "cooperative" and "competitive" were not opposites, and that a category of behaviour, "individualistic" must be added, in the sense that collective as opposed to individualistic refers to behaviour, and not to goals. Similarly, in summarizing the findings, one of the headings for character formation is "ego development", in the sense in which Freud used it to contrast ego drives with sexual drives. Her co-worker states (p. 48):[44]

> "In the use of our term ego needs is implied the self-protective, self-maximising tendencies so often described under the caption of the self-preservation instinct."

This perpetrates the greatest confusion of all. "Self-preservation" and "self-maximising" describe entirely different acts. None of the cultures classified as "weak" in ego development has any trouble in self-preservation, though they often regard self-maximising behaviour as socially abhorrent. In the West, people seem quite unable to distinguish between the "sense of self" and "selfishness", and many indigenous peoples and foreigners alike are said to lack a sense of self simply because they do not value material possessions and do not act selfishly. Mead had to conclude from the studies that strong ego development can occur in individualist, competitive or cooperative societies.

Of the 13 cultures examined, six were identified as cooperative, four individualistic, and only three were competitive. Of the four cultures classified as individualistic, the Arapesh are a peaceable, good-natured people, helpful to a fault. They minimize blood relations, fixed membership in any given group or rigid association with any piece of land, and would travel for miles to help friends harvest their crops. There is complete individual freedom of choice in association with any group, and the groups are extremely fluid, changing freely with the particular collective task at hand. Their economic affiliations therefore cut across all boundaries of geography and kinship, and are based upon personal ties and friendship between individuals.

The Arapesh, however, are not considered to be "cooperative", but are classified as "individualistic", although they engage in cooperative collective tasks almost all of the time, simply because *mere helpfulness without any personal gain or end in sight is considered other than cooperation*. Helpfulness is an expression of sociality for its own sake, which is more fundamental than cooperation towards a common end. In many respects, *the Arapesh culture exhibits a coherent society where individual and collective are maximally expressed, and there is no conflict between the two*. Mead attributed that to the elimination of the distinction between the self and the good of others, which is also achieved in all the societies classified as cooperative. To me, that is also a concrete demonstration of empathy: satisfying the needs of others (and thereby sharing their satisfaction) is itself intrinsically rewarding.

The most suggestive generalization from the studies are that cooperative societies are those in which personal property is consistently undervalued and which have a strong sense of security afforded by kin group and other extension groups. There is no attempt to exercise power over other persons *and all share a belief in an ordered universe.* Competitive societies, on the other hand, place a high valuation on property for individual ends; they have a low sense of security correlated with a strong will to dominate over others; and they believe in an arbitrary, disordered domain of the supernatural which is prevailingly antagonistic to them.

In social structure, Mead suggested that competition was prevented by three features. First, a rigid hierarchical social system such that rank interposes between would-be competitors; second, a social system through which the desired end is converted from an individual to a group end; and third, a "cultural phrasing" which displaces the emphasis from objective situation to some other sphere in which competition is not so possible (for example, the Zuni and Arapesh transform the scarcity of land into a perception of the scarcity of labour, and hence encourage cooperative labour). Inherent in Mead's suggestion is the widely held underlying assumption that competition is a pre-existing or inevitable condition, originating with Hobbs (see above), that must be mitigated or ameliorated by some means.

One might easily gain the impression from the foregoing description that cooperative societies are closed and hierarchical, and rigidly controlled by rituals. That is not the case. Although the Maori have a system in which status is inalienable, the Samoans have a system in which status is movable; the Zuni have no status at all; and the Dakota are quite egalitarian. So Mead's conclusion did not accord with her findings. What is more significant is that these cultures share an emphasis on natural kin relationship, which can be quite extended. On account of that, I suggest they are best able to retain and express their natural sociality and sense of security on which cooperation depends. In this light, the competitive societies may be such because they have lost the means to express their natural sociality, resulting in a pervasive sense of insecurity and hence, in competition.

Perhaps the most significant feature of cooperative societies, for the purpose of our discussion, is that they all have a view of an ordered universe. Inherent in this belief is that nature is knowable and hence it is possible to live within her. By contrast, a view of antagonistic, disordered nature, such as pervades our Western industrialized society, gives rise to the idea that nature is ultimately unknowable, being governed solely by chance, and must hence be dominated, conquered and kept under control, much as Hobbs had envisaged.

The concept of a coherent society goes further than a cooperative society. It suggests a relationship between the local and global, or the individual and collective that has previously been deemed contradictory or impossible, basically because all social theories are still under the influence of a mechanistic perspective. The coherent society maximizes both the individual and the cooperative, both local freedom and global cohesion, that is the hallmark of quantum coherence of the organic whole. Conflict between the individual and the collective, and between private and public interests, is not at all inevitable.

From our present perspective, culture is the creation of meaning and knowledge in partnership with nature, in which every social being participates. The coherent society is the society of natural beings living in harmony with nature's creative process. It is time social theorists and our politicians transcend the mechanistic worldview to adopt a new set of premises.

References and Notes

1. Heilpern, J. *Conference of the Birds, the Story of Peter Brook in Africa.* Methuen, London, 1989.
2. Ho, M.W. "Toward an Indigenous Western Science". In *Reassessing the Metaphysical Foundations of Science* (W. Harman, ed.), pp. 179–213, Noetic Sciences Institute Publications, Sausalito, California, 1994a.
3. Ho, M.W. "Genetic Fitness and Natural Selection, Myth or Metaphor?" In *Evolution of Social Behavior and Integrative Levels*

(E. Tobach and G. Greenberg, eds.), pp. 87–112, Erlbaum, New Jersey, 1988a.
4. Gilbert, W.S. "Princess Ida". In *The Savoy Operas*, p. 321f, Oxford University Press, Oxford, 1962.
5. Young, R.M. *Darwin's Metaphor*, Cambridge University Press, Cambridge, UK, 1985.
6. Darwin, C. *On the Origin of Species by Means of Natural Selection, or the Preservation of Favoured Races in the Struggle for Life*. John Murray, London, 1859.
7. Barzun, J. *Darwin, Marx, Wagner*. Doubleday Anchor, New York, 1958.
8. Huxley, T.H. "The Struggle for Existence: A Programme". *Popular Science Monthly*, Volume 32, April 1888.
9. Brown, J.A.C. *Freud and the Post-Freudians*, p. 13. Pelican, Harmondsworth, 1961.
10. Ghiselin, M. *The Economy of Nature and the Evolution of Sex*, p. 247. University of California Press, Berkeley, 1974.
11. Wilson, E.O. *Sociobiology*, p. 3. Belknap Press, Cambridge, Massachusetts, 1975.
12. See Ho, M.W. and Saunders, P.T. "Beyond Neo-Darwinism: An Epigenetic Approach to Evolution". *J Theor Biol* 78 (1979): 573–591; also Ho, 1998, Ho, M.W. "Organism and Psyche in a Participatory Universe". In *The Evolutionary Outrider* (D. Loye, ed.), pp. 49–65, Adamantine Press, Twickenham, 1998.
13. Clairborne, R. "How Homo Sapiens Learned to Be Good". *Horizon*, Volume 16, pp. 30–35, American Heritage Publishing Co., 1975.
14. Kropotkin, P. *Mutual Aid: A Factor of Evolution*, p. x. Extending Horizon Books, Boston, Massachusetts, 1914.
15. Csanyi, V. Personal communication, 1995.
16. An explanation offered by among others, British ethologist Richard Dawkins, a populariser of Darwinism; see Dawkins, R. *The Selfish Gene*. Oxford University Press, Oxford, 1976.
17. Allee, W.C. *The Social Life of Animals*. The London Book Club, London, 1951.
18. Bateson, P. "The Biological Evolution of Cooperation and Trust". In *Trust: Making and Breaking Cooperative Relations* (D. Gambetta, ed.), pp. 14–30, Basil Blackwell, Oxford, 1988.
19. Suttie, I. *The Origins of Love and Hate*. Penguin, Harmondsworth, 1989.

20. Winnicott, D.W. *Playing and Reality*. Pelican, Harmondsworth, 1974.
21. See Sherry, J.F., Jr. "Gift Giving in Anthropological Perspective". *J Consum Res* 10 (1983): 157–168.
22. See Ho, M.W. "Evolution in Action, and Action in Evolution". In *Gaia and Evolution: Implications of the Gaia Thesis* (P. Bunyard and E. Goldsmith, eds.), pp. 14–28, Wadebridge Ecological Press, Wadebridge, 1989a.
23. Ho, M.W. *The Rainbow and the Worm: The Physics of Organisms*. World Scientific, Singapore, 1993; also Ho, M.W. "Toward an Indigenous Western Science." In *Reassessing the Metaphysical Foundations of Science* (W. Harman, ed.), pp. 179–213, Noetic Sciences Institute Publications, Sausalito, California, 1994a.
24. There is no recognized authority on the meaning of *tao*, but a good description is given here in Wikipedia, accessed 10 August 2015, https://en.wikipedia.org/wiki/Tao.
25. See Barfield, O. *Saving the Appearances: A Study in Idolatry*. Wesleyan University Press, Connecticut, 1956.
26. Ho, M.W. "Heredity as Process: Towards a Radical Reformulation of Heredity". *Rivista di Biologia* 79 (1986): 407–444.
27. Cairns, J., Overbaugh, J. and Miller, S. "The Origin of Mutants". *Nature* 335 (1988): 142–145; also Hall, B.G. "Spontaneous Point Mutations That Occur More Often When Advantageous Than When Neutral". *Genetics* 126 (1990): 5–16.
28. Ho, M.W. "How Rational Can Rational Taxonomy Be: A Post-Darwinian Rational Taxonomy Based on a Structuralism of Process". *Rivista di Biologia* 81 (1988b): 11–55.
29. Lovelock, J.E. *A New Look at Gaia*. Oxford University Press, Oxford, 1979.
30. Saunders, P.T. "Evolution without Natural Selection: Further Implications of the Daisyworld Parable". *J Theor Biol* 166 (1994): 365–373.
31. Bohm, D., Hiley, B.J. and Kaloyerou, P.N. "An Ontological Basis for Quantum Theory". *Phys Rep* 144 (1987): 323–348.
32. Ho, M.W. "A Quest for Total Understanding". Transcript of Saros Seminar on the Dilemma of Knowledge, Saros Book Club, Bristol, 1990.

33. Brown, F.A. "Extrinsic Rhythmicality: A Reference Frame for Biological Rhythms Under So-Called Constant Conditions". *Ann N Y Acad Sci* 98(1962): 775–787.
34. Shulman, S. "Cancer Risks Seen in Electro-Magnetic Fields". *Nature* 345 (1990): 463.
35. Ho, M.W. "Coherent Excitations and the Physical Foundations of Life". In *Theoretical Biology: Epigenetic and Evolutionary Order from Complex Systems* (B.C. Goodwin and P.T. Saunders, eds.), Edinburgh University Press, Edinburgh, 1989b.
36. Szent-Györgyi, A. *Introduction to a Sub-Molecular Biology*. Academic Press, New York, 1960.
37. Fröhlich, H. "Long-Range Coherence and Energy Storage in Biological Systems". *Int J Quant Chem* 2 (1968): 641–649.
38. Popp, F.A., Li, K.H., Mei, W.P., Galle, M. and Neurohr, R. "Physical Aspects of Biophotons". *Experientia* 44 (1988): 576–585.
39. See Ho, M.W., Xu, X., Ross, S. and Saunders, P.T. "Light Emission and Re-Scattering in Synchronously Developing Populations of Early *Drosophila* Embryos: Evidence for Coherence of the Embryonic Field and Long Range Cooperativity. In *Advances in Biophotons Research* (F.A. Popp, K.H. Li and Q. Gu, eds.), pp. 387–406, World Scientific, Singapore,1992.
40. Glauber, R.J. "Quantum Theory of Coherence". In *Quantum Optics* (S.M. Kay and A. Maitland, eds.), Academic Press, London, 1970; see also Ho, M.W. (2008). *The Rainbow and the Worm: The Physics of Organisms* (3rd edn), World Scientific, Singapore.
41. Jones, A.K. "Social Symbiosis: A Gaian Critique of Contemporary Social Theory". *The Ecologist*, Volume 20, pp. 108–113, 1990.
42. Korton, D.C. *The Post-Corporate World*. Berrett-Koehler, San Francisco, 1999 (which the author has most generously dedicated to me and Elizabet Sahtouris).
43. See Ho, M.W. (1994b). "The Physics of Organisms and the Naturalistic Ethic of Organic Wholeness". *Ethical Record*, January, pp. 8–12.
44. Mead, M. *Cooperation and Competition among Primitive Peoples*. Beacon Press, Boston, 1961. I thank Edward Goldsmith for directing me to this book.

2

Towards an Indigenous Western Science: Causality in the Universe of Coherent Space-Time Structures

A reformulation of Western philosophy based on quantum physics that reconnects to nature and to indigenous knowledge systems all over the world.

Magister Ludi and the Rainmaker: Two Ways of Knowing

Hermann Hesse (1877–1962) was a German-born Swiss poet, novelist and painter who received a Nobel Prize in Literature. His famous last novel describes the life of Joseph Knecht, Magister Ludi, or the supreme master of the Glass Bead Game.[1] The game is one of pure intellect directed at the synthesis of the spiritual and aesthetic abstractions in diverse disciplines of all ages, and it is the prerogative and *raison d'etre* of an entire spiritual institution, Castalia. Isolated within its enclaves and unsullied by reality, the chosen elite undertake arduous scholastic studies, the sole purpose of which is to create ever more intricate themes and variations of the game. Castalia and the Glass Bead Game developed as an antithesis to the philistine, superficial bourgeois society intent on its pursuit of conventional, establishment values. In the end, however, Joseph Knecht turns his back on Castalia, disillusioned with a life consecrated exclusively to the mind, recognizing not only its utter futility, but also its inherent danger and irresponsibility.

As an addendum to the biography of Joseph Knecht, Hesse included Knecht's posthumous writings: some poems and an account of three lives each set in a different age, but all dedicated to knowledge in different ways. The first interests me most of all. It is that of "the Rainmaker" also called Knecht, also a man of knowledge in his time. But there, the parallel ends. In contrast to the master of the Glass Bead Game who excels in scholastic, intellectual knowledge, the Rainmaker's knowledge comes directly from nature. Nature is the primary and only text: the waxing and waning of the moon, the disposition of the stars, the scent in the air, the wind in the trees, the call of birds, the chirring of insects, the spoor of animals and bits of fur left on their trails, are so many signs woven into the fabric of a mutually permeating, mutually defining reality that is nature. His predecessor and mentor taught him to experience nature until he too knew her with the same intimacy and sensitivity, until he could be one with her (pp. 433–434):[1]

> "He concentrated the very vibrations of the weather within himself, holding them within him in such a way that he could command the clouds and the winds — not, to be sure, just as he pleased, but out of the *very* intimacy and attachment he had with them, which totally erased the difference between him and the world, between inside and outside. At such times he could stand rapt, listening, or crouch rapt, with all his pores open, and not only feel the life of the winds and clouds within his own self, but also direct and engender it, somewhat in the way we can awaken and reproduce within ourselves a phrase of music that *we* know by heart. Then he needed only to hold his breath — and the wind or the thunder stopped; he needed only to nod or shake his head — and the hail pelted down or ceased; he needed only to express by a smile the balance of the conflicting forces within himself — and the billows of clouds would part, revealing the thin, bright blueness. There were many times of unusually pure harmony and composure in his soul when he carried the weather of the next few days within himself with infallible foreknowledge, as if the whole score were already written in his blood."

This remarkable passage captures the intensely aesthetic and spiritual, and at the same time resolutely practical, orientation of a direct, participatory knowledge. The indigenous consciousness of undivided body and soul, feeling and intellect, permeates the whole of nature as nature holds him enthralled. This is the state of grace in which knower and known become mutually transparent and coherent; knowledge is both sacred and authentic because it comes from the source which is nature herself.

To those indigenous to nature — and I use "indigenous" in that sense because it is more fundamental than being indigenous to any particular culture or place — knowledge is also power; it is a matter of life and death, for a misreading of nature could end in famine and starvation. Indigenous knowledge is therefore effective where the knowledge of the Glass Bead Game is effete; divorced, as it were, from reality and from action. The Rainmaker, by his knowledge, integrates the social life of his tribe into the order of nature's ways. Nature and culture are thus inseparable and in harmony one with the other.

We can see that indigenous knowledge is the *active cause* of things. By reading nature accurately and sensitively, the knower can engage in nature's process, and create in partnership with her. It is the way of the *tao* in indigenous Chinese philosophy, the effortless power of being and becoming with nature that was brought home to me after my decades of wandering in the wilderness of modernist, Newtonian Western science. The term "Newtonian" derives from the English physicist and mathematician Isaac Newton (1642–1727) who laid the foundations of classical mechanics in his book, *Philosophiae Naturalis Principia Mathematica* (Mathematical Principles of Natural Philosophy), published in 1687, which formulated the laws of motion and universal gravitation that has dominated the development of Western science ever since.[2] Within the Newtonian framework, there is no real action, effortless or otherwise; and cause is a blind extraneous force acting on a dumb indifferent object.

Traditional indigenous knowledge was closely and jealously guarded, not so much because it gave power, but because that power could be misused in the wrong hands. Along with power also came responsibility. In the end, the Rainmaker offered himself as sacrifice in an ultimate attempt to make rain.

What relevance has all this for us here and now when the days of the Rainmaker are irretrievably lost in the mists of time? Like many of his contemporaries and predecessors, Hesse was concerned with the alienation of the human spirit in a society increasingly dominated by a reductionist, positivist science that interpreted and validated only those human experiences expressible in terms of logical and mathematical laws, in which life no longer has meaning.[1] The only alternative seemed to be an outright rejection of that society for a life of intellectual contemplation, a retreat into spiritual mysticism.

Today, we live in the crisis of a global ecological devastation that is the direct legacy of the same positivist science and technology, further spurred on by the Darwinian creed of ruthless exploitation of nature and of our fellow human beings (derived from the extremely influential theory of evolution by natural selection propounded by the English naturalist Charles Darwin (1809-1882)).[3] And yet, out of the ruins of destruction, we see some glimmer of hope.

Western science was premised on the separation of the observer as disembodied mind from an objective nature observed. This was also the beginning of the demise of our natural being,[3] which I take to be the unfragmented consciousness wholly connected to nature. The mind-body or mind-matter dualism, which goes back to French philosopher and mathematician René Descartes (1596–1650) presupposes that we can know nature without experiencing her. More than that, our subjective experiences are deemed to be unreliable, and must be denied at all costs in order to preserve the objectivity of science. Hence it is that generations of Western philosophers remain perplexed as to how we can know when the knowing being has been abstracted away from a reality which must be experienced to be understood.[4] Consequently, we can have no basis for linking cause and effect, nor for distinguishing good from bad, and the authentic from mere simulacrum.[5]

The history of Western culture since then is one of increasing alienation from nature and hence from our natural being, while nature becomes fragmented into atoms and fundamental particles. And yet when this reductionist, atomistic science is pursued to its logical conclusion and pushed to its very limits, it can only undermine the basis on which it was built. For everywhere, it reaffirms the unity of nature in which the knowing being is inextricably embedded,[4] compelling us towards a new knowledge system and a new way of knowing that is at the same time very old. It harks back to the universal indigenous wisdom that sees nature as she really is: the evolving plenitude that affords the existence of all things, the ultimate inspiration for the consciousness and striving to know.[4,5]

The new science will come to resemble traditional indigenous science in many fundamental respects. To make the affinities explicit, I propose to call it the indigenous Western science; "indigenous" because it is premised on a direct experience of nature. In this sense, indigenous knowledge is indifferent to time and place, but is the prerogative of the natural being of all ages. The natural being is the vehicle to authentic knowledge. Knowing involves the undivided indigenous consciousness — the natural being — communicating with nature, engaging her wholeheartedly without bounds or boundaries. There is hence no mismatch between knowledge and our experience of reality. Reality is not a flat impenetrable surface of common-sensible literalness. It has breadths and depths beyond our wildest imagination. The quality of our vision depends entirely on the extent to which our consciousness permeates and resonates within nature's magical realm. In this respect, there is complete symmetry between science and art. Both are creative acts of the most intimate communion with nature.[6]

As consistent with the character of indigenous science, it can be approached through many avenues: art, psychology, biology, ecology, sociology, linguistics, anthropology and philosophy, to name but a few. My approach here is through contemporary physics and chemistry, especially of the organism, linking up with the thoughts of English mathematician and philosopher Alfred North Whitehead (1861–1947) and French philosopher Henri Bergson

(1859–1941) — though not claiming to be true to either — in order to outline what I hope to be a coherent and contemporary indigenous Western narrative on the nature of reality.[3]

The Fall from Grace

The biblical account of the fall from grace — of Adam and Eve being driven out of paradise because Eve has bitten the apple from the Tree of Knowledge — is a parable of our exile as knowing beings from the magic kingdom of nature; it is a radical severing of form from content in that knowledge can no longer mean anything because it is unconnected to nature. In isolating the human consciousness from the reality that must be experienced to be understood, the content of knowledge is renounced at the outset. In turn, the imposition of an inappropriate form derived from the mechanical, materialistic worldview led to the progressive erosion of the content of experience, resulting in the disenchantment of nature and of life.

A disjunction is imposed between the quality of authentic experience and the description of reality in science. No one has written more vividly and poignantly on the subject than Bergson.[7] He invites us to step into the rich flowing stream of our consciousness to recover the authentic experience of reality for which we have substituted a flat literal simulacrum given in language, in particular, the language of mechanistic science.

In that language, words that represent our feelings — love and hate, joy and pain — emptied of their experiential content, are taken for the feelings themselves. They are then defined as individual psychic entities (or psychological states), each uniform for every occasion across all individuals, differing only in magnitude or intensity. But should we, as Bergson urges, connect our mind to our inner feelings, we will discover that what we actually experience is not a quantitative increase in intensity of some psychological state, but a succession of qualitative changes which "melt into and permeate one another" with no definite localizations or boundaries, each occupying the whole of our being within a span of feeling that Bergson refers to as "pure duration".

Pure duration is our intuitive experience of inner process, which is also inner time with its dynamic, heterogeneous multiplicity of succession without separateness. Each moment is implicated in all other moments. Newtonian time, in which separate moments, mutually external to one another, are juxtaposed in linear progression, arises from our attempt to externalize pure duration, an indivisible heterogeneous quality, to an infinitely divisible homogeneous quantity. In effect we have reduced *time* to Newtonian *space*, an equally homogeneous medium in which isolated objects, mutually opaque, confront one another in frozen immobility.

Bergson emphasizes the need for introspection in order to recover the quality of experience. Yet introspection divorced from external reality will only give a partial insight into the indigenous experience of our Rainmaker. Thus, Bergson opposes an inner "succession without mutual externality" to an outer "mutual externality without succession". He distinguishes two different selves, of which one is the external projection of the other, inner self, into its spatial or social representation. The inner self is reached "by deep introspection, which leads us to grasp our inner states as living things, constantly becoming". He writes (p. 231):[7]

> "But the moments at which we thus grasp ourselves are rare, and that is just why we are rarely free. The greater part of the time we live outside ourselves, hardly perceiving anything of ourselves but our own ghost, a colorless shadow which pure duration projects into homogeneous space. Hence our life unfolds in space rather than in time; we live for the external world rather than for ourselves; we speak rather than think; we 'are acted' rather than act ourselves."

This passage anticipates the sentiment of the existentialist writers such as French authors and philosophers Albert Camus (1913–1960) and Jean Paul Sartre (1905–1980) *who* emphasize individual freedom and choice. They believe it is for humans to define their meaning in life and try to make rational decisions despite existing in an irrational universe.[8] A similar sentiment pervades the poetry of

T.S. Eliot (1888–1965), one of the 20th century's major poets. The opening lines from "The Hollow Men"[9] evoke strong reminiscences of Bergson's projected being outside oneself:

We are the hollow men
We are the stuffed men
Leaning together
Headpiece filled with straw. Alas!
Our dried voices, when
We whisper together
Are quiet and meaningless
As wind in dried grass
Or rats' feet over broken glass
In our dried cellar
Shape without form, shade without color,
Paralyzed force, gesture without motion ...

The Fallacy of Misplaced Concreteness

Bergson's protestations[7] are really directed against one of the most fundamental assumptions underlying the modernist, scientistic culture of the West, which claims to express the most concrete, common-sensible aspect of nature: that material objects have simple locations in space and time. Yet space and time are not symmetrical. A material object is supposed to have extension in space in such a way that dividing the space it occupies will divide the material accordingly. On the other hand, if the object endures within a period of time, then it is assumed to exist equally in any portion of that period. In other words, dividing the time does nothing to the material because it is always assumed to be immobile. Hence the lapse of time is a mere accident, the material being indifferent to it. The world is simply made of a succession of instantaneous immobile configurations of matter, each instant bearing no inherent reference to any other instant of time. How, then, is it possible to link cause and effect? How are we justified in inferring by observation the great "laws of nature"? This was essentially the problem of induction — whether it is legitimate to generalize from repeated observations — raised by Scottish philosopher David Hume (1711–1776).

This problem is created ultimately because the abstraction of reality in the "laws of nature" are mistaken for reality, a case of the fallacy of misplaced concreteness, says Whitehead in a thorough critique of mechanical materialism.[10] Ideal Newtonian objects are devoid of attributes except for location, mass or momentum. Yet the objects of our experience are saturated with qualities such as colours, scents, and sounds. How can this square with the science of physical reality? English philosopher John Locke (1632–1704) had attempted to solve the problem by elaborating a theory of primary and secondary qualities. Primary qualities are the essential qualities of substances whose spatiotemporal relationships constitute nature. The occurrences of nature are apprehended by minds associated in some ways with living bodies. The mind, in apprehending, also experiences sensations that are qualities of the mind alone, but are projected by the mind to "clothe" the appropriate bodies in external nature. From henceforth, mind and nature (as inert matter) confront each other across an impermeable divide, and philosophers align themselves in each of three possible camps: the dualists who insist on the division, and the two varieties of monists — "those who put mind inside matter and those who put matter inside mind". The most famous philosopher in the last category is Anglican-Irish philosopher Bishop George Berkeley (1685–1753) who contended that natural entities exist only in the perception of the mind.

The Organism as Agent of Prehensive Unity

In order to transcend the philosophical ruin left in the wake of mechanical materialism, Whitehead[10] attempts to return to a kind of native realism, not unlike the pan-psychism or pan-animism usually attributed to the so-called primitive mind by Western anthropologists. Whitehead, however, cites English philosopher and scientist Francis Bacon (1561–1626) as his source of inspiration. Bacon not only attributed consciousness or perception to all bodies, including the "inanimate", but also endowed them with distinctive qualities. Whitehead cites Bacon (p.87):

> "It is certain that all bodies whatsoever, though they have no sense, yet they have perception ... and whether the body be alterant or altered, evermore a perception precedes operation; for else all bodies would be alike one to another."

What follows from this in his own thought is not completely clear, and with due apologies to many excellent Whitehead scholars,[11] in my exposition below I shall be interpreting his ideas to suit my own purpose. A major difficulty is that our language is quite unequal to the task of the organicist process philosophy Whitehead is developing; the same difficulty frustrates my own attempt here.

Whitehead rejects the existence of inert objects or things with simple locations in space and time. As all nature is process, there is only the progressive realization of natural occurrences. For mind, he substitutes a process of "prehensive unification". The realization of an occurrence is thus (p. 88):[10]

> "the gathering of things into the unity of a prehension ... This unity of a prehension defines itself as a here and a now, and the things so gathered into the grasped unity have essential references to other places and other times."

The focus of prehensive unification is the "event" which has been individuated by the act of prehensive unification. (We shall see later that an organism is an event with enduring pattern.) Seen in this way, neither the event nor the prehensive unification of "things" gathered into a unity is subject to simple location, for (p. 90)[10]

> "each volume of space, or each lapse of time includes in its essence aspects of all volumes of space, or of all lapses of time."

The event, therefore, has a space-time structure of its own (although Whitehead did not use that term).

Each event also possesses its own special qualities, which Whitehead refers to as "eternal objects", whereby the "interfusion of events" is accomplished. Furthermore (pp. 128–129),[10]

> "There is a reciprocity of aspects and there are patterns of aspects. Each event corresponds to two such patterns: namely the pattern of aspects of other *events* which it grasps into its own unity, and the pattern of Its aspects which other events severally grasp into their unities. Accordingly, a non-materialist philosophy of nature will identify a primary organism as being the emergence of some particular pattern as grasped in the unity of a real event."

In other words, the event, or the organism, and its environment of other events are mutually implicated and mutually constitutive. Realization depends on the act of prehension, which enfolds the environment consisting of others into a unity residing in a "self", while aspects of the self are communicated to others for their enfoldments. The realizations of "self" and "other" are thus completely intertwined; the one does not occur without the other.

Towards a Theory of the Organism

Whitehead refers to a "primary organism" as the emergence of some particular pattern as grasped in the unity of a real event (see above). Defined in this way, the question is left open as to whether the fundamental particles of physics such as protons and electrons, or, at the other extreme, entire planets such as Earth, or a galaxy like our own Milky Way, are also organisms. As an event, the concept of an organism includes the concept of the interaction of organisms. The organism in its own prehension is an achievement in its own right and for its own sake, because it has enfolded diverse entities into an enduring pattern that is uniquely itself. In its own realization, it also contributes to the realization of other organisms. But though each organism is necessary for the community of organisms, its "value" or contribution depends on something intrinsic to it: its endurance, and its ability to recover its self-identity or specific organization in the midst of the flux of reality. This does not mean it necessarily maintains some constant shape or form. It may realize itself in the guise of an individual enduring entity that nevertheless undergoes a sequence of transformations constituting its life-history.[12]

An organism comes into being by its own activity out of a substrate that Whitehead refers to as the underlying, eternal "energy of realization". The analogy between Whitehead's view of nature and many traditional indigenous philosophies is notable. The Ufaina Indians in the Colombian Amazon, for example, believe in a vital force called *fufaka* present in and circulating among all living things. The source of this vital force is the sun.[13] Traditional Chinese Taoist philosophy, similarly, speaks of the *qi*, the undifferentiated primal energy pervading the universe, the substrate for realizing the multiplicity of things.[14] In turn, these philosophies are not so far removed from the biochemical ecology of the biosphere in Western science, in which the energy from the sun, trapped by green plants, is channelled into a cyclic web of biosynthesis and degradation involving practically all organisms. Material and energy flow link the entire planet together. We shall see later on that organisms are also linked by information flow.

Though consciousness is constitutive of the universe, located in organisms that are the foci of prehensive unification, Whitehead perceived that organisms differ in "value" or purposiveness in terms of their endurance of pattern in the flux of process. (This is related to the quality of organization that I shall presently introduce as its space-time structure.) Some form no more than a ripple on the surface of the energy substrate; the vast majority in the middle apparently "obey" the laws of physics; while at the other extreme are those who rise to conscious thought and are capable of exercising abstract judgments (pp. 131–132):[10]

> "The individual perception arising from enduring objects will vary in its individual depth and width according to the way in which the pattern dominates its own route. It may represent the faintest ripple differentiating the general substrate energy; or in the other extreme, it may rise to conscious thought, which includes

> poising before self-conscious judgment the abstract possibilities of value inherent in various situations of ideal togetherness. The intermediate cases will group round the individual perception as envisaging (without self-consciousness) that one immediate possibility of attainment which represents the closest analogy to its own immediate past, having regard to the actual aspects which are there for prehension."

An individual consists of a distinctive enfoldment of its environment; therefore each individual is simultaneously delocalized over all individuals. There is another sense in which individuality is relative. An individual (p. 132)[10]

> "whose own life-history is part within the life-history of some larger, deeper, more complete pattern, is liable to have aspects of that larger pattern dominating its own being, and so to experience modifications of that larger pattern reflected in itself as modifications of its own being. This is the theory of organic mechanism."

An obvious situation where nested individualities occur is in a species or a society. Societies evolve slowly, their slowly changing variables defining parameters for the evolution of individuals within them. The picture of nested individuality and constitutive mutuality is also consistent with British space scientist Jim Lovelock's Gaia hypothesis,[15,16] which proposes that Earth is one cybernetic system maintained far from thermodynamic equilibrium in conditions suitable for life by the actions of the "organisms" (both physical and biological) within it.

In these examples, not only are individual organisms parts of a larger organism, but the substance, the very essence of each individual is constitutive of every other. There is a mutual enfoldment and unfoldment, of implicate and explicate, between organism and environment. The organism on individuating enfolds the environment into itself simultaneously as it unfolds to the environment. Explicate order in the environment is implicate in the organism just as the organism's explicate pattern is implicate in the environment. This anticipates the account of the quantum universe[17] by American-born theoretical physicist David Bohm (1917–1992).

In analogy to the hologram, the implicate order of an object is contained in an interference pattern of light distributed throughout space, in which it can be said to be enfolded. By an act of unfoldment, however, the original form of the object could once again be made explicate (p. 43):[17]

> "Each separate and extended form in the explicate order is enfolded in the whole and ... in turn, the whole is enfolded in this form."

The account[17] is the latest version of Bohm's theory presented before his death. The universe is pictured as a continuous field with quantized values for energy, momentum and angular momentum. Such a field will manifest as both particles and as waves emanating and converging on the regions where particles are detected. This field is organized and maintained by the "superquantum potential", a function of the entire universe. There is a universal process of constant creation and annihilation, determined through the superquantum potential so as to give a world of form and structure in which all manifest features are only relatively constant, recurrent and stable aspects of the whole.

In summary, organisms are enduring patterns of activity and concrescences of such activities emerging from the underlying energy substrate. They are thus dynamic structures that evolve in the process of continual realization or becoming, ever enfolding from the environment to unfold to the next novelty of being.[12] In this light, causality is *immanent* to being and not external to it. *Organisms are their own cause as well as the "cause" of other events or organisms in their environment.*

Whitehead's organicist philosophy is in many ways a logical progression from the demise of mechanical materialism that began towards the end of the 19th century. The rise of thermodynamics introduced a new kind of conservation law, of energy in place of mass. Mass was no longer the pre-eminent permanent quality. Instead, the notion of energy became fundamental, especially after German-born theoretical physicist Albert Einstein (1879–1955) worked out the famous mass-energy equation, and nuclear fission in the atomic bomb proved him right, with devastating consequences. Scottish mathematical physicist James Clerk Maxwell (1831–1879) formulated a theory of electromagnetism that demanded there should indeed be energy, in the form of electromagnetic fields pervading throughout all space, which is not immediately dependent on matter. Finally, the development of quantum theory reveals that even the atoms of solid matter are composed of vibrations that can radiate out into space under certain circumstances. Matter loses material solidity more and more under the steady scrutiny of relentless rationality.

Meanwhile, the Newtonian picture of homogeneous absolute time and space gives way to Einstein's theory of relativity. Each inertial frame of reference (associated with its own observer or prehensive organism) must be considered as having a distinct space-time metric. The organism has no simple location in space-time. It can alter its space-time by its own motion or activity. It is possible that some organisms will no longer "endure" under changes of space-time. Thus, the organism's space-time metric, and perforce its internal space-time, cannot be regarded as a given, but arises out of its own activities. The organisation of these activities is also its internal space-time structure, its implicatedness of individuation and its "value", which is not a quantity but a quality. Bergson's "pure duration" is a quality of the same cloth, and is akin to temporal organization in living systems proposed in the 1960s by Canadian-born British mathematical biologist Brian Goodwin (1931–2009).[18]

"Internal space-time structure" introduced in the following sections refers to that which is generated by the organism's own activities of prehension or enfoldment.

The Organism as a Space-Time Structure

The Space-Time Catenation of Living Process

It is obviously true that organisms possess both spatial and temporal organization. Spatial organization is the focus of investigation for developmental biologists, anatomists and histologist alike; how spatial organization comes into being in the course of development has remained perhaps the deepest mystery in science.[19] Temporal organization, in terms of biological rhythms, has exercised physiologists for almost half an century with but little progress in their search for the biochemical mechanism of the "biological clock".[20] What I address here is a somewhat different perspective on the "implicatedness", the "complicity" or "depth" of the organism's space-time structure.

The organism is a dissipative structure[21] in the sense that it is maintained in a steady slate by a flow of energy and chemicals. As soon as that flow is interrupted, disintegration and death begin. However, that steady state is not a static bulk phase in a rigid unvarying container. Even a single cell has its characteristic shape and anatomy, all parts of which are in constant motion; its electrical potentials and mechanical properties similarly are subject to cyclic and non-cyclic changes as it responds to and counteracts environmental fluctuations.[6] Spatially, the cell is partitioned into numerous compartments, each with its own steady state of processes that can respond directly to external stimuli and relay signals to other compartments of the cell. Within each compartment, microdomains can be separately energized to give local circuits, and complexes of two or more molecules can function autonomously as "molecular machines". In other words, the "steady state" is not a state at all but a conglomerate of processes that are spatiotemporally organized; it has a deep space-time structure, and cannot be represented as an instantaneous state or even a configuration of states. Relaxation times of processes range from $< 10^{-14}$ second for resonant energy transfer between molecules, to 10^7 seconds for circannual rhythms. The spatial extent of processes, similarly, spans at least ten orders of magnitude from 10^{-10} metre for intermolecular interactions to several metres for nerve conduction and the general coordination of movements in larger animals.

The problem immediately arises as to how we can describe such a space-time structure, and, in particular, whether the thermodynamics of ordinary equilibrium processes have any relevance for the living system. Faced with this problem, English physical chemist R.J.P. Williams (1926–2015) advocated a shift from the conventional thermodynamic approach to a dynamic approach.[22] He proposed that the changes are catenated in both time and space: the extremely rapid transient flows (very short-lived pulses of chemical or of energy) are propagated to longer and longer time domains of minutes, hours, days, and so on into the "future" via interlocking processes that ultimately straddle generations. These processes include the familiar enzyme activation cascades that lead to the expression of different genes and morphological differentiation as cells respond to changes in their immediate environment; nerve cells on electrical or neurotransmitter stimulation; and liver, pancreatic or gonadal cells on being activated by hormones. These are in effect projections into the future, consisting of an enfoldment of the present environment subject to past history. In total, they constitute that which is usually referred to as the system's memory, determining to some extent how the system unfolds and enfolds in future. Simultaneously, the locus of change propagates spatially to the rest of the cell or organism, stimulating other processes to take place that feedback on the earlier process to dampen or amplify its effects.[6] (It may lead, for example, to the organism altering its environment.) Physiologically, therefore, every volume of space has reference to every other volume, just as each lapse of time is implicated in all other lapses of time, as Whitehead has said.[10]

The dynamism of the living system is such that each single cell is simultaneously crisscrossed by many circuits of flow, each with its own time domain and vectorial direction specified by local pumping, gating and chemical transformation, and classical equilibrium constants are quite irrelevant. Williams wrote (p. 331):[22]

> "In such a system there are no fixed constants such as equilibrium constants or solubility products, as each such constant, defined relative to a conventional standard state, is a continuous function of variables."

The variables that make up the "constants" include the flow rates, the electrical and mechanical field strengths and so on. Furthermore, as the products of reactions alter the chemical potentials of all the components by altering the variables, the equilibrium "constants" will also be functions of time. All is flux: there is indeed no holding nature still and looking at it, as Whitehead says in his book *Concept of Nature*.[23]

The above biochemical description also gives considerable substance to Bergson's intuition of "pure duration" as a succession of qualitative changes melting into and permeating one another (see above); for that is the essence of living process.

The Thermodynamics of Living Systems

(A more complete version of the thermodynamics of organisms and sustainable systems is presented in the third edition of the Rainbow Worm;[24] *the description below gives a more detailed account of its origins.)*

The question alluded to above is whether living systems violate the laws of physics. Our present knowledge of biochemistry already shows how inadequate the laws of thermodynamics are; at least, as most scientists understand them to be. But there is a deeper problem.

The physical world runs down, according to the second law of thermodynamics, such that useful energy continually degrades into heat or random molecular motion. Concomitantly, order dissolves into disorder, the measure of which is *entropy*. The biological world, by contrast, seems capable of going in the opposite direction, of increasing organization by a flow of energy and matter. Many physicists and chemists feel that as all biological processes require either chemical energy or light energy and involve real chemical reactions, both the first and second law of thermodynamics must apply to living systems (the first law of thermodynamics is the law of conservation of energy). Yet chemical reactions within organisms have efficiencies that greatly exceed those occurring outside. For example, motorcars convert only 10–20% of the energy available in petrol into work. Energy conversion in animal muscles, by contrast, is up to 98% efficient;[25] in other words, little of the available energy is degraded into entropy.

So what is the secret of the organism? One answer is that because living systems are open, they can create a local decrease in entropy at the expense of the rest of the universe, so that the entropy of living systems plus the universe always increases in all real processes.[26] But a more fundamental reason may have to do with the formulation of the second law itself.

The second law of thermodynamics, as usually formulated, is a statistical law that can only apply to a system consisting of a large number of particles, that is, a bulk phase system. As is clear from the previous description, living systems do not have a bulk phase. Instead, they consist of compartments and microdomains, each with its own steady state, and complexes of a few molecules can act as efficient cyclic molecular machines with no immediate reference to the "steady state" of its surroundings. This implies that if thermodynamics is to apply to living systems, it must apply to single individual molecules, as British physiologist Colin McClare (1937–1977) contends.[27,28]

In order to formulate the second law of thermodynamics so that it applies to single molecules, McClare introduces the key notion of a characteristic time interval. Consider a system at equilibrium at temperature θ within an interval of time τ. The energies contained in this system can be partitioned into stored energies *versus* thermal energies. Thermal energies are those that exchange with each other and reach equilibrium in a time less than τ (so technically they give the typical Boltzmann distribution characterized by the temperature θ). Stored energies are those that remain in a non-equilibrium distribution for a time greater than τ, either as characterized by a higher temperature, or such that states of higher energy are more populated than states of lower energy. So, stored energy is any form that does not thermalize or degrade into heat in the interval τ.

McClare goes on to restate the second law as follows: useful work is only done by a molecular system when one form of stored energy is converted into another. In other words, thermalized energy is unavailable for work and it is impossible to convert thermalized energy into stored energy. But in my view, this is unnecessarily restrictive, and possibly untrue, for thermal energy *can be* harvested to do useful work in a cooperative system (see below). It is only necessary to recognize that useful work can be done by a molecular system by a transfer of stored energy in a time less than τ.

The major consequence of McClare's formulation arises from the explicit introduction of time. For there are now two quite distinct ways of doing useful work: slowly according to conventional thermodynamics theory, and quickly, in which case conventional thermodynamics does not apply. Both of these are reversible and operate at maximum efficiency as no entropy is generated. (This is implicit in the classical formulation, $dS \geq 0$, for which the limiting case is $dS = 0$.)

Let us consider the slow process first. A slow process is one that occurs at or near equilibrium. By taking explicit account of characteristic time, a reversible thermodynamic process merely needs to be slow enough for all thermally exchanging energies to equilibrate, that is, to spread evenly throughout the system. In other words, it needs to be slower than τ, which can in reality be a very short period of time. So, high efficiencies of energy conversion can still be attained in thermodynamic processes that occur quite rapidly, provided that equilibration is fast enough.

This may be where spatial partitioning and microdomains are crucial for restricting the volume within which equilibration occurs, thus reducing the equilibration time. This means that local equilibrium may be achieved at least for some biochemical reactions in the living system.

At the other extreme, there is the quick process, where an exchange of energy occurs so fast that it, too, is reversible. In other words, provided the exchanging energies are not thermal energies in the first place, but remain stored, then the process is limited only by the speed of light. Resonant energy transfer between molecules is an example of a fast process. As is well known, chemical bonds, when excited, will vibrate at characteristic frequencies, and any two (or more) bonds that have the same intrinsic frequency of vibration will resonate with one another. More important, the energy of vibration can be transferred through large distances (theoretically infinite, if the energy is radiated, as electromagnetic radiations travel through space at the speed of light, though in practice it may be limited by non-specific absorption in the intervening medium). Resonant energy transfer occurs typically in 10^{-14} second, whereas the vibrations themselves die down, or thermalize, in 10^{-9} second to 10^{1} second. The process is 100 % efficient and highly specific, being determined by the frequency of the vibration itself, and resonating molecules (like people) can attract one another. By contrast, conventional chemical reactions depend on energy transfer that occurs only at collision; they are inefficient because a lot of the energy is dissipated as heat, and specificity is low, for non-reactive species could collide with each other as often as reactive species.

Does resonant energy transfer occur in the living system? McClare suggests it occurs in muscle contraction,[28] where it has already been shown that the energy released in the hydrolysis of ATP (the immediate source of chemical energy supply for the muscle) is almost completely converted into mechanical energy. This is the molecular machine, which can cycle autonomously without equilibration with its environment. The reaction has been reinvestigated[29] using much more sophisticated techniques to monitor the chemical-mechanical energy transduction. The results suggest that the formation of the myosin-actin complex (the major proteins in muscle) is coupled to the release of inorganic phosphate from ATP in a reaction that is readily reversible; in other words, this is a reaction that generates no entropy.

Similarly, in photosynthesis, whereby green plants convert light energy into chemical energy, charge separation occurs when a quantum of light is absorbed by the chlorophyll molecules in the antenna complex. More recent work[30] shows that the first step of the charge separation is a readily reversible reaction that takes place in less than 10^{-13} second, again implying a fast process that generates no entropy.

Thus, the living system may use both means of efficient energy transfer — slow and quick reactions — always with respect to the relaxation time, which is itself a variable according to the processes and the spatial extents involved. This insight is offered by taking into account the space-time structure of living systems explicitly.

Another important insight is the fundamental quantum nature of important biological processes. McClare defines a "molecular energy machine" as one in which the energy stored in single molecules is released in a specific molecular form and then converted into another specific form so quickly that it never has time to become heat.[27] It is also a quantum machine because it sums up the effects produced by single molecules. Muscle contraction is the most obvious example. A muscle is a sum of many fibres, each of which is in turn a sum of many individual molecules; the action of all in concert produces a muscle contraction.[6] Even in conventional enzyme kinetics, more and more quantum mechanical effects are being recognized. Electron tunnelling is already well known. Now it appears that hydrogen transfer reactions also involve tunnelling under energy barriers via an overlap of quantum mechanical wave functions between substrates and products,[31] and not by thermal activation as conventionally conceived.

The existence of molecular quantum machines immediately raises the question as to how the astronomical numbers of individual quantum machines can be coordinated over the macroscopic distances characteristic of biological functioning. This is the fundamental problem of biological organization. Just as bulk phase thermodynamics is inapplicable to the living system, which consists of quantum molecular machines, so, perforce, some new principle is required for the coordination of quantum molecular machines. This principle is *quantum coherence*, which will be developed later.

For now, we continue to examine the nature of space and time. In particular, we would like to give substance to the idea, as consistent with a process ontology, that they are constructed by the actions of organisms, organisms interpreted widely as in Whitehead's philosophy.[10]

Space-Time Structure and Quantum Theory

The nature of space and time is fundamental to our theory of reality. The mismatch between the Newtonian universe and our intuitive experience of reality hinges on space and time. In fact, all developments subsequent to Descartes and Newton in Western science may be seen as a struggle to reinstate our intuitive, indigenous notions of space and time, which deep within our soul we feel to be more consonant with authentic experience. But there has only been limited success so far.

Einstein's theory of special relativity substitutes for absolute space and absolute time a four-dimensional space-time continuum different for each observer in its own inertial frame. Space and time have become symmetrical to each other, but they remain definite quantities. In quantum theory — which originated with German theoretical physicist Max Planck (1858–1947) though many others contributed to its development — on the other hand, space coordinates lose definiteness in becoming operators, and hence statistical quantities, but time remains a simple parameter as in classical mechanics.

Another problem in connection with time is that the laws of physics in both classical and quantum mechanics, as well as in relativity, are time symmetric; that is, they do not distinguish between past and future. Yet real processes seem to have an "arrow of time". So time ought to be related to real processes. And it would have the quality that Bergson refers to as pure duration. In other words, it would have a structure, and not just a simple parameter.

Theoretical physicist Wolfram Schommers at Karlruhe University in Germany, like both Whitehead and Bergson, argues for the primacy of process, and in a significant reformulation of quantum theory shows how time and space are tied to real processes.[32] He begins from a consideration of Mach's principle (after Austrian physicist and philosopher Ernst Mach (1838–1916)), which eliminates absolute space and time from mechanics. According to Mach, particles do not move relative to space, but to the centre of all the other masses in the universe. In other words, absolute space and time coordinates cannot be determined empirically. Any change in position of masses is not due to the interaction between coordinates and masses, but entirely between the masses. However, neither relativity nor quantum mechanics has incorporated Mach's principle in their formulation.

If one takes account of Mach's principle, space-time must be considered an auxiliary element for the geometrical description of real processes. In other words, real processes are projected to space-time or "(r,t)-space" from perhaps a more fundamental space — that which represents reality more authentically in terms of the parameters of interactions, that is, momentum and energy, the "(p,E)-space". The two spaces are equivalent descriptions and are connected by the mathematical device of a Fourier transformation, which decomposes a function into a series of sinusoidal or oscillatory component functions. (Intermediate spaces can also be formed that are similarly connected.) The result is that time takes the form of an operator and become statistical in (p,E)-space where momentum and energy are parameters and take on definite values.

Hence the usual wave function for space-time ψ (r,t) leads to probability distributions for *both* space and time. Processes consisting of matter interacting — that is in (p,E)-space — generate space-rime structures. In other words, space-time structures are *caused* by action, and in the limiting cases of a stationary process and a free particle, that is, the wave function ψ (r,t) = 0, no time- or space-structures are defined.

In Schommer's scheme, energy and time representations are complementary, and for non-stationary processes an uncertainty relationship exists between *them of the same form as that between position and momentum in conventional quantum theory*, which is referred to as the Heisenberg uncertainty relationship, due to German theoretical physicist and a pioneer of quantum mechanics Werner Heisenberg (1901–1976). The consequence is that both energy structure and the internal time structure are different for different systems compared to an external reference time structure such as a clock.

In fact, space-time forms a non-absolute continuum depending on the reference mass. In turn, mass is nothing but a quality of action, represented perhaps somewhat inadequately as energy and momentum.

So far, nothing has been said concerning the interaction between space-time structures, such as occurs in perception or prehension between "subject" and "object", which may have very different space-time structures. Before I venture into this territory, it is necessary to introduce the idea of coherence, which I believe to be the key to biological organization.

Coherence and Biological Organization

What is the basis of the remarkable spatiotemporal organization in all living systems, a pattern that is stably maintained in the face of a constant flux of energy and matter? How does this organization enable them to transform energy so rapidly, and with such high efficiency? How do organisms react so promptly and sensitively to specific cues from the environment? Despite great advances in molecular biology within the past 50 years, we have as yet no satisfactory explanation for any of these distinguishing characteristics of living systems. Nobel laureate biochemist Albert Szent-Györgyi (1893–1986) was one of the first to suggest that we can only begin to understand the characteristics of living systems if we take into account the collective properties of the molecular aggregates, such as those observed under special conditions in solid-state physics.[33] (Actually, Erwin Schrodinger (1887–1961) had alluded to the problem of coherence and the irrelevance of statistical thermodynamics in describing living systems.[6,26])

The molecules in most physical matter have a high degree of uncoordinated or random thermal motion. But when the temperature is lowered to below a critical level, all the molecules may condense into a collective state, and exhibit the unusual properties of superfluidity or superconductivity. In other words, all the molecules of the system move as one, or conduct electricity with zero resistance (by a coordinated arrangement of conducting electrons). Liquid helium, at temperatures close to absolute zero, was the first and only superfluid substance known, and various pure metals and alloys are superconducting at liquid helium temperatures. Recently, superconducting materials have been found that can work at much higher temperatures above absolute zero.

German-born British solid-state physicist Herbert Fröhlich (1905–1991) proposes that something like a condensation into a collective mode of activity may be occurring in living systems such that they are in effect superconductors working at physiological temperatures.[34] He suggests that metabolic energy, instead of being lost as heat, is actually stored in the form of collective modes of electromechanical and electromagnetic vibrations extending over macroscopic distances within the organisms. He calls these collective modes "coherent excitations". The collective modes can vary from a stable or metastable highly polarized state (resulting from mode softening of interacting frequencies towards a collective frequency of zero), to limit cycle oscillations, to much higher frequencies when the energy supply exceeds a certain threshold. Each collective "mode" can be a band of frequencies, with varying spatial extent as consistent with the spatiotemporal structure of the living system. Nevertheless, the frequencies are coupled together so that energy fed into any specific frequency is readily communicated to other frequencies. (Conceptually, Fröhlich achieved energy exchange via a "heat-bath", though, as we shall see later, this energy coupling may have a more fundamental origin.)

Coherent excitations are responsible for long-range order in the living system, as well as for efficient energy transfer. Under those conditions, the organism will also be very sensitive to external electromagnetic fields, and weak signals will be greatly amplified. Fröhlich's theory of coherent excitations involves classical mechanisms and offers a plausible explanation for the biological effects of electromagnetic fields in the environment, which have increasingly become the focus of public attention.

Russian biophysicist Alexander Presman was among the first to review observations suggesting that diverse organisms are sensitive to electromagnetic fields of extremely low intensities, of magnitudes similar to those occurring in nature.[35] The natural electromagnetic sources, such as Earth's magnetic field, provide information for navigation and growth in a wide variety of organisms, while major biological rhythms are closely attuned to the natural electromagnetic rhythms of Earth, which are in turn tied to periodic variations in solar and lunar activities. In many cases, the sensitivity of the organisms to electromagnetic fields is such that they detect signals below the level of thermal noise (energy level of individual molecules vibrating incoherently as temperature increases). This indicates that the electromagnetic field cannot be acting on the biological system by conventional energy transfer, but by informational transfer. It points to the existence of amplifying mechanisms in the organisms receiving the information (and acting on it). Specifically, the living system itself must also be organized by intrinsic electrodynamical fields, capable of receiving (and transmitting) electromagnetic information in a wide range of frequencies rather like an extraordinarily efficient and sensitive, and extremely broad-band radio receiver and transmitter, much as Fröhlich has suggested.[34]

In my own laboratory, we have just completed a study showing that brief exposures of early fruit fly *(Drosophila)* embryos to weak magnetic fields result in a high proportion of characteristic body pattern abnormalities in the larvae hatching 24 hours later.[36] As the energies involved are below the thermal threshold, there can be no significant effect unless there is a high degree of cooperativity or coherence in the pattern determination processes reacting to the external field. A wide-ranging review of many similar observations since the 1970s is in a recent book by Becker.[37]

Fröhlich's hypothesis of coherent excitations in living systems[34] has received independent support from research on light emission carried out by German quantum physicist and biophysicist Fritz-Albert Popp and his colleagues.[38] They show how the characteristics of light emitted from diverse organisms point to the existence of a coherent photon field underlying living organization. In a very real sense, we are all beings of light. "Light" here refers to electromagnetic radiation in the entire spectrum, from the optical region (that which we normally call light) to electromagnetic fields of one hertz and below. Not only are organisms fields of coherent light waves that organize their activities, but they are literally immersed in a sea of light consisting of the cosmic radiation background, the sun's radiant energy and Earth's fields, to which they are attuned and through which they receive electromagnetic signals from and transmit signals to other organisms, possibly to all other organisms. This vision is completely consonant with that presented by Whitehead[10] and Bohm[17] described earlier. But I am jumping ahead.

Quantum Coherence

A key notion in the new perspective of living organization is coherence. So I shall start from first principles and try to explain what it is.

Coherence in ordinary language means correlation, a sticking together, or connectedness; also, a consistency in the system. So we refer to people's speech or thought as coherent if the parts fit together well, and incoherent if they are uttering meaningless nonsense, or presenting ideas that don't make sense as a whole. Thus, coherence always refers to wholeness. However, in order to appreciate its full meaning, it is necessary to make incursions into its quantum physical description, which gives us some insights that are otherwise not accessible from a common-sensible, literal description.

Let us begin with Young's two-slit experiment[39] (named after English polymath Thomas Young (1773–1829) who first performed it at the beginning of the 19th century. The experiment played a major role in the general acceptance of the wave theory of light. In fact, it is also a key experiment demonstrating the wave-particle duality of light — light being simultaneously both wave and particle — and provides a rigorous definition of quantum coherence.[6]

The experiment is illustrated in Fig. 2.1. A source of monochromatic light is placed in front of a screen that has two narrow slits *t* and *b*. The light passes through the slits and on to a photographic plate.

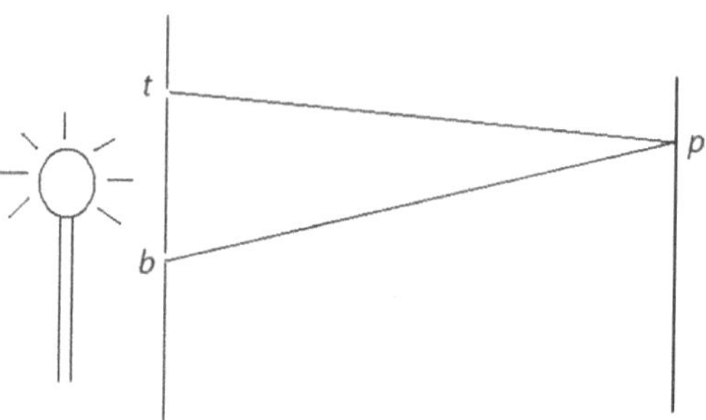

Figure 2.1 *The two-slit experiment.*

As is now well known, light behaves as either particle or wave according as to whether one or both slits are opened. When both slits are opened, even single photons behave as waves in that they pass through both slits at once, and, falling upon a photographic plate, form a characteristic interference pattern. The intensity or brightness of the pattern at each point depends on a "probability" that light falls on to that point.

The "probability" is placed between quotation marks because it is not probability in the ordinary sense. One way of representing those probabilities is as correlation functions consisting of the product of two complex amplitudes.

Light arriving at the point *p* on the photographic plate has taken different paths, *tp* and *bp*. The intensity at *p* is given as the sum of four such correlation functions:

$$I = G(t, t) + G(b, b) + G(t, b) + G(b, t) \tag{1}$$

where $G(t, t)$ is the intensity with only top slit opened, $G(b, b)$ the intensity with only bottom slit opened, and $G(t, b) + G(b, t) = 2G(t, b)$ is the additional intensity when both slits are opened.

At different points on the photographic plate, the intensity is

$$I = G(t, t) + G(b, b) + 2G(t, b) \cos\theta \tag{2}$$

where θ is the angle of the phase difference between the two light waves.

The fringe contrast in the interference pattern formed on the photographic plate depends on the magnitude *of* $G(t, b)$. If this correlation function vanishes, it means that the light coming out of t and b are uncorrelated; and if there is no correlation, we say that the light at t and b is incoherent. On the other hand, increase in coherence results in an increase in fringe contrast, that is, the brightness of the bands. As $\cos\theta$ is never greater than one (when the two beams are perfectly in phase), then the fringe contrast is maximized by making $G(t, b)$ as large as possible and that signifies maximum coherence. But there is an upper bound to how large $G(t, b)$ can be. It is given by the so-called Schwarz inequality:

$$G(t, t)\, G(b, b) \geq |G(t, b)|^2 \tag{3}$$

The maximum of $G(t, b)$ is obviously obtained when the two sides are equal:

$$G(t, t)\, G(b, b) = |G(t, b)|^2 \tag{4}$$

It is this equation that gives us the criterion of quantum coherence. A field is coherent at two space-time points, say, t and b, if the above equation is true. Furthermore, we have a coherent field if this equality holds for all space-time points, X_1 and X_2. This coherence is referred to as first-order coherence because it refers to correlation between two space-time points, and we write it more generally as

$$G_{(1)}(X_1, X_1)\, G_{(1)}(X_2, X_2) = |G_{(1)}(X_1, X_2)|^2 \tag{5}$$

The above equation tells us that paradoxically the correlation between two space-time points in a coherent field factorizes, or decomposes, neatly into the self-correlations at the two points separately, and that this *factorizability* or decomposability is a sufficient condition for coherence. What it means is that any two points in a coherent field behave statistically independently of each other. If we put two photon detectors in this field, they will register photons independently of each other. (Note that although they appear independent, there is strong synergy due to positive or constructive interference; the central bright fringe is four times the intensity of a single source.)

Coherence can be generalized to arbitrarily higher orders, say, to n approaching ∞, in which case we shall be talking about a fully coherent field. If n^{th} order coherence holds, then all of the correlation functions which represent joint counting rates for m-fold coincidence experiments (where $m < n$) factorize as the product of the self-correlations at the individual space-time points. In other words, if we put n different counters in the field, they will each record photons in a way that is statistically independent of all the others with no special tendency towards coincidences, or correlations. The above description is simplified from an account[40] given by American quantum physicist Roy Glauber (who was subsequently awarded the 2005 Nobel Prize in physics for this work).

A coherent state is thus one of maximum global cohesion and also maximum local freedom! Nature presents us a deep riddle that compels us to accommodate seemingly polar opposites. What she is telling us is that coherence does not mean uniformity, where everybody must be doing the same thing all the time. An intuitive way to think about it is in terms of a symphony orchestra or a grand ballet. Or, much better yet, a jazz band where every individual is doing his or her own thing, but keeping in tune or in step with the whole. This is precisely the biochemical picture we now have of the living state: micro-compartments and microdomains, right down to molecular machines, all functioning autonomously, doing very different things at different rates, yet in step with the whole organism.

As Popp pointed out to me,[41] factorizability optimizes communication by providing an apparently uncorrelated (yet actually correlated) network of space-time points that can be modulated instantaneously by specific signals. Furthermore, it provides the highest possible fringe contrast (or visibility) for pattern recognition, which may account for the great specificities in the response of organisms to diverse stimuli. The factorizability of coherent fields may also underlie the efficiency of bioenergetic processes in two respects. First, it concentrates the highest amount of energy of the same field (by constructive interference) as well as creating effectively field-free zones within the field (by destructive interference). Second, as the higher order correlations (correlations of many points) are the smallest in a completely coherent field, so the tiniest possible amount of energy is subject to correlated transfer between an arbitrarily large numbers of space-time points in the field with minimum loss.

One of the ways of achieving coherence is to occupy only one mode, which is precisely the sort of thing that happens in superconductivity and superfluidity. That "mode" does not have to be a single frequency; it can be a broadband of frequencies, such as are found in biological organisms. But as mentioned earlier, in order to constitute one "mode", these frequencies have to be coupled together or intercommunicating. It means energy fed into one frequency must be capable of being propagated to all the frequencies. Another characteristic of the coherent field is that it is fluctuationless or noiseless; the sort that any communication engineer working at radio-frequencies, for example, would say is coherent.

In summary, quantum coherence is capable of explaining many of the most distinctive properties of living systems. Next, I review some of the evidence for coherence from work on biophotons and other areas of research.

Biophotons and Coherence in Living Systems

Practically all organisms emit light at a steady rate that varies from a few photons per cell per day to several photons per organism per second.[42] The emission of biophotons, as they are called, is strongly correlated with the functional states of the organisms, and responds to many external stimuli or stresses (see below). It shows a highly nonlinear response to temperature characteristic of many physiological processes. Spectral analyses of the emitted light reveal that it typically covers the entire range of optical frequencies, with approximately equal distribution of photons throughout the range. Such a distribution deviates markedly from the equilibrium (random) Boltzmann distribution, and is never found in non-living systems. It also suggests that the different frequencies are in fact intercommunicating, so that energy fed into any frequency is rapidly distributed to all other frequencies, precisely as predicted for a coherent state.

Biophotons can also be studied as rescattered emission after a brief exposure to light of different spectral compositions. It has been found without exception that the rescattered emission decays not according to an exponential function characteristic of non-coherent light, but according to a hyperbolic function which, Popp asserts, is a sufficient condition for a coherent light-field.[43] The hyperbolic function takes the general form,

$$x = A\,(t + t_0)^{-1/d} \tag{6}$$

where x is the light intensity, A and d are constants and t is time after light exposure. The light intensity is inversely proportional to time, which points to the existence of memory in the system and, hence, of time-measure.

This phenomenon can be intuitively understood as follows. In a system consisting of non-interacting molecules emitting at random, the energy of the emitted photons is lost completely to the outside, or converted into heat, which is the ultimate non-coherent energy. If the molecules are emitting coherently, however, the energy of the emitted photons is not completely lost. Instead, part of it is coherently coupled back or restored to the system. The consequence is that the decay is delayed, and follows typically a hyperbolic curve with a long tail. Other nonlinear forms of delayed decay kinetics can be predicted from a coherent field, such as oscillations, which are also frequently observed.

The typical hyperbolic and nonlinear decay kinetics is uniform throughout the visible spectrum, as evidenced both by the rescattering of monochromatic light, or light of restricted spectral compositions, and by the spectral analysis of the rescattered emission.[43] The rescattered emission also covers the same broad range of frequencies, and its spectral distribution may be maintained even when the system is perturbed to such an extent that the emission intensity changes over several orders of magnitude, suggesting that all the frequencies are in effect coupled together. Such observations are consistent with the idea that the living system is one coherent photon field far from equilibrium, with coherence simultaneously in the whole range of frequencies that are nonetheless coupled together to give a single degree of freedom as a *statistical* average.

As is made clear in the description of quantum coherence above, coherence does not mean uniformity, or that every part of the organism must be doing the *same* thing or vibrating with the same frequencies. There can indeed be domains of local autonomy such as that we know to exist in the organism. Furthermore, as argued in preceding sections, organisms have a space-time structure; hence any measurement of the degree of freedom performed within a finite time interval will deviate from the ideal of one. Another source of variation may arise because some parts of the system are in fact temporarily decoupled from the whole, and hence the degree of coherence will also reflect changes in the functional states of the system. Such a variation in the degree of coherence appears to be associated with the development of malignancy in cells (see below).

Evidence of Coherence from Other Areas of Research

Evidence for various aspects of coherence come from many areas of biological and biochemical research, although the researchers themselves do not recognize them as such. Frequency coupling has long been observed for biological rhythms,[44] which often show harmonic relationships with one another, as for example, the relationship between respiratory rhythm and heartbeat frequency. Similarly, the phenomenon of sub-harmonic resonance is well known in metabolic oscillations, where entrainment of the metabolic oscillators is obtained to external driving frequencies that are approximately integer multiples of the fundamental.[45] Recently, a gene has been isolated in *Drosophila,* mutations of which alter the circadian period. Remarkably, the wing-beat frequency of its love song is correspondingly speeded up or slowed down according to whether the circadian period is shortened or lengthened.[46]

Many organisms, tissues and cells show spontaneous oscillatory contractile activities that are coherent over large spatial domains with periods ranging from 10^{-1} second to minutes. Similarly, spontaneous oscillations in membrane potentials can occur in a wide range of "non-excitable" cells as well as in cells conventionally regarded as excitable, that is, neurons, and these range in frequencies from 10^{1} second to minutes, again involving entire cells or tissues (such as the stomach and the intestine).[47] Recent applications of supersensitive SQUID (superconducting quantum interference device) magnetometers to monitor electrical activities of the brain have revealed an astonishing repertoire of rapid coherent changes (in milliseconds) which sweep over large areas of the brain.[48] These and the observations on synchronous firing patterns (40 to 60 hertz) in widely separated areas of the brain recorded by conventional electrodes[49] are compelling neurobiologists to consider mechanisms that can account for such long-range coherence. The authors suggest that the synchronization of oscillatory response in spatially separate regions may serve as "a mechanism for the extraction and representation of global and coherent features of a pattern", and for "establishing cell assemblies that are characterized by the phase and frequency of their coherent oscillations" (pp. 336–337).

At the molecular level, muscle contraction is shown to occur in definite quantal steps that are synchronous over entire muscle fibres, and measurements with high-speed ultrasensitive instrumentation suggest that the contraction is essentially fluctuationless[50,51] (as characteristic of a coherent quantum field, see above). Similarly, the beating of cilia in mussels and other organisms also occurs in synchronized quantal steps with little or no fluctuation.[52]

One consequence of coherence is energy storage. Coherence is associated with a time and a volume over which phase correlation is maintained. The coherence time for a quantum molecular process is just the characteristic time interval over which energy remains stored in McClare's formulation of the second law[27] referred to above. Thus, in conformity with the second law of thermodynamics, the longer the coherence time τ, the more extended is the timescale over which efficient energy transfer processes can take place, provided that they are much less than τ, or in the quasi-equilibrium approximation if they take place slowly with respect to τ. In other words, efficient energy transfer processes can in principle occur over a wide range of timescales, depending on the coherence times in the system, which we shall return to later.

Long-Range Communication between Cells and Organisms

In considering the possibility that cells and organisms may communicate at long range by means of electromagnetic signals, Presman[35] points to some of the perennial mysteries of the living world: how do birds in a flock, or fish in a shoal, move so effortlessly and simultaneously as one? During emotional mobilization, the speed and strength of action of the organism are much greater than the normal working level. The motor nerve to the muscle conducts at 100 times the speed of the vegetative nerves, which are responsible for the activation of processes leading to the enhancement of the contractile activity of the muscles required in a crisis: adrenaline release, dilatation of muscular vessels and increase in the heart rate. Thus, it appears that the muscle receives the signals for enhanced coordinated action long before the signals arrive at the organs responsible for the enhancement of muscle activity! This suggests that there may be a system of communication that sends emergency messages simultaneously to all organs, including those not directly connected with the nerve network; the speed with which the system operates rules out all conventional mechanisms (see a similar assessment of the visual system presented elsewhere[53]). Furthermore, there are grounds for believing that electromagnetic signals are involved. Animals are highly sensitive to electromagnetic fields, which can also act as conditional stimuli for the elaboration of conditioned reflexes. Special electromagnetic receptors are present in animals, including humans. And electromagnetic signals of various frequencies, other than visible biophotons, can be recorded in the vicinity of isolated organs and cells, as well as close to entire organisms. In my laboratory, we have indeed recorded profuse electrical signals from fruit fly embryos (about 1 hertz to 30 hertz) during the earliest stages of development.[54] Thus, long-range communication between cells and organisms is a distinct possibility, given that they both emit and are sensitive to weak electromagnetic fields.

In a pioneering experiment investigating the photon-emission characteristics of normal and malignant cells, Dutch scientists D.H. Schamhart and Roland van Wijk found that while normal cells exhibit decreasing light re-emission with increasing cell density, malignant cells show a highly nonlinear increase with increasing cell density, suggesting long-range interactions between the cells are responsible for their differing social behaviour: the tendency of disaggregation in the malignant tumour cells as opposed to attractive long-range forces between normal cells.[55] The difference between cancer cells and normal cells may lie in their communicative capability, which in turn depends on their degree of coherence. The parameter *1/d* in the hyperbolic decay function (Eq. 6) can be taken as a measure of incoherence, as it is directly correlated with the inability of the system to reabsorb emitted energy coherently. This value was shown to increase with increasing cell density in the malignant cells, whereas that of normal cells decreased.

Similar long-range interactions between organisms have been demonstrated in *Daphnia*[56] where the emission rate varies periodically with cell number in such a way as to suggest a relationship to average separation distances that are harmonics of the body size. In synchronously developing populations of early *Drosophila* embryos, we have recently discovered the remarkable phenomenon of super-delayed luminescence in which intense prolonged flashes of light are re-emitted with delay times of 20 minutes to eight hours after a single brief light exposure. These may result from cooperative interactions among embryos within the entire population, such that all the embryos re-emit in synchrony.[57] The long delays in re-emission also imply that coherence times for the process range from minutes up to eight hours.

Organisms are Coherent Space-Time Structures

We can now see that organisms are coherent space-time structures, and furthermore, that this is the basis of biological organization. Such an understanding also provides a solution to one of the perpetual riddles of life: what constitutes an individual or a "self"? The answer is that an individual is simply a field of coherent activities. Defined in this way, it readily opens the way to Whitehead's nested individualities,[10] but with the added insight that individualities are spatially and temporally fluid entities, in accordance with the extent of the coherence established. Thus, in long-range communication between cells and organisms, the entire community may become one when coherence is established and communication occurs without obstruction or delay. This is the ideal coherent society[3] which maximizes both global cohesion and individual freedom. Within the coherence time there is no space separation; that is, the usual spatial neighbourhood relationship becomes irrelevant. Similarly, within the coherence volume, there is no time separation, hence instantaneous (faster-than-light) communication can occur. Feelings can indeed spread "like wildfire", and people everywhere can get caught up simultaneously in a sudden fervour.

Quantum coherence is also a solution to the Einstein-Rosen-Podolsky paradox in which two originally correlated particles become spatially widely separated and yet seem able to communicate instantaneously. This solution is already implicit in Einstein's remarks (cited by American physicists Robert Jahn and Brenda Dunne (p. 746):[58]

> "If the partial systems A and B form a total system which is described by its psi-function ψ (*AB*), there is no reason why any mutually independent existence (state of reality) should be ascribed to the partial systems A and B viewed separately, *not even if the partial systems are spatially separated* from each other

> *at the particular time under consideration.* The assertion that, in this latter case, the real situation of *B* could not be (directly) influenced by any measurement taken on *A* is therefore, within the framework of quantum theory, unfounded and (as the paradox shows) unacceptable."

In other words, if the particles remain coherent, then they must be considered effectively as one system, despite spatial separation.

Within the living system, coherence times and coherence volumes are themselves determined by the relaxation times and volumes of the processes involved. We may envisage biological rhythms as manifesting a hierarchy of coherence times that define the time frames of the processes within the organism. This fits with Bergson's concept of pure duration, which we may now identify as the time taken for the completion of a process. A heartbeat requires a full cycle of contraction and relaxation before we recognize it as such, that is, the duration of a heartbeat, which is about one second in external reference time. In the brain, neurobiologists have recently discovered an endogenous 40-hertz rhythm that is coherent over the entire brain.[48] It is possible that this may define the duration of primary perception. Within that duration, which we can regard as the coherence time at that level of the nested hierarchy of time structure, processes coherent with it will generate no time at all. A similar argument should apply to the corresponding coherence volume; i.e., coherent processes will have no space separation wherever they may be occurring.

The representation of individuals as coherent space-time structures implies that space and time, in terms of separation and passage, are both generated, perhaps in proportion to the incoherencies of action. Thus, a coherent sage may well be living in a truly timeless-spaceless-state beyond ordinary comprehension. I believe some of us get glimpses of that in a particularly inspired moment, or during an aesthetic or religious experience, not unlike that achieved by Hesse's Rainmaker.[1]

In ordinary perception, on the other hand, the organism interacts with the environmental object and a perturbation propagates within the organism and is registered or assimilated in its physiology (enfolded). This results in time generation. The greater the wave function ψ (r, t) changes perhaps, the more time is generated. Conversely the more match or transparency there is between object and subject, the less time is generated.

Another consequence is that coherent states encompass regimes of simultaneity within which instantaneous communication can take place and spatial separation becomes irrelevant. This can occur within individuals, and, as we have seen, also between members of a population. Whole societies can, in principle, match coherent space-time structures and achieve instantaneous communication over long distances, transcending ordinary, common-sensible space-time. Perhaps unusual states of consciousness such as clairvoyance, telepathy and so-called extrasensory perception are due to such matching of coherent space-time structures.

But something further may be required for the emergence of what I have referred to as a coherent society, that is, a society that maximizes both global cohesion and local freedom, in accordance with the definition of quantum coherence. From the foregoing discussion, it is obviously a society where social space-time structure matches both natural space-time and individual private space-time. This has considerable relevance for the idea of convivial scales of machinery as well as communities and institutions due to Austrian philosopher Ivan Illich (1926–2002).[59]

Causality in a Universe of Coherent Space-Time Structures

The primary implication of the organicist view as represented by both Bergson[7] and Whitehead[10] is that causation is immanent to process, embodied in organisms. Causation is coextensive with being; it is both local to the organism and distributed in the community of other organisms in a mutuality of enfoldment and creative unfolding. It is not mediated by so many external forces acting on so many indifferent objects. This has deep implications for the issue of free will as opposed to determinism, which I can only touch upon here.

First of all, the positing of "self" as a domain of coherent space-time structure implies the existence of active agents who are free; "free" in this context means being true to self, in other words, being coherent. A free act is thus a coherent act. Of course not all acts are free, as one is seldom coherent. Yet the mere possibility of being unfree affirms the opposite: that freedom is real. As Bergson puts it (p. 172),[7]

> "... we are free when our acts spring from our whole personality, when they express it, when they have that indefinable resemblance to it which one sometimes finds between the artist and his work."

As the self is distributed — being implicated in a community of other entities — to be true to self does *not* imply acting against others. On the contrary, sustaining others sustains oneself, so being true to others is also being true to self. It is only within a mechanistic Darwinian perspective that freedom becomes perverted into acts against others.[3]

According to British philosopher and economist John Stuart Mill (1806–1873), to be free "must mean to be conscious, before I have decided that I am able to decide either way."[60] So defenders of free will claim that when we act freely, some other action would have been equally possible. Conversely, proponents of determinism assert that given certain antecedent conditions, only one resultant action was possible.

The problem itself is posed on the mechanistic assumption of immobility and mutual externality of events. This gives rise to two equally unacceptable alternatives: either that an immobile configuration of antecedents "determines" another immobile configuration of resultants, or that, at any frozen instant, to be or not to be are equally likely choices for a consciousness that is external to itself, a process that immediately leads us back to Cartesian mind-matter dualism, which makes us strangers to ourselves.

In the reality of process, where the self is ever becoming, it does not pass like an automaton from one frozen instant to the next — instead, the quality of experience permeates the whole being in a succession without separateness, as Bergson[7] says in "a self which *lives* and develops by means of its very hesitations, until the free action drops from it like an overripe fruit" (p. 176).

One might represent consciousness as a wave function that evolves, constantly being transformed by experience and overt acts. The issue of quantum indeterminism is a very deep one, but the picture of a wave function — a pure state — consisting of a total interfusion of feelings, each of which occupies the whole being, is precisely what Bergson[7] describes. Such a pure state cannot be resolved or factorized into a mixture of states, except under certain ideal conditions. Thus, the overt act, or choice, does follow from the antecedent, but it cannot be predicted in advance. One can at best retrace the abstract "steps" and represent the evolution of the consciousness as having followed a "trajectory". In truth, the so-called trajectory was traced out by one's own actions, both overt and covert up to that point.

Moreover, when one reinstates the full quality of our consciousness, we can see that there can be no identical or repeatable states, which, when presented again at any time, will bring about identical resultant states. The "wave function" of consciousness is always changing and always unique, as it is coloured by all the tones of our personality. Thus each of us loves and hates in peculiar ways different on every occasion; yet language makes no distinction from one to the other. Only by the efforts of the great novelists and artists can we recover the plenum that is life and reality.

American psychologist James Gibson (1904–1979), a chief exponent of process ontology in perception, has this to say on consciousness (pp. 134–138):[61]

> The stream of consciousness does not consist of an instantaneous present and a linear past receding into the distance; it is not a 'travelling razor's edge' dividing the past from the future. Perhaps the present has a certain duration. If so, it should be possible to find out when perceiving stops and remembering begins. But it has not been possible ... A perception in fact, does not have an end. Perceiving goes on."

Nature is ever-present to us, as we are to ourselves. This ever-presence is structured, as we have seen. Our experience consists of the catenation of events of different durations, which propagate and reverberate in and around our being, constantly being registered and recreated. What constitutes memory of some event is the continuing present for the overarching process of which the event is part.[62]

The universe of coherent space-time structures is a nested hierarchy of individualities and communities that come into being through acts of prehensive unification. Just as the organism is ever-present to itself during its entire life history, and all other individualities are ever-present to it, the universe is ever-present to itself in the universal duration where creation never ceases by the convocation of individual acts, now surfacing from the energy substrate, now condensing to new patterns, now submerging to re-emerge in another guise.

Reality is thus a shimmering presence of infinite planes, a luminous labyrinth of the active now connecting "past" and "future", "real" with "ideal", where potential unfolds into actual and actual enfolds to further potential through the free action and intentions of the organism. It is a sea awash with significations, dreams and desires. This reality we carry with us, an ever-present straining towards the future. The act is the cause; it is none other than the creation of meaning, the realization of the ideal and the consummation of desire.[6]

References and Notes

1. Hesse, H. *Magister Ludi: The Glass Bead Game,* translated by R. and C. Winston. Bantam Books, 1943.
2. "Isaac Newton", Wikipedia, accessed 13 August 2015, https://en.wikipedia.org/wiki/Isaac_Newton.
3. Ho, M.W. "Natural Being and Coherent Society", 1992. This paper was intended to be part of *Social and Natural Complexity* (special issue of *The Journal of Social and Biological Structures*), edited by E.L. Khalil and K.E. Boulding, but was rejected by the co-editor

after Kenneth Boulding's death in 1993. The article went through several versions before it was finally published under the same title in *Gaia in Action* (P. Bunyard, ed.), pp. 286–307, Edinburgh, Floris Books, 1996.

4. Ho, M.W. "Reanimating Nature: The Integration of Science with Human Experience". *Beshara Magazine* 8 (1989a): 16–25; reprinted with minor modifications in *Leonardo* 24(5)(1991): 607–615.
5. Ho, M.W. "A Quest for Total Understanding". In *Learning without Limits: The Dilemma of Knowledge, transcript*, pp. 47–66, Saros *Seminars*, Saros Publications, 1990.
6. See Ho, M.W. *The Rainbow and the Worm: The Physics of Organisms*, World Scientific, Singapore, 1993.
7. Bergson, H. *Time and Free Will: An Essay on the Immediate Data of Consciousness*, translated by F.L. Pogson, George Allen & Unwin, London, 1916.
8. Existentialism, The Basics of Philosophy, accessed 21 August 2015, http://www.philosophybasics.com/branch_existentialism.html.
9. Eliot, T.S. "The Hollow Men", http://allpoetry.com/The-Hollow-Men.
10. Whitehead, A.N. *Science and the Modern World*, p. 611. Fontana Books, London, 1925.
11. For example, Emmet, D. *The Effectiveness of Causes*, State University of New York Press, New York, 1984.
12. See Ho, M.W. "How Rational Can Rational Morphology Be? A Post-Darwinian Rational Taxonomy Based on a Structuralism of Process". *Rivista di Biologica* 81 (1988): 11–55.
13. Bunyard, P. *The Colombian Amazon: Policies for the Protection of the Indigenous Peoples and Their Environment*, p. 68. The Ecological Press, London, 1989.
14. "Qi", Wikipedia, accessed 19 December 2015, https://en.wikipedia.org/wiki/Qi.
15. Lovelock, J.E. *Gaia: A New Look at Life on Earth*. Oxford University Press, Oxford, 1979.
16. Lovelock, J.E. *The Ages of Gaia*, Oxford University Press, Oxford, 1988.
17. Bohm, D. "Hidden Variables and the Implicate Order". In *Quantum Implications: Essays in Honor of David Bohm* (B.J. Hiley

and F.D. Peat, eds.), pp. 33–45, Routledge and Kegan Paul, London, 1987.
18. Goodwin, B.C. *Temporal Organization in Cells.* Academic Press, London, 1963.
19. See Bard, J. *Morphogenesis: The Cellular and Molecular Processes of Developmental Anatomy.* Cambridge University Press, Cambridge, 1990.
20. See Morse, D.S., Fritz, L. and Hastings, J.W. "What Is the Clock? Translational Regulation of Circadian Bioluminescence". *TIBS* 15 (1990): 262–265.
21. Prigogine, I. *Introduction to Thermodynamics of Irreversible Processes.* Wiley, London, 1967.
22. Williams, R.J.P. "On First Looking into Nature's Chemistry. Part I: The Role of Small Molecules and Ions; the Transport of the Elements. Part II: The Role of Large Molecules, Especially Proteins". *Chem Soc Rev* 9(3) (1980): 281–324, 325–364.
23. Whitehead, A.N. *Concept of Nature.* Cambridge University Press, Cambridge, 1920.
24. Ho, M.W. *The Rainbow and the Worm: The Physics of Organisms*, 3rd edition. World Scientific, Singapore and London, 2008.
25. Kushmerick, M.H., Larson, R.E. and Davies, R.E. "The Chemical Energetics of Muscle Contraction. I. Activation Heat, Heat of Shortening and ATP Utilization for Activation-Relaxation Processes". *Proc Roy Soc Lond* B174 (1969): 293–313.
26. Schrödinger, D. *What Is Life?* Cambridge University Press, Cambridge, 1944.
27. McClare, C.W.F. "Chemical Machines, Maxwell's Demon and Living Organisms". *J Theor Biol* 30 (1971): 1–34.
28. McClare, C.W.F. "A 'Molecular Energy' Muscle Model". *J Theor Biol* 35 (1972): 569–575.
29. Hibbard, M.G., Dantzig, J.A., Trentham, D.R. and Goldman, V.E. "Phosphate Release and Force Generation in Skeletal Muscle Fibres". *Science* 228 (1985): 1317–1319.
30. Fleming, G.E., Martin, J.L. and Breton, J. "Rate of Primary Electron Transfer in Photosynthetic Reaction Centers and Their Mechanistic Implications". *Nature* 333 (1988): 190–192.

31. Klinman, J.Y. "Quantum Mechanical Effects in Enzyme: Catalyzed Hydrogen Transfer Reactions". *TIBS* 14 (1989): 368–373.
32. Schommers, W. "Space-Time and Quantum Phenomena". In *Quantum Theory and Pictures of Reality* (W. Schommers, ed.), pp. 217–277, Springer·Verlag, Berlin, 1989.
33. Szent-Györgyi, A. *Introduction to a Submolecular Biology*, Academic Press, London, 1960.
34. Fröhlich, H. "The Biological Effect of Microwaves and Related Questions". *Adv Electron El Phys* 53 (1980): 85–152.
35. Presman, A.S. *Electromagnetic Fields and Life.* Plenum Press, New York, 1970.
36. Ho, M.W., Stone, T.A., Jerman, I., Bolton, J., Bolton, H., Goodwin, B.C., Saunders, P.T. and Robertson, F. "Brief Exposure to Weak Static Magnetic Fields During Early Embryogenesis Cause Cuticular Pattern Abnormalities in Drosophila Larvae". *Phys Med Biol* 37 (1992): 1171–1179.
37. Becker, R.O. *Cross Current: The Promise of Electromedicine, The Perils of Electropollution.* Jeremy P. Tarcher, Inc., New York, 1990.
38. Popp, F.A., Li, K.H., Mel, W.P., Galle, M. and Neuroh, R. "Physical Aspects of Biophotons". *Experientia* 44 (1988): 1576–1585.
39. "Young's Interference Experiment", Wikipedia, accessed 31 May 2015, https://en.wikipedia.org/wiki/Young%27s_interference_experiment.
40. Glauber, R.J. "Coherence and Quantum Detection". In *Quantum Optics* (R.J. Glauber, ed.), pp. 15–60, Academic Press, London, 1969.
41. This Insight is due to Fritz-Albert Popp in Popp, F.A. and Ho, M.W. *Light and Life*, 1991 (unpublished manuscript).
42. For a fuller description of the characteristics of biophotons, see Popp, F.A. "On the Coherence of Ultraweak Photoemission from Living Tissues". In *Disequilibrium and Self-Organization* (C.W. Kilmister, ed.), pp. 207–230, D. Reidel, Lancaster, 1986.
43. For a detailed analysis of hyperbolic decay kinetics in different parts of the light spectrum, see Musumeci, M., Godlewski, M., Popp, F.A. and Ho, M.W. "Time Behavior of Delayed Luminescence in *Acetabularia acetetabulum*". In *Recent Advances in Biophoton*

Research and Its Applications (F.A. Popp, K.H. Li and Q. Gu, eds.), pp. 327–344, World Scientific, Singapore, 1992.

44. Breithaupt, H. "Biological Rhythms and Communications". In *Electromagnetic Bioinformation,* 2nd edition (F.A. Popp, H. Warnke, H.L. König and W. Peschka, eds.), pp. 18–41, Urban & Schwartzenberg, Vienna and Berlin, 1989.
45. Hess, G. "The Glycolytic Oscillator". *J Exp Biol* 81 (1979): 7–14.
46. Kyriacou, C.B. "The Molecular Ethology of the Period Gene". *Behavioral Genetics* 20 (1990): 191–211.
47. Berridge, M.J., Rapp, P.E. and Treherne, J.E., eds. *Cellular Oscillators.* Cambridge University Press, Cambridge, 1979.
48. Ribary, U., Ionnides, A.A., Singh, K.D., Hasson, R., Boiton, J.P.R., Lido, F., Mogilner, A. and Llinas, R. "Magnetic Field Tomography (MFT) of Coherent Thalamocortical 40hz Oscillations in Humans. *Proc Natl Acad Sci* 88 (1991): 11037–11041.
49. Gray, C.M., Konig, P., Engel, A.K. and Singer, W. "Oscillatory Responses in Cat Visual Cortex Exhibit Inter-Columnar Synchronization Which Reflects Global Stimulus Properties". *Nature* 338 (1989): 334–337.
50. Granzier, H.L.M., Myers, J.A. and Pollack, G.H. "Stepwise Shortening of Muscle Fibre Segments". *J Muscle Res Cell Motil* 8 (1987): 241–251.
51. Iwazumi, T. "High Speed Ultrasensitive Instrumentation for Myofibrll Mechanics Measurements". *Am J Physiol* 252 (1987): 253–262.
52. Baba, S.A. "Regular Steps in Bending Cilia During the Effective Stroke". *Nature* 282 (1979): 717–772.
53. Ho, M.W. "Coherent Excitations and the Physical Foundations of Life". In *Theoretical Biology* (B.C. Goodwin and P.T. Saunders, eds.), pp. 162–176, Edinburgh University Press, Edinburgh, 1989b.
54. Ho, M.W., Ross, S., Bolton, H., Popp, F.A. and Li, K.H. "Electrodynamic Activities and Their Role in the Organization of Body Pattern". *JSE* 6 (1992): 59–77.
55. Schamhart, S. and van Wijk, R. "Photon Emission and Degree of Differentiation". In *Photon Emission from Biological Systems*

(B. Jezowska-Trzebiatowki, B. Kochel, J. Slawinski and W. Strek, eds.), pp. 137–150, World Scientific, Singapore, 1986.

56. Galle, M., Neurohr, R., Allman, G. and Nagl, W. "Biophoton EmIssion from *Daphnia magna*; A Possible Factor in the Self-Regulation of Swarming". *Experientia* 47 (1991): 457–460.
57. Ho, M.W., Popp, F.A., Xu, X., Ross, S. and Saunders, P.T. "LIght Emission and Rescattering in Synchronously Developing Populations of Early *Drosophila* Embryos: Evidence for Coherence of the Embryonic Field and Long Range Cooperativity". In *Biophotons Research and Coherence in Biological Systems* (F.A. Popp, ed.), pp. 287–306, World Scientific, Singapore, 1992.
58. Cited in Jahn, R.B. and Dunne, B.J. "On the Quantum Mechanics of Consciousness, with Application to Anomalous Phenomena". *Found Phys* 16 (1986): 721–772.
59. Illich, E. *Tools for Conviviality.* Fontana, Holland, 1973.
60. Mill, J.S. *Examination of Sir W. Hamilton's Philosophy,* 5th edition, pp. 580–583, 1878; cited in Bergson, H. *Time and Free Will: An Essay on the Immediate Data of Consciousness,* translated by F.L. Pogson, p. 174, George Allen & Unwin, London, 1916.
61. Gibson, J.J. *The Ecological Approach to Visual Perception.* MIT Press, Cambridge, Massachusetts, 1966.
62. Ho, M.W. "The Role of Action in Evolution: Evolution by Process and the Ecological Approach to Perception". In *Evolutionary Models in the Social Sciences: Cultural Dynamics* (T. Ingold, ed.), pp. 336–354, Brill, Leiden, 1991.

3

Organism and Psyche in a Participatory Universe

The organism is the most universal Jungian archetype in which love, wholeness and creativity feature prominently; a theory of the organism based on quantum coherence is in some respects a microcosm of the quantum holographic universe, and it has profound implications for our global future.

Organism, the Universal Archetype and Love

In the summer of 1991, I saw something in Mexico City that haunted me for months afterwards. It was a thick round slab of sculpted rock, about 3.25 metres in diameter. The official guide book says it depicts the Aztec moon goddess, embodying the powers of night, who was killed and gruesomely dismembered by her brother the sun god, an act so terrible that the world itself was torn asunder. Yet, the beautifully executed symmetries of the form evoke a sense of the dismembered parts drawing together again to make a whole, counteracting the violent severance of head and limbs. Musician and composer Mazatl Galindo,[1] who teaches indigenous American cultures and is himself of Aztec Indian descent, has since explained to me that this sculpted disc, which has the same dimensions as the much better known and widely reproduced calendar stone, is actually also a calendar: the 13 main joints of her dismembered body representing the 13 divisions of the year. The alternating disintegration and reintegration it evokes signify the cycles of death and rebirth that mark the passage of time.

I came upon the sculpture while accompanying a group of university undergraduates travelling around the world on an intensive, year-long education programme: Global Ecology — Integrating Nature and Culture (of which I was a founding faculty member). In the course of the year and throughout the Third World, we had experienced the same distressing disintegration of the environment and indigenous communities brought on by industrial developments; and yet there remains everywhere an indestructible, irrepressible spirit to make things whole again. It was not just a survival instinct, but a genuine lust for life — the psychic energy that created the calendar stone is at work, initiating the healing process even as disintegration is continuing apace. The meaning of that year's journey and the journey of my life as symbolic of life itself came to me like an avalanche. I have died several deaths since my encounter with that symbol. I found myself standing at the gates of the underworld, as Orpheus[2] must have done, torn between the fear of impending hell and the overriding need to recover a lost love. Eventually, it transformed my life, in much the way that German psychologist and founder of analytical psychology Carl Jung (1875–1961) has envisaged the transforming power of symbols.[3]

Love rules our lives on many planes. Scottish psychologist Ian Suttie (1898–1935), a critic of German psychoanalyst Sigmund Freud (1856–1939), whose theory accorded sexual desire as the main drive to psychic development,[4] proposed instead that love, as distinct from sex, is primary for all social organisms.[5] Love comes from the nurturing ministrations of the mother or caretaker during infancy. From this arises a feeling of tenderness that regards all people to be possible companions, to be enjoyed and loved, and from whom approval is sought. On another plane, the successful separation of child from mother creates a field of attraction, a "virtual space" of love which we fill with our social and creative activities, as English paediatrician and psychoanalyst Donald Winnicott (1896–1971) wrote.[6]

Love and Wholeness

Love is a desire for wholeness. It is a desire for resonance, for intimacy, a longing to embrace and complete a larger whole. And it is that which motivates our social and creative acts and our knowledge of nature on the most universal plane.[7] At its most restricted and personal, love is our affection for specific human beings. But it is also one's own process of individuation, of remaking one's "self" out of the fabric of experiences, transcending the well-worn archetypes to become a unique whole person. The whole person is one whose sense of uniqueness is premised on her relationship with all of nature. Thus, the personal and universal are inextricably intertwined. The most intimate knowledge of oneself is at the same time, the most profound knowledge of nature.

The true love of self is inextricably the love of humanity and of all nature. That is why we feel obliged to serve, to help, to alleviate suffering and pain just as they were our own. I suggest that great scientists like David Bohm (1917–1992), Ervin Laszlo, and others are indeed trying to recover that lost love, the universal wholeness and entanglement that enables us to empathize and to be compassionate.

The whole is never static; it is constantly dying and rebirthing, decaying and renewing, breaking down to build up again. The same cycles of disintegration and reintegration occur whether one is looking at the energy metabolism of our body or the stream of consciousness out of which we individuate our psyche. During the normal "steady state" of our existence, the multitudes of infinitesimal deaths and rebirths are intricately balanced so that the old changes imperceptibly into the new. However, whenever the attracting centre of the new is radically different from the old, a larger, and at times, complete, disintegration may be needed before the new can individuate. It is like the caterpillar that must completely dissolve so that the beautiful butterfly can emerge. That is our hope for the approaching millennium.

Organism and Psyche

The psyche has so much in common with the organism that many of the most perceptive biologists and psychologists have proposed a complete continuity and identity between the two. They are impressed with the "directiveness" of all vital processes, whether developmental, physiological or psychical. In development, the fertilized egg goes through a series of morphogenetic changes directed towards producing the adult organism, and is remarkably resistant to disturbing influences. Similarly, the organism is able to maintain its internal physiology in a constant state despite large changes in the external environment. So it is with the purposiveness of all living things. One has only to try to stop a cat from doing what it wants to do. The mark of a living being is that it always has its own way of doing things, its own directed purpose in life that resists what is imposed on it. It is not at the mercy of its surroundings. It is so even for the simplest unicellular organism. American biologist Herbert Spencer Jennings (1868–1947) took a lifetime to study the ciliate protozoa *Paramecium*, and became convinced of its purposiveness, its autonomy at the very least.[8] For example, it will swim towards the light, or not, according as to whether it is hungry or fully fed. Geneticist and botanist Edmund Ware Sinnott (1888–1968), also American, argued in his book *Cell and Psyche*[9] that biological organization, concerned with development and physiology, and psychical activity, concerned with behaviour and leading to mind, are fundamentally the same thing (p. 10):

> "In some unexplained fashion, there seems to reside in every living thing ... an inner subjective relation to its bodily organization. This has finally evolved into what is called consciousness ... through this same inner relationship, the mechanism which guides and controls vital activities towards specific ends, the pattern or tension set up in protoplasm, which so sensitively regulates its growth and behaviour, can also be experienced, and this is the genesis of desire, purpose, and all other mental activities."

To me, the Jungian ideal of the whole person is also one whose cell and psyche, body and mind, inner and outer, are fully integrated, and hence completely in tune with nature. That may be the secret of Jung's "golden flower",[10] the immortal spirit-body created out of the resolution of opposites, the intertwining of darkness and light (moon goddess and sun god) that is the essence of life itself. The encounter with the Aztec calendar stone is the immediate prelude to my work towards a theory of the organism, much of which is in *The Rainbow and The Worm* written almost a year later.[11] This was after months of extraordinarily vivid lucid dreams of flying through the air to ancient cities, being transported down into the depths of pyramids, riding on the back of the Jaguar and a most beautiful erotic encounter with the dazzling white Plumed Serpent.

Jung's ideas on psychical development show many parallels to those relating to living organization, and have since been borrowed back into biology. "Individuation", for example, has been used by British embryologist/geneticist C.H. Waddington (1905–1975) to describe the process of forming a whole, or a whole organ, such as a limb from the global morphogenetic field.[12] Jung himself was not unaware of these parallels when he presented the psyche as a dynamic, self-regulating system, motivated by psychic energy or *libido*, a general sense of desire or longing, an urge that flows between opposite poles, so that the stronger the opposition the greater the tension.[10] The allusion to the living system and energy flow is unmistakable. Jung's theory of the psyche, drawn largely from his own experiences and imagination, is also a theory of the organism. The organism is the most universal Jungian archetype. Similarly, Laszlo's theory of the quantum holographic universe views the universe effectively as a kind of super-organism, constantly becoming, being created through the activities of its constituent organisms at every level.[13,14] These activities leave traces (quantum interferences) in the universal vacuum field that feedback on the future evolution of the organisms themselves. The universal quantum holographic field is the collective consciousness (including the unconscious) of *all* organisms. My theory of the organism is in some respects, a microcosm of Laszlo's quantum holographic universe.

The Organism's Irrepressible Tendency Towards the Whole

What is it to be an organism? It is, at bottom, the irrepressible tendency towards being whole. It is that which underlies both the directedness of vital activities and the love we express on many planes. In biological development, the most characteristic feature of the embryo is not so much its directedness towards producing an adult organism or any archetype; rather it is its tendency to maintain and develop into an organized *whole*, however it is disturbed. Sometimes, this organized whole is so altered that it is no longer recognizable as the same organism, but it is nonetheless an organism in the sense of being an organized whole.

More significantly, there is a special relationship between part and whole in the organism. The egg starts to develop by cell division. At a sufficiently early stage, the cells in the embryo are typically *totipotent*, in that they have the potential to develop into any part of the whole. When they are separated, each cell can develop into a whole organism, albeit a much smaller one than the original. Similarly, if a part of the early embryo is removed, that part can be regenerated from the remaining so that the whole is again recovered. German biologist Hans Driesch (1867–1941) discovered these phenomena in his extensive experiments in embryology, from which he proposed his neo-vitalist philosophy of entelechy, the tendency towards completing a whole that living things possess.[15]

Regeneration can also occur in adult organisms of some species such as the salamander. It is part and parcel of the healing process that enables all organisms to recover from illnesses and injuries. Whole and part are therefore mutually implicated in the organism. This quality of organic wholeness has eluded mechanistic science right from the beginning, and has been the main sticking point of the debate between the mechanists, who believe that life can in principle be explained in terms of mechanical physics and chemistry, and their opponents the vitalists who held that living things contain a vital principle irreducible to chemistry and physics. As I shall show, they are both mistaken; the answer lies in an organic materialist understanding based on quantum physics that transcends both.

Organic Space-Time versus Mechanical Space and Time

The mechanistic framework broke down at the turn of the 20th century, giving way to quantum theory at the very small scale of elementary particles and to general relativity at the large scales of planetary motion. In place of the static, eternal universe of absolute space and time, there is a multitude of contingent, observer-dependent space-time frames. Instead of solid objects with simple locations in space and time, one finds delocalized, mutually entangled quantum entities evolving like organisms. The opposition between the mechanistic and the organic worldview hinges on the fundamental nature of space and time.

Mechanical space and time are linear, homogeneous, separate and local. In other words, both are infinitely divisible, and every bit of space or of time is the same as every other bit. A billiard ball *here* cannot affect another one *there*, unless someone pushes the one here to collide with the one there. Mechanical space-time also happens to be the space and time of the "common-sensible" world in our mundane, everyday existence. It is the space-time of frozen instantaneity abstracted from the fullness of real process, rather like a still frame taken from a bad movie-film, which is itself a flat simulation of life. The passage of time is an accident, having no connection with the change in the configuration of solid matter located in space. Thus, space and time are merely coordinates for locating objects. One can go forwards or backwards in time to locate the precise objects at those particular points. In reality, we know that we can as much retrace our space-time to locate the person that was 30 or 50 years younger as we can undo the wrongs we have committed then. As so clearly argued by English mathematician and philosopher Alfred North Whitehead (1861–1947), there is no simple location in space and time.[16]

British psychoanalyst-artist Marion Milner (1900–1998) described her experience of "not being able to paint" as the fear of losing control, of no longer seeing the mechanical common-sensible separateness of things.[17] It is really a fear of being alive, of entanglement and process in the organic reality that ever eludes mechanistic description. And yet, it is in overcoming the imposed illusion of the separateness of things that the artist/scientist enters into the realm of creativity and real understanding, the realm of organic space-time. Mechanical physics has banished organic space-time from our collective public consciousness, though it never ceases to flourish in the subterranean orphic universe of our collective unconscious and our subjective aesthetic experience. In a way, all developments in Western science since Descartes and Newton may be seen as a struggle to reclaim our intuitive, indigenous notions of organic space-time, which, deep within our soul, we feel to be more consonant with authentic experience.

Organism versus Mechanism

The mechanistic worldview indeed officially ended at the beginning of this century. But the profound implications of this decisive break with the intellectual tradition of previous centuries were recognized by a mere handful of visionaries, especially by Whitehead mentioned earlier and by the French philosopher Henri Bergson (1859–1941).[18] Between them, they articulated an organicist philosophy in place of the mechanistic. Let me summarize some of what I see to be the major contrasts between the mechanical universe and the universe of organisms (see Box 3.1).

The contrasts are brought into sharper relief by considering the differences between mechanism and organism, or, more accurately, the opposition between a mechanical system and an organic system. First of all, a mechanical system is an object *in* space and time, whereas an organism is, in essence, *of* space-time. An organism creates its own space-times by its activities, so it has control over its space-time, which is not the same as external clock

Box 3.1 The mechanical and organic universes.

Mechanical Universe	**Organic Universe**
Static, deterministic	Dynamic, evolving
Separate, absolute space and absolute time for all observers	space-time inseparable, contingent observer(process)-dependent space-time frames
Inert objects with simple locations in space and time	Delocalized organisms with mutually entangled space-times
Linear, homogeneous space and time	Nonlinear, heterogeneous, multidimensional space-times
Local causation	Non-local causation
Given, non-participatory and hence, impotent observer	Creative, participatory; entanglement of observer and observed

time. Second, a mechanical system has a stability that belongs to a *closed* equilibrium, depending on controllers, buffers and buttresses to return the system to set, or fixed, points. It works like a non-democratic institution, by a hierarchy of control: a boss who sits in his office doing nothing (bosses are still predominantly male) except giving out orders to line managers, who in turn coerce the workers to do whatever needs to be done. An organism, by contrast, has a dynamic stability, which is attained in open systems far away from equilibrium. It has no bosses, no controllers and no set points. It is radically democratic; everyone participates in making decisions and in working by intercommunication and mutual responsiveness. Finally, a mechanical system is built of isolatable parts, each external and independent of all the others. An organism, however, is an irreducible whole, where part and whole, global and local are mutually implicated.

An even more significant change in worldview is the dissolution of the Cartesian barrier separating the observer from the observed.

In the quantum universe, observer and observed are mutually entangled, each act of observation determining the evolution of *both*. Knowledge, therefore, involves the full participation of the knower in the known. As the knower is an organism, she is also an actor who participates in constructing and shaping the universe, and *she does so knowingly*. There is, thus, no escaping from the responsibility of a participatory universe and the moral imperative of one's mutual entanglement, ultimately with all of nature. But let us begin with the central percept of being an organism.

A Theory of the Organism

There are about 75 trillion cells in our body, made up of astronomical numbers of molecules of many different kinds. How can this huge conglomerate of disparate cells and molecules function so perfectly as a coherent whole? How can we summon energy at will to do whatever we want? And most of all how is it possible for there to be a singular "I" that we all feel ourselves to be amid this diverse multiplicity?

To give an idea of the coordination of activities involved, imagine an immensely huge super-orchestra playing with instruments spanning an incredible spectrum of sizes from a piccolo of 10^{-9} metre up to a bassoon or a bass viol of a metre or more, and a musical range of perhaps *70 octaves*. The amazing thing about this super-orchestra is that it never ceases to play out our individual song lines, with a certain recurring rhythm and beat, but in endless variations that never repeat exactly. Always, there is something new, something made up as it goes along. It can change key, change tempo, change tune perfectly, as it feels like it, or as the situation demands, spontaneously and without hesitation. Furthermore, each and every player, however small, can enjoy maximum freedom of expression, improvising from moment to moment, while maintaining in step and in tune with the whole.

I have just described a theory of the *quantum coherence* that underlies the radical wholeness of the organism, which involves total participation, maximizing *both* local freedom and global cohesion. It involves the mutual implication of global and local, of part and whole, from moment to moment. It is on that basis that we can have a sense of ourselves as a singular being, despite the diverse multiplicity of parts. That is also how we can perceive the unity of the here and now, in an act of "prehensive unification" according to Whitehead.[16] Artists, like scientists, depend on the same exquisite sense of prehensive unification, to see patterns that connect apparently disparate phenomena. To add corroborative details to the theory of the organism, I shall give a more scientific narrative beginning with energy relationships.

Thermodynamics of Organized Complexity

Textbooks tell us that living systems are open systems dependent on energy flow. Energy flows in together with materials, and waste products are exported, as well as the *spent* energy that goes to make up *entropy*. And that is how living systems can, in principle, escape from the second law of thermodynamics. The second law, as you may know, encapsulates the fact that all physical systems run down, ultimately decaying to homogeneous disorganization when all useful energy is spent, or converted into entropy. But how do living systems manage their anti-entropic existence?

I have suggested[19] that the key to understanding how the organism overcomes the immediate constraints of thermodynamics is in its capacity to *store* the incoming energy, and in somehow closing the energy loop within to give a reproducing, regenerating life cycle. The energy, in effect, circulates among complex cascades of coupled cyclic processes within the system before it is allowed to dissipate to the outside. These cascades of cycles span the entire gamut of space-times from slow to fast, from local to global, that all together, constitutes the life-cycle. Each cycle is a domain of *coherent* energy storage; coherent energy is simply energy that can do work because it is all coming and going together, as opposed to incoherent energy that goes in all directions at once and cancel out, and is therefore, quite unable to do work.

Coupling between the cycles ensures that the energy is transferred directly from where it is captured or produced, to where it is used. In thermodynamic language, those activities going thermodynamically *down*-hill, and therefore yielding energy, are coupled to those that require energy and go thermodynamically *up*hill. This coupling also ensures that *positive* entropy generated in some space-time elements is compensated by *negative* entropy in other space-time elements. There is, in effect, internal energy conservation as well as internal entropy compensation. The whole system works by reciprocity, a cooperative give and take that balances out over the system as a whole, and within a sufficiently long time.[19–21] Thus, there is always coherent energy available in the system, which can be readily shared throughout the system, from local to global and *vice versa*, from global to local. That is why, in principle, we can have energy at will, whenever and wherever it is needed. The organism has succeeded in gathering all the necessary vital processes into a unity of coupled non-dissipative cycles spanning the entire gamut of space-times up to and including the life-cycle itself, which effectively feeds off the dissipative irreversible energy flow. In thermodynamic terms, the living system can be represented as a superposition of cyclic non-dissipative processes, for which entropy production balances out to zero, $\Sigma\Delta S = 0$, and dissipative, irreversible processes, for which net entropy production is positive, $\Sigma\Delta S > 0$. (I later show that the net entropy production is also minimum, i.e., $\Sigma\Delta S > 0$, see "Chap. 14 — Sustainable Cities As Organisms: A Circular Thermodynamics perspective" in this book.)

But how can energy mobilization be so perfectly coordinated? That is a direct consequence of the energy stored, which makes the whole system *excitable*, or highly sensitive to specific weak signals. It does not have to be pushed and dragged into action like a mechanical system. Weak signals originating anywhere within or outside the system will propagate throughout the system and become automatically amplified by the energy stored, often into macroscopic action. Intercommunication can proceed very rapidly, especially because organisms are completely *liquid crystalline*.

The Liquid Crystalline Organism

In 1992, we discovered an optical technique that enables us to see living organisms in brilliant interference colours generated by the liquid crystallinity of their internal anatomy.[22,23] We found that all live organisms are completely liquid crystalline in their cells as well as the extracellular matrix, or connective tissues. Liquid crystals are states of matter between solid crystals and liquids. Like solid crystals, they possess long-range orientation order, and often, also varying degrees of translational order (or order of motion). In contrast to solid crystals, however, they are mobile and flexible and highly responsive. They undergo rapid changes in orientation or phase transitions when exposed to weak electric (or magnetic) fields, to subtle changes in pressure, temperature, hydration, acidity or pH, concentrations of inorganic molecules or other small molecules. These properties happen to be ideal for making organisms, as they provide for the rapid intercommunication required for the organism to function as a coherent whole.

The imaging technique enables us to literally see the whole organism at once, from its macroscopic activities down to the long-range order of the molecules that make up its tissues. The colours generated depend on the structure of the particular molecules — which differ for each tissue — and their degree of coherent order. (We have provided a mathematical derivation showing how, for weakly birefringent material like biological liquid crystals, the colour intensity is approximately linearly related to both intrinsic birefringence and the order parameter[23]). The principle is exactly the same as that used in detecting mineral crystals in geology; but with the important difference that the living liquid crystals are *dynamic* through and through. The molecules are all moving about busily transforming energy and material in the meantime, and yet they still appear crystalline. The reason is because visible light vibrates much faster than the molecules can move, the tissues will appear indistinguishable from static crystals to the light transmitted, *so long as the movements of the constituent molecules are sufficiently coherent*. In fact, the most actively moving parts of the organism are always the brightest, implying that their molecules are moving all the more coherently. With our optical technique, therefore, one can see that the organism is thick with coherent activities at all levels, which are coordinated in a continuum from the macroscopic to the molecular. That is the essence of the organic whole, where local and global, part and whole are mutually implicated at any time and for all times.

Those images draw attention to the wholeness of the organism in another respect. Without exception, all organisms — from protozoa to vertebrates — are polarized along the anterior-posterior axis, or the oral-adoral axis, such that all the colours in the different tissues of the body are at a maximum when the axis is appropriately aligned in the optical system, and they change in concert as the axis is rotated from that position.

Knowledge as Intercommunication in a Participatory Universe

The images demonstrate something profound about the nature of knowledge. Are the colours really in the organisms? Yes and no. They are dependent on the particular organism and its physiological state, but no colours would be produced unless we set up the observation in a certain way. Therefore, the observation, and hence the knowledge gained, is always dependent on both the observer and the observed. It is an act of intercommunication, which, in the ideal, is just like that between different parts of the organism (see below). The authenticity of the knowledge gained depends on this delicate balance of obtaining information while respecting the object of one's interrogation. That is why one uses minimally invasive, non-destructive techniques for investigating living organization, which allows organisms to be organisms.[11] Crude destructive methods of interrogation will invariably yield misleading information of the most mechanistic kind, reinforcing a mechanistic view of organisms and of the universe.

In the same way, as we participate in universal wholeness, in Laszlo's quantum holographic field,[15] we do so with the requisite sensitivity and respect. Knowledge is always a gift one accepts with responsiveness and responsibility. Let us look at how intercommunication takes place within the organism.

The Quantum Holographic Body Field of the Organism

There is no doubt that if we could look inside our bodies the same way we have done for the small creatures, we would see our living body as an incredibly colourful, liquid crystalline continuum, with all parts rapidly intercommunicating and colours flashing, so that it can act as a coherent whole. (That may be why we say we are off-colour when we don't feel well.)

One has been led to believe that intercommunication in large animals like us depends on the nervous system controlled by the brain. However, that may be only half the story, as nerves do not reach all parts of the body, and animals without a nervous system nevertheless have no problems acting as a coherent whole.

The clue to the other half of the story is in the connective tissues which make up the bulk of most animals including ourselves. These are the skin, the bones, cartilage, tendons, ligaments and other tissues that fill up the spaces between the usual organs. Most people still think that these tissues fulfil mechanical functions of protection and support, like packing material. However, we now know they are all liquid crystalline, and have much more exotic properties.

The connective tissues are further connected to the intracellular matrices of all individual cells that are also liquid crystalline. There is thus an excitable, liquid crystalline continuum for rapid intercommunication permeating the entire organism, enabling it to function as a coherent whole, as we have directly demonstrated with our non-invasive optical imaging technique. This continuum constitutes a "body consciousness" that precedes the nervous system in evolution; and I suggest, it still works in tandem with, and to some extent, independently of, the nervous system.[24] Body consciousness is the prerequisite for conscious experience that involves the participation of the intercommunicating whole. *When the body is fully coherent,* intercommunication is instantaneous and nonlocal. By nonlocal, I mean that distant sites say my left hand and my right hand take no time at all to reach agreement as to what to do next, so it is impossible to know where the "signal" originated. This is the *quantum* coherent state.

The quantum coherent state is a very special state of being whole, which maximizes *both* local freedom and global cohesion.[11] This arises from the *factorizability* of the quantum coherent state[25] in which the parts are so perfectly coordinated that the correlation between them resolves neatly into products of the self-correlations of the parts, so the parts behave as though they are independent of one another. Remember the huge super-orchestra I mentioned earlier? Factorizability of the quantum coherent state explains why the body can be performing all sorts of different but coordinated functions simultaneously. As I am writing this paper, my metabolism is working in all the cells of my body, my trunk and leg muscles are keeping in tone so I don't collapse into a heap, while the muscles in my arms and fingers are working together in just the right way to make the appropriate taps on the keyboard, and my eyes are tracking the words on the monitor screen; and hopefully, the nerve cells in my brain are firing coherently; all that is possible also because noiseless and instantaneous intercommunication can occur throughout the quantum coherent system. In practice, quantum coherence occurs to different degrees, and factorizability is never perfect except in the ideal, otherwise we would never age and never die. Nevertheless, our body approaches that ideal, which also tends to be restored after decohering interactions.[21]

The Coherence of Brain and Body Consciousness

It is generally assumed that the brain's primary function is to mediate coherent coupling of all subsystems, so the more highly differentiated or complex the system, the bigger the brain required. Substantial parts of the brain are indeed involved in integrating inputs from all over the body, and over long time scales. But it is also clear that not all the coordination required is provided by the brain, for this coordination seems instantaneous by all accounts.

During an olfactory experience, slow oscillations in the olfactory bulb (in the brain) are in phase with the movement of the lungs, as discovered by neurobiologist Walter Jackson Freeman III and his team at University of California Berkeley.[26,27] Similarly, the coordinated movement of the four limbs (or all the hundreds of limbs in the millipede) in locomotion is accompanied by patterns of activity in the motor centres of the brain, which are in phase with those of the limbs.[28,29] Those are remarkable phenomena, which physiologists and neuroscientists alike have taken too much for granted. The reason macroscopic organs, such as the four limbs, can be precisely coordinated is that each is individually a coherent whole, so that a definite phase relationship can be maintained among them. The hand-eye coordination required for the accomplished pianist is extremely impressive, but depends on the same inherent coherence of the subsystems which, I suggest, enables instantaneous intercommunication to occur. There simply isn't time enough, from one musical phrase to the next, for inputs to be sent to the brain, there to be integrated, and coordinated outputs to be sent back to the hands.

I raised the possibility that a "body consciousness" works in tandem with the "brain consciousness" of the nervous system, and suggested that instantaneous coordination of body functions is mediated, not so much by the nervous system, but by the body consciousness inhering in the liquid crystalline continuum of the body.[24,30] This liquid crystalline continuum not only possesses all the properties required for a body consciousness that can register tissue memory of previous experiences; but may also contain the anatomical basis of the acupuncture meridians of traditional Chinese medicine.[30]

Up to 70% of the proteins in the connective tissues consist of collagens that exhibit constant patterns of alignment, as characteristic of liquid crystals. Collagens have distinctive mechanical and dielectric properties that make them very sensitive to mechanical pressures, changes in pH, inorganic ions and electromagnetic fields.

In particular, a cylinder of water surrounds the collagen molecule, giving rise to an ordered array of bound water on the surface of the collagen network that supports rapid "jump conduction" of protons, or positive electric charges through the ordered array of bound water. Proteins in liquid crystals have coherent motions, and will readily transmit weak signals by proton conduction, or as coherent electric waves. Thus, extremely weak electromagnetic signals or mechanical disturbances may be sufficient to set off a flow of protons that will propagate throughout the body, making it ideal for intercommunication, which is what the acupuncture meridians of traditional Chinese medicine accomplish.

The liquid crystalline nature of the continuum also enables it to function as a distributed memory store. The water bound on the surfaces of proteins is known to be altered when the proteins change their shape. Proteins undergo a hierarchy of shape changes over a range of time scales and of different energies. The shapes are clustered in groups that have nearly the same energies, with very low energetic barriers between them. Thus, global shape changes in a liquid crystalline network can easily be triggered, that will in turn alter the structure of bound water. As the bound water forms a global network in association with the collagen, it will have a certain degree of stability, or resistance to change. By the same token, it will also retain tissue memory of previous experiences. Additional chemical modifications of the collagen network could also contribute to this memory.

A yet more interesting possibility is that the liquid crystalline continuum may function as a quantum holographic medium, recording the interference patterns arising from interactions between local activities and a globally coherent field.[24] This is analogous to Laszlo's suggestion that the "zero-point field" of the universe functions as a universal holographic medium, recording the experiences of all the particles, each of which is subject to influences from the rest of the universe as well as feedback from the particle's own activities on the universal medium.[15] If the organism is coherent as I have suggested, then the conditions are there for a quantum holographic memory store in the liquid crystalline continuum of the body itself. Holographic memory is unique in that it is distributed globally, and yet can be accessed and recovered locally. It captures an aspect of the organic whole in developmental biology that has completely eluded mechanistic understanding. It is that which can give rise to the subjective self, or psyche, that guides and regulates all vital activities towards a specific end. It is possible that biological development is based on the same holographic memory so that the entire organism can be engendered locally in a germ cell, from which the organism is, in turn, recoverable.

Thus, consciousness is distributed throughout the entire body; "brain consciousness", associated with the nervous system, being embedded in "body consciousness". Brain and body consciousness mutually inform and condition each other. The singularity of purpose of the individual is based on a complete coherence of brain and body. The implications for holistic and psychic health are clear. A stressful situation will affect body consciousness through subtle ways in which mechanical pressures build up in the body to block intercommunication. That acts on the nervous system to give a diminished self-image of the body, which feeds back on the body in a vicious cycle that further undermines the individual's physical well-being. By contrast, a supple body is a responsive body that moves and responds with the greatest of ease. It leads to a buoyant self-image that again feeds back to further enhance all bodily functions.

Quantum Coherence and Brain Consciousness

Many recent studies of brain activities are revealing impressive large-scale spatiotemporal coherence that suggests the brain also functions with a high degree of quantum coherence.[11,21,24] Measurements carried out with the ultrasensitive, non-invasive SQUID magnetometer, also referred to as magnetoencephalography (MEG) as well as conventional electroencehalography (EEG) over the past five years have revealed 40 Hz activities that are coherent at both deep and superficial layers of the brain, for which no obvious "sources" could be identified.[26,27] This is a strong indication that a quantum coherent field exists over the entire body, of which the brain is part.

In the same way that body consciousness associated with the liquid crystalline continuum registers memory of its experience, brain consciousness registers memory of sensory images. The idea that brain memory is distributed and holographic has been suggested by a number of neurobiologists over the past 40 years. Holographic memory storage is orders of magnitude more efficient than any model that makes use of "representations", as conventionally assumed.

As suggested earlier, the liquid crystalline continuum supporting the body field may also take part in memory storage, although this possibility has never been seriously considered. Laszlo goes even further to propose that much of memory may be stored in an ambient, collective holographic memory field delocalized from the individual; and that memories are only accessed by the brain from the ambient field.[15] This ambient field may well be our collective unconscious according to Jung.[3] One can begin to see the organism with its own local quantum holographic field as a microcosm of the universal field in which it participates.

The Organism's Macroscopic Wave Function and Universal Entanglement

If quantum coherence is characteristic of organism and psyche, as argued here, the organism will possess something like a macroscopic wave-function. This wave function is ever evolving, entangling its environment, transforming and creating itself, and no "collapse of the wave function" need ever occur, in contrast to conventional quantum theory.[11] British quantum theorist David Bohm (1917–1992) and his long-time colleague and collaborator Basil Hiley at Birkbeck College, London, also reject collapse of the universal wave function in their scheme of super-quantum potential.[31] Moreover, when quantum systems interact, they become mutually entangled, and there may be no resolution of their respective wave functions afterwards. So one may remain entangled and indeed, delocalized over past experiences (i.e., in Laszlo's ambient field).

The whole organism is thus a domain of coherent activities, constituting an autonomous free entity[19] *not* because it is separate and isolated from its environment, but precisely *by virtue of its unique entanglement of other organisms* in its environment. In this way, one can see that organic wholes are nested as well as entangled individualities. Each can be part of a larger whole, depending on the extent over which coherence can be established. When many individuals in a society have a certain rapport with one another, they may constitute a coherent whole, and ideas and feelings can indeed spread like wildfire within that community. In the same way, an ecological community, and by extension, the global ecosystem may also be envisaged as a super-organism within which coherence can be established in ecological relationships over global, geological space-times.[32]

There is an important debate going on concerning "globalization", the idea that the greater part of our life is determined by global processes in which national or local cultures, economies and borders are dissolving. While some are questioning the reality of globalization,[33] others like American author and former professor of Harvard Business School David Korten sees the globalized economy as the greatest threat to the survival of the global community.[34,35] The problem with the globalized economy under the current terms is that it does not respect the autonomy of individual persons, local communities or nation states, nor does it enable universal participation of all the parties concerned. Local autonomy and universal participation are some of the prerequisites for a coherent, sustainable global society,[11,36] in which the players must also be sensitive and responsive, or responsible and accountable. Instead, "unaccountable corporate powers"[34,35] effectively rule the world, depleting the planet's natural resources with impunity, degrading the environment and creating poverty on a massive scale. The challenge of globalization is, indeed, to create a fully participatory global society, served by an appropriate global economy that maximizes *both* local autonomy and global cohesion, as consistent with the quantum coherent organism and psyche.

References and Notes

1. Unfortunately I lost touch with Mazatl Galindo after having shared several platforms with him at conferences and workshops around the world. But here is a website that contains a good biography of him and some of his latest musical works: "Traditional Aztec Musician Releases New LP 'Red'", Botheyesshut, 23 April 2013, https://sonicsmashmusic.wordpress.com/2013/04/23/sean-mccauley/.
2. Orpheus was a musician, poet and prophet in ancient Greek myth. He was able to charm all living things with his music, and was able to visit the underworld and return to life. In one myth about Orpheus, he went into the underworld to try to bring his dead wife back to life. See "Orpheus", Wikipedia, accessed 2 August 2015, https://en.wikipedia.org/wiki/Orpheus.
3. Jung, C.G. *Man and His Symbols*. Aldus Books, London, 1964.
4. Jones, E., abridged by Trilling, L. and Marcus, S. *Sigmund Freud: Life and Work*. Basic Books, New York, 1961.
5. Suttie, I. *The Origins of Love and Hate*. Penguin Books, Harmondsworth, 1924, 1989.
6. Winnicott, D.W. *Playing and Reality*. Pelican, Harmondsworth, 1974.
7. Ho, M.W. "In Search of the Sublime". *Metanoia* (Introductory issue) (1994): 9–16.
8. Jennings, H.S. *The Universe and Life*. Yale University Press, New Haven, 1933.
9. Sinnott, E.W. *Cell and Psyche: The Biology of Purpose*. The University of North Carolina Press, Chapel Hill, 1950.
10. Fordham, F. *An Introduction to Jung's Psychology*, Pelican Books, Harmondsworth, 1966.
11. Ho, M.W. *The Rainbow and the Worm: The Physics of Organisms*. World Scientific, Singapore, 1993.

12. Waddington, C.H. *Principles of Embryology*, Allen & Unwin, London, 1956.
13. Laszlo, E. *The Interconnected Universe*, World Scientific, Singapore, 1995.
14. Laszlo, E. *The Whispering Pond*, Element, Rockport, Massachusetts, 1996.
15. Driesch, H. *The Science and Philosophy of the Organism: The Gifford Lectures Delivered before the University of Aberdeen in the Year 1907 and 1908* (2 vols.). Adam and Charles Black, London, 1908.
16. Whitehead, A.N. *Science and the Modern World*, Penguin Books, Harmondsworth, 1925.
17. Milner, M. *On Not Being Able to Paint*. International Universities Press, Inc., Madison, Connecticut, 1950.
18. Bergson, H., translated by Pogson, F.L. *Time and Free Will: An Essay on the Immediate Data of Consciousness*. George Allen & Unwin, Ltd., New York, 1916.
19. Ho, M.W. "The Biology of Free Will". *J Conscious Stud* 3 (1996a): 231–244.
20. Ho, M.W. "Bioenergetics and Biocommunication." In *Computation in Cellular and Molecular Biological Systems* (R. Cuthbertson, M. Holcombe and R. Paton, eds.), pp. 251–262, World Scientific, Singapore, 1996b.
21. Ho, M.W. "Towards a Theory of the Organism." *Integr Physiol Behav Sci* 32 (1997a): 343–363.
22. Ho, M.W., Haffegee, J., Newton, R., Zhou, Y.M., Bolton, J.S. and Ross, S. "Organisms are Polyphasic Liquid Crystals." *Bioelectrochemistry and Bioenergetics* 41 (1996): 81–91.
23. Ross, S., Newton, R., Zhou, Y.M., Haffegee, J., Ho, M.W., Bolton, J.P. and Knigh, D. "Quantitative Image Analysis of Birefringent Biological Material". *J Microscopy* 187 (1997): 62–67.
24. Ho, M.W. "Quantum Coherence and Conscious Experience". *Kybernetes* 26 (1997b): 265–276.
25. Glauber, R.J. "Coherence and Quantum Detection". In *Quantum Optics* (R.J. Glauber, ed.), Academic Press, New York, 1970.
26. Freeman, W.J. *Societies of Brains: A Study in the Neuroscience of Love and Hate*, Lawrence Erlbaum Associates, New Jersey, 1995.

27. Freeman, W.J. and Barrie, J.M. "Chaotic Oscillations and the Genesis of Meaning in Cerebral Cortex". In *Temporal Coding in the Brain* (G. Bizsaki, ed.), pp. 13–37, Springer-Verlag, Berlin, 1994.
28. Collins, J.J. and Stewart, I.N. "Symmetry-Breaking Bifurcation: A Possible Mechanism for 2:1 Frequency-Locking in Animal Locomotion". *J Math Biol* 30 (1992): 827–838.
29. Kelso, J.A.S. "Behavioral and Neural Pattern Generation: The Concept of Neurobehavioral Dynamical Systems". In *Cardiorespiratory and Motor Coordination* (H.P. Koepchen and T. Huopaniemi, eds.), pp. 224–234, Springer-Verlag, Berlin, 1991.
30. Ho, M.W. and Knight, D. "Liquid Crystalline Meridians". *Am J Chin Med* 26 (1998): 251–263.
31. Bohm, D. and Hiley, B.J. *The Undivided Universe*. Routledge, London, 1993.
32. Ho, M.W. "Natural Being and Coherent Society". In *Gaia in Action: Science of the Living Earth* (P. Bunyard, ed.), pp. 286–307, Floris Press, Edinburgh, 1996c.
33. Hirst, P. and Thompson, G. *Globalization in Question*. Polity Press, Cambridge, 1996.
34. Korton, D.C. *When Corporations Rule the World*. Kumarian Press, West Hartford, Connecticut, 1995.
35. Korten, D.C. "The Responsibility of Business to the Whole". A People-Centred Development Forum paper, 1997.
36. Ho, M.W. *Genetic Engineering Dream or Nightmare: The Brave New World of Bad Science and Big Business*. Third World Network, Penang, 1997c.

4

Evolution, Genetics and Epigenetics

4

Epigenetics and Generative Dynamics

The intrinsic dynamics of developmental processes is the source of non-random variations that directs evolutionary change in the face of new environmental challenges; and the resulting evolutionary novelties are reinforced in successive generations through epigenetic mechanisms, independently of natural selection.

In 1979, I and British bio-mathematician Peter Saunders proposed the then outrageous idea that the intrinsic dynamics of developmental processes is the source of non-random variations that *directs* evolutionary change in the face of new environmental challenges; and the resulting evolutionary novelties are reinforced in successive generations through epigenetic mechanisms, *independently of natural selection.*[1]

The proposal has held up well against subsequent research findings, and is all the more relevant in view of the numerous molecular mechanisms discovered in epigenetic inheritance[2,3] that could transmit developmental novelties to subsequent generations.

We have demonstrated how the nonlinear dynamics of living processes predicts the major features of macroevolution such as "punctuated equilibria" (long period of stasis interrupted by abrupt changes); large changes from small critical disturbances, and discontinuous changes from continuously varying parameters; and why macroevolution of form and function is decoupled from the microevolution of gene sequences. We showed that the same (non-random) developmental changes are repeatedly produced by specific environmental stimuli. Furthermore, we demonstrated how general mathematical models can account for all the developmental transformations experimentally produced, which can make strong evolutionary predictions, and offer a natural taxonomy based on the predicted transformations.

However, neither the epigenetic mechanisms nor the dynamics of developmental processes are taken into account in the recent studies on evolution and development.

The totality of research findings gives no support to the neo-Darwinian theory of evolution by the natural selection of random genetic mutations, nor to any theory ascribing putative differences in human attributes predominantly to genes. The overwhelming determinants of health and behaviour are social and environmental. Heredity is distributed over the seamless web of nested organism-environment interrelationships extending from the social and ecological to the genetic and epigenetic. Consequently, there is no separation between development and evolution, and the organism actively participates in shaping its own development as well as the evolutionary future of the entire ecological community of which it is part.

"Epigenetic" Then and Now

The term "epigenetic" as used today in epigenetic inheritance refers to effects that do not involve DNA base sequence changes, but only the chemical modifications of DNA or histone proteins in chromatin (complex of DNA and protein that make up chromosomes in the nucleus of cells), which alter gene expression states. Epigenetic inheritance has been defined in a review (p. 398)[4] as "the structural adaptation of chromosomal regions so as to register signal or perpetuate altered activity states". But these definitions are rapidly becoming obsolete.[2,3,5–9] In reality, epigenetic modifications encompass a great variety of mechanisms. They act during and after transcription, and at translation of genetic messages; they can even rewrite genomic DNA.[2] Hence the distinction between genetic and epigenetic is increasingly blurred.

"Epigenetic" as originally used, was derived from *epigenesis*, the theory that organisms are not *preformed* in the germ cells, but come into being through a process of development in which the environment plays a formative role. Most evolutionists have used "epigenetic" to mean hereditary influences arising from environmental effects in the course of development.

Evolution: Lamarck versus Darwin

Evolution refers to the natural (as opposed to supernatural) origin and transformation of organisms on earth throughout geological history to the present day. The first comprehensive *general* theory of evolution — that evolution has occurred — was proposed by French scientist Jean Baptiste de Lamarck (1944–1829) in his book published more than 200 years ago (1809).[10–12] It was a *uniformitarian* theory in that causes proposed to be operating in the past are the same as those that can be observed at present. The theory postulated the spontaneous generation of the living from the non-living and unlimited transformation over time, which gave rise to whole kingdoms of organisms beginning from a single origin of life. In addition, Lamarck proposed special mechanisms whereby new species could evolve through changes in how the organism relates to its environment in *pursuing its basic needs*, which produce new characteristics that become inherited after many generations. These special mechanisms are "use and disuse" (use enhances and reinforces the development of the organs or tissues while disuse results in atrophy) and the "inheritance of acquired characters", the transmission to subsequent generations the tendency to develop certain new characteristics that the organism has acquired in its own development.

Thus, Lamarck also proposed the first epigenetic theory of evolution, with development playing a key role in initiating evolutionary change while specific epigenetic mechanisms transmit the change and reinforce it in subsequent generations.[12–14]

British naturalist Charles Darwin (1809–1882) should be credited only with the special theory of evolution by natural selection (1859),[15] which states that, given the organisms' capability to reproduce more of their numbers than the environment can support, and there are heritable variations, then, within a population, individuals with the more favourable variations would survive to reproduce their kind at the expense of those with less favourable variations. The ensuing competition and "struggle for life" results in the "survival of the fittest", so the species will become better adapted to its environment. And if the environment changes in time there will be a gradual but definite "transmutation" of species. Thus, nature effectively "selects" the fittest in the same way that artificial selection by plant and animal breeders ensures that the best or the most desirable characters are bred and preserved. In both cases, new varieties are created after some generations.

In *addition* to natural selection, Darwin invoked the effects of use and disuse, and the inheritance of acquired characters in the transmutation of species. However, those Lamarckian ideas do not fit into the theory of natural selection, and Darwin's followers all regard the lack of a theory of heredity and variation as the weakest link in the argument for natural selection.[16,17]

In *Genetic Engineering Dream or Nightmare? The Brave New World of Bad Science and Big Business.*[18] I described in detail how Darwin's followers created the "neo-Darwinian synthesis" by expurgating Darwin's Lamarckian tendencies, including his theory of pangenesis. Darwin's theory of pangenesis had actually received a great deal of support.[19] The rediscovery of Mendel at the turn of the last century provided evidence that particulate genes controlling the characteristics of organisms are passed on unchanged, except for rare random mutations. This fits in perfectly with the discovery of German biologist August Weismann (1834–1914) that the material basis of heredity was the "germplasm" in germ cells that becomes separate from the rest of the animal's body early in development to ensure it would be protected from environmental influences. Development is therefore irrelevant to evolution. We now know that Weissman's theory is wrong and there are numerous exceptions to Mendelian inheritance. Nevertheless, Darwinism was promptly reinterpreted according to the gene theory in the "neo-Darwinian synthesis" from the 1930s up to the 1950s and 60s.

As the result of the neo-Darwinian synthesis, evolution is supposed to occur strictly by the natural selection of *random* gene mutations, or changes in base sequence of DNA; those that happen to increase reproductive fitness are selected at the expense of the others that do not.

Evolution, Development and Heredity

The theories of evolution, development and heredity are closely intertwined. Just as evolutionists needed a theory of heredity, so plant breeders in the 18th century who inspired Mendel's discovery of genetics were motivated by the question as to whether new species could evolve from existing ones. In accounting for change or transformation, it is also necessary to locate where constancy or stability resides, which constitutes heredity. In order to explain the evolution of form and function, development (epigenesis) is central, as Lamarck clearly grasped. In contrast, Darwin, and neo-Darwinists see new variations arising at *random* in the sense that they bear no direct relationship to the environment, those that happen to be adaptive are selected, while the rest are eliminated. The theory of natural selection is essentially preformist, development playing little or no role in determining evolutionary change.[13,14,20]

There are a number of different epigenetic theories of evolution since Lamarck; some predating the neo-Darwinian synthesis. A common starting point for all epigenetic theories is the developmental flexibility of all organisms. In particular, it has been observed that artificially induced developmental modifications often resemble (*phenocopy*) those existing naturally in related geographical races or species that appear to be genetically determined. Thus, it seemed reasonable to assume that evolutionary novelties first arose as developmental modifications, which somehow became stably inherited (or not, as the case may be) in subsequent generations.

Epigenetic Reorganization Initiates Evolutionary Change

Early proponents of epigenetic theory included American philosopher James Mark Baldwin (1861–1934), who suggested in 1896[21] that modifications arising in organisms developing in a new environment produce "organic selection" forces internal to the organism, which stabilize the modification in subsequent generations. Another notable figure was German-born American geneticist Richard Goldschmidt (1878–1958) who proposed in 1940[22] that evolutionary novelties arise through *macromutations* producing "hopeful monsters" that can initiate new species. In his defence, he pointed out that monsters could be hopeful because of the inherent *organization* of the biological system that tends to "make sense" of the mutation. Following Goldschmidt, Søren Løvtrup (1922–2002) advocated a similar theory of macromutations in 1974[23] for the origin of major taxonomic groups of organisms such as phyla.

But random mutations — changes in the DNA — that generate hopeful monsters must be hopelessly rare, and to make things worse, major taxonomic groups tend to appear suddenly in clusters, "adaptive radiations", rather than isolated at different geological times.

The extraordinarily rich fossil finds of the Cambrian "explosion" responsible for most of the major animal phyla is a prime example of evolution occurring in bursts of "adaptive radiation" followed by relatively long periods of stasis, or "punctuated equilibria" proposed by American palaeontologists Stephen Jay Gould (1941–2002) and Niles Eldredge in 1972.[24] Furthermore, evolution does seem to proceed top-down, from phyla, to subphyla, classes, orders and so on according to James Valentine[25] (now at University of California Berkeley) rather than the converse, as predicted by Darwin and neo-Darwinian natural selection of small random mutations. And crucially, all the evidence indicates that macroevolution is decoupled from molecular or microevolution (more below).

These considerations suggest that "adaptive radiations" involve major novelties arising from *epigenetic reorganisation* provoked by large environmental changes or changes in the organisms' way of life, which also seem to coincide with adaptive radiations. For example, oxygen is very important for the evolution of complex organisms, and the Cambrian "explosion" is believed to have been triggered by the rapid increase in atmospheric O_2 levels from a low of ~15% to the current level of ~20% between 1 billion to 0.5 billion years ago.[26]

"Evo-devo" Still Blinded by "Genetic Programme" of Development

In a sense, there is nothing new about the current revival of "evo-devo".[27–31] It is still dominated by the idea going back at least 20 years that genes control development in a "genetic programme" of gene regulation and interaction;[31] and that large evolutionary changes in body pattern are the result of changes in gene regulation due to natural selection. There is still no recognition that the *patterns* themselves, and biological *form* need to be explained in their own right, independently of whether natural selection operates or not, and independently of the action of specific genes.[1,16,32,33] Not surprisingly, there is still little or no recognition that epigenetic and non-genetic environmental influences can give rise to large alterations in form and function.

In a brilliant critique of the genetic determinist approach to behaviour, American development psychologist Gilbert Gottlieb (1929–2006) deconstructed the idea that genes determine body pattern by pointing to the very different expression patterns of the same *Hox* genes in the fruit fly, the centipede, and the Onychophora.[34] *Hox* (homeotic) genes are supposed to control segmental patterning during development; instead, the same genes appear to be simply responding to different patterning processes in the different animals. There is decidedly no homology of genes corresponding to homology of biological structures.

The same theme emerged in a comprehensive review of segmentation in arthropods,[35] which showed that different groups have distinct modes of segmentation and divergent genetic mechanisms.

In a study of left-right asymmetry in development,[36] the genes determining vertebrate heart asymmetry known as *the nodal signalling cascade* were analysed. Like the molecular mechanisms that define the anteroposterior and dorsoventral axes, the nodal signalling cascade includes a "curious mix" of conserved and divergent elements. But non-conserved elements greatly outnumber the conserved elements. The idea that natural selection of developmental genes accounts for the evolution of form is obsolete; it is time to move on.

Some researchers now despair of trying to explain pattern formation with complicated computational networks of genes that pass for "systems biology". Japanese researchers Shigeru Kondo and Takashi Miura at Osaka University stated in a review that[37] "the behaviour of such systems often defies immediate or intuitive understanding" and "it becomes almost impossible to make a meaningful prediction". For that reason, they are turning back to mathematical reaction-diffusion models (more later).

Waddington's Theory of Canalization and Genetic Assimilation

The most influential figure among the "epigenetic evolutionists" was British developmental and evolution biologist Conrad H. Waddington (1905–1975), who attempted to accommodate "pseudo-Lamarckian" phenomena within neo-Darwinism in his theory of genetic assimilation. Like all Darwinian and neo-Darwinian evolutionists, he wanted to explain the origin of *adaptive* characters, i.e., characters that seem to fit the functions they serve.

In his most influential book *The Strategy of the Genes*[38] published in 1957, Waddington conceptualized the flexibility and plasticity of development, as well as its capacity for regulating against disturbances, in his famous "epigenetic landscape", a general metaphor for the nonlinear dynamics of the developmental process.[39] The developmental paths of tissues and cells are constrained or *canalized* to "flow" along certain valleys due to the "pull" or force exerted on the landscape by the various gene products that define the fluid topography of the landscape (Fig. 4.1). Thus, certain paths along valley floors will branch off from one another to be separated by hills (thresholds) so that different developmental results (alternative attractors) can be reached from the same starting point.

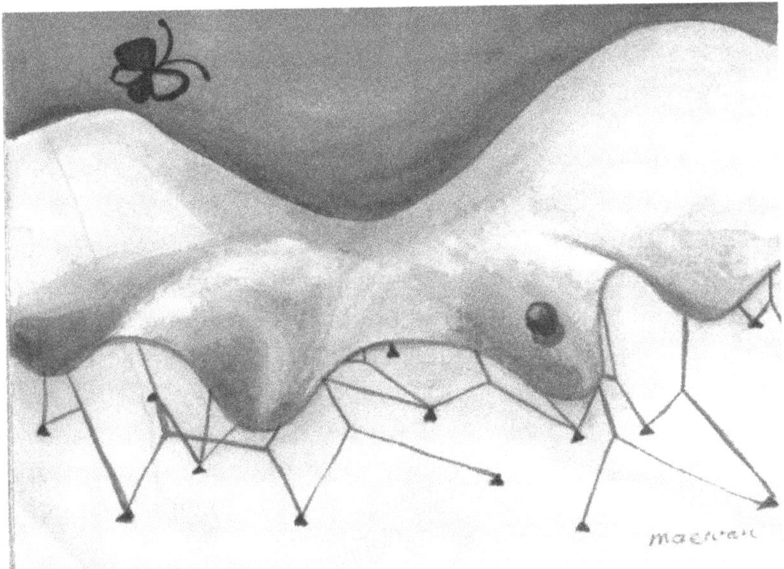

Figure 4.1 Waddington's epigenetic landscape.

However, some branches may rejoin further on, so that different paths will nevertheless lead to the same developmental result. Genetic or environmental disturbances tend to "push" development from its normal pathway across the threshold to another pathway. Alternatively, other valleys (developmental pathways) or hills (thresholds) may be formed due to changes in the topography of the epigenetic landscape itself.

The significance of the conceptual epigenetic landscape is that its topography is determined by *all* of the genes whose actions are inextricably interlinked, and is not immediately dependent on specific alleles of particular genes.[1] That is in accord with what we know about metabolism and the epigenetic system, particularly as revealed by the new genetics (see later). It also effectively decouples the evolution of the organism, of form and function, from alleles of specific genes, and explains the notable lack of correlation between morphological and genetic differences between species.[40]

Waddington proposed that a new phenotype arises when the environment changes so that development proceeds to a new pathway in the epigenetic landscape, or else a remodelling of the epigenetic landscape itself takes place (both of which are possible from what we now know about epigenetic processes at the molecular level). Thereafter, the new phenotype becomes reinforced or "canalized" through natural selection for modifier genes so that a more or less uniform phenotype results from a range of environmental stimulus, and later, the phenotype is "genetically assimilated", so it occurs in the absence of the original environmental stimulus.

Waddington and colleagues carried out experiments showing that artificial selection for the bithorax phenocopy in *Drosophila* induced by ether exposure during early embryogenesis resulted in canalization and genetic assimilation.

Ho and Saunders' Epigenetic Theory of Evolution

The first distinctive feature of our epigenetic theory of evolution[1,41,42] is that neo-Darwinian natural selection plays little or no role, based on evidence suggesting on the one hand that most genetic changes are irrelevant to the evolution of organisms, and on the other, that a relative *lack* of natural selection may be the prerequisite for major evolutionary change.

The second feature is that the intrinsic dynamics of the epigenetic system — developmental dynamics — is determined not so much by gene interactions as by *physical and chemical forces* of nonlinear complex systems in general, which are amenable to mathematical description.[32,43] That is why, contrary to the neo-Darwinian view, variations of the phenotype that arise during development in response to new environments are *non-random* and *repeatable*.

We proposed, therefore, that the intrinsic dynamical structure of the epigenetic system is the source of non-random variations that *direct* evolutionary change in the face of new environmental challenges. These evolutionary novelties are reinforced (canalized) in subsequent generations through cytoplasmic/epigenetic mechanisms, *independently of natural selection.*

When a population of organisms experience a new environment, or *adopt a new behaviour*, the following sequence of events is envisaged:

a. A novel response arises during development in *a large proportion, if not all of* the organisms in a population experiencing a new environment, due to the intrinsic dynamics of the epigenetic system. In the case of a new behaviour initiated by a single individual in a social group, the behaviour can also spread quite rapidly. For example, the new habit of washing sweet potatoes in the sea initiated by a young female had spread to the entire troop of wild monkeys on Koshima Island in Japan within nine years.[44] This behaviour may also have triggered developmental changes in the monkey's brain.
b. This response is "canalized" in successive generations through epigenetic mechanisms *independent of natural selection*, and this has been demonstrated in experiments in our laboratory subsequently (see later).
c. After some generations, the response *may* become "genetically assimilated", in that it arises even in the absence of the stimulus. As in Waddington's epigenetic landscape, this could entail a change in the topography to bias the original branch point in favour of the new pathway, so that the new phenotype will persist in the absence of the environmental stimulus. Random genetic mutations could also be involved.

Corroborations of Ho and Saunders' Epigenetic Theory

Since our theory was proposed, we have obtained important empirical and theoretical corroboration. We questioned Waddington's assumption that selection of (modifier) genes is necessary for canalization and genetic assimilation, and in a series of experiments[45] demonstrated that canalization occurred in the *absence* of selection *for* the new character. We showed that successive generations of ether treatment during early embryonic development in *Drosophila* increased the frequency of the bithorax phenocopy in the adults, without selecting *for* the phenocopy. If anything, the phenocopy was almost certainly selected *against*, as it obviously interfered with flight and other normal functions. We had identified a case of "epigenetic inheritance" of a maladaptive character, consistent with recent findings in "epigenetic toxicology", in which toxic effects of exposure to environmental pollutants are transmitted to grandchildren.[7]

At least one study of the fossil record[36] provided evidence that left-right asymmetry in animals and plants may have originated as phenotypic novelties that became genetically assimilated subsequently.

We stipulated that genetic assimilation is not a necessary part of the response to change,[1] as it would preserve the important property of developmental flexibility or "adaptability". In retrospect, this has proved correct. We now know that maternal behaviour, long regarded as genetically inherited and instinctive, is actually associated with epigenetic gene markings made during early infancy that are erased at every generation, yet perpetuated indefinitely from mother to daughter, parent to offspring.[5,46] More than that, it is an epigenetic system that enables foster parents to literally influence their child biologically.

An important motivation for focussing on development for evolutionary change is that developmental changes are far from random or arbitrary;[1,32,33,41,42] but are determined by dynamical processes, independently of the action of specific genes.

Physicochemical Forces and Flows in Growth and Form

Patterns are generated everywhere in the physical world where no genes are involved, and many of the patterns closely resemble those found in the living world. It is the dynamics of physical and chemical forces and flows that generate patterns and forms, as Scottish biologist and mathematician D'Arcy Thompson (1860–1948) so beautifully argued in his classic book, *On Growth and Form*.[47] Closer to our time, Alan Turing (1912–1954), English mathematician, logician, code breaker and computer pioneer, is also well known for his work on morphogenesis.[48] Turing's reaction-diffusion model shows, for the first time, how patterns can arise *spontaneously* in an initially homogeneous domain, precisely the problem of how patterns can form in a featureless egg in development.[48–50]

The Turing model inspired much work on pattern formation in biological systems before it got lost in the proliferating thicket of genes that "control pattern formation".

The complex nonlinear dynamics of the developmental process has been explored mathematically in greater detail,[32,39,43,51] and its evolutionary consequences made explicit. For example, it accounts for "punctuated equilibria".[24] It also shows how large organized changes can occur with a relatively small disturbance, and how continuously varying environmental parameters can nevertheless precipitate discontinuous phenotypic change.[39]

The physical and chemical forces that organize living systems were the subject of my book, *The Rainbow and the Worm: The Physics of Organisms*[52] now in its third enlarged edition. The book presents evidence that cells and organisms are liquid crystalline, with water the most important constituent of the liquid crystalline matrix. I pointed out that electrical polarities determine the alignment of the liquid crystals and hence the major body axes. Furthermore, electrodynamical forces acting on liquid crystal mesophases may play a key role in pattern formation and morphogenesis. As consistent with this hypothesis, we demonstrated dramatic effects with brief exposures of early *Drosophila* embryos to very weak static magnetic fields; the segmental body patterns of the larva that emerged 24 hours later were transformed into helices.[53]

Recently, there has been a revival of interest in electrodynamical processes in development, as changes in membrane potential and the establishment of ionic currents and endogenous electric fields appear to determine polarities long before the relevant genes are expressed.[54] These and other evidence suggest that electrodynamical processes are involved in pattern formation via the liquid crystalline cortex of the egg and epithelial cells in regeneration. The connection to genes is hence indirect and nonlinear.

In contrast, developmental geneticists generally assume that diffusion gradients of special "morphogens" determine body pattern by providing "positional information" for particular genes to "interpret". For example, in *Drosophila*, where the most complete genetic analysis of development has been carried out,[55] the maternal gene product Bicoid is identified as the morphogen; its anteroposterior gradient serving to initiate the cascade of "combinatorial regulation" of genes that eventually gives rise to the complete body pattern. The difficulty is that very few molecules, if any, diffuse freely in the liquid crystalline matrix, and Bicoid protein is no exception. If anything, it appears that a gradient of transcription/translation and degradation is actively maintained in the embryo during several cycles of synchronous nuclear divisions[56,57] by an as yet unknown patterning process, which may well be electrodynamical.

Status of Ho and Saunders' Epigenetic Theory

To summarize the present status of Ho and Saunders' epigenetic theory, I want to stress again that natural selection plays little or no role in evolution, especially in the evolution of major novelties or macroevolution (except in the negative sense of eliminating deleterious mutations with large effects) for three reasons. First, the epigenetic (developmental and non-genetic) novelties produced in response to new environments are common to most, if not all, individuals in a population, and would swamp out residual effects due to genetic variation. Second, the fluidity of the genome — the constant interaction between genome and environment, the epigenetic markings of genes, and the blurring between genetic and epigenetic — makes it clear that organism and environment are inseparable; hence there can be no selection of any static, preformed variant that is independent, or random, with respect to the selective environment. Third, the physical and chemical forces and flows that *generate* biological patterns and forms are independent of natural selection, and require their own explanations.[54]

Our theory has been corroborated since it was first proposed, as follows:

- Experimental demonstration of canalization of bithorax phenocopy independent of natural selection.[45]
- Epigenetic toxicology show that harmful effects are transmitted across generations.[7]
- Evidence for genetic assimilation in the evolution of left-right asymmetry and other morphologies.[36]
- Genetic assimilation may not occur, for example, stable epigenetic inheritance of maternal behaviour nevertheless reversed in cross-adoption studies in rats.[6,46]
- Complex nonlinear dynamics do exhibit properties of the epigenetic landscape and account for major features of macroevolution.[39]
- Electrodynamical forces are involved in pattern formation, with indirect and nonlinear connection to genes.[52–54]

I shall show why the dynamics that generate patterns and forms are much more than weak "developmental constraints" to natural selection, and then address the "neutral mutation hypothesis", the proposal that most, if not all, DNA base changes during evolution are due to random genetic drift decoupled from the evolution of organisms.

Rational Taxonomy Based on the Generative Dynamics of Biological Form

The dynamics of developmental (epigenetic) processes, being amenable to mathematical description, provide a powerful perspective for understanding the development and the evolution of form. That is the basis of "structuralism in biology";[33,58] or more accurately in our view, "process structuralism".[14,32,39,42,43,51,59]

The developmental dynamics define a set of possible transformations that is highly constrained, so that particular transformations may be *predictably* linked to specific environmental stimuli. The fundamental importance of development for evolution is that evolutionary transformations can ultimately be understood in terms of developmental transformations that can be empirically investigated and that this in turn provides us with the criteria for a rational taxonomy, a natural system of classification based on the generative dynamics of form. I shall describe two examples: the segmentation defects in *Drosophila* larva produced by exposing early embryos to ether vapour, and phyllotaxis, the arrangement of leaves around the stem.

The segmentation pattern of the first instar *Drosophila* larva is determined during early embryogenesis. In the course of our studies on the bithorax phenocopy, we discovered that brief exposures to ether vapour also produced characteristic defects in the segmental pattern reflecting a dynamic process arrested at different stages.[60] These defects phenocopy *all* the major genetic mutants identified. And the most general model of successive bifurcation could produce all the observed defects, giving a rational taxonomy of both the observed and yet-to-be-observed forms.[61–62] This rational taxonomy based on generative dynamics differs from one based on genealogy or similarity of DNA, and interestingly, also differs significantly from one based on cladistic analysis.[61] We subsequently produced a mathematical model of reliable segmentation based on successive bifurcation.[63]

Figure 4.2[61,62] is a transformational "tree" of the range of segmental patterns obtained *during development*. The main sequence, going up the trunk of the tree, is the normal transformational pathway, which progressively divides up the body into domains, ending up with 16 body segments of the normal larva. All the rest (with solid outlines) are transformations in which the process of dividing up the body has been arrested at different positions in the body. The patterns with dotted outlines are hypothetical forms, not yet observed, connecting actual transformations.

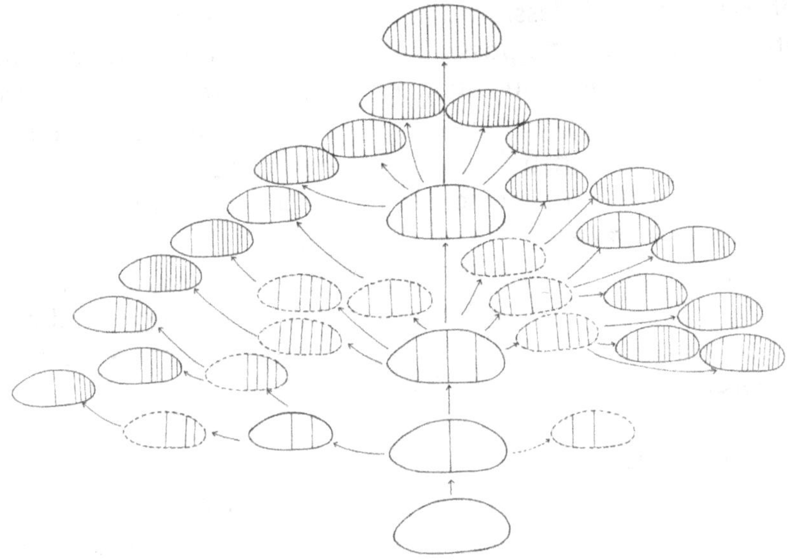

Figure 4.2 Transformation tree of body patterns in fruit fly larvae based on a model of successive bifurcation.

This transformational tree reveals how different forms are related to one another; how superficially similar forms are far apart on the tree, while forms that look most different are neighbours. It is the most parsimonious tree relating all the forms.

More importantly, the ontogenetic transformation tree predicts the possible forms that can be obtained in evolution (phylogeny), mostly likely by going up the sequence of successive bifurcations, but occasional reversals to simpler forms could also take place. This is why phylogeny appears to recapitulate ontogeny.[64] But actually *it does not*; ontogeny and phylogeny are simply related through the dynamics of the generic processes generating form.

A *natural* system of classification results from the tree. The 24 actual forms or species are classified hierarchically into one "family" with two "orders", the first order containing three "genera", and the second order, eight genera. The forms not yet found (depicted in dotted lines in Fig. 4.2), would also fit neatly in the natural system of classification should they be discovered in the future. There are 676 possible forms according to the dynamic model of successive bifurcation. If all the body segments were free to vary independently, the number of possible forms would have been 2^{16}, or more than 60,000. This demonstrates how highly the generative dynamics can constrain the possible forms, and why, incidentally, parallelisms are rife in evolution.[16,41]

Obviously, the scheme proposed is an oversimplification, which is why eight hypothetical intermediates (represented in dotted outlines) were not actually observed. The actual process itself may well predict many fewer possible forms.

In the second example, we produced a transformation tree for all possible ways leaves are arranged around the stem in plants (Fig. 4.3)[65] based on the generic and robust dynamics that generate the patterns, discovered by French mathematical physicists S. Douady and Y. Couder.[66] The discovery caused quite a stir in France, as leaf arrangement, or *phyllotaxis*, has been a long-standing problem in biology, ever since Alan Turing drew attention to how the spiral patterns of leaves around the stem conform to the Fibonacci sequence.[32,49–51] Many neo-Darwinian "just-so stories"

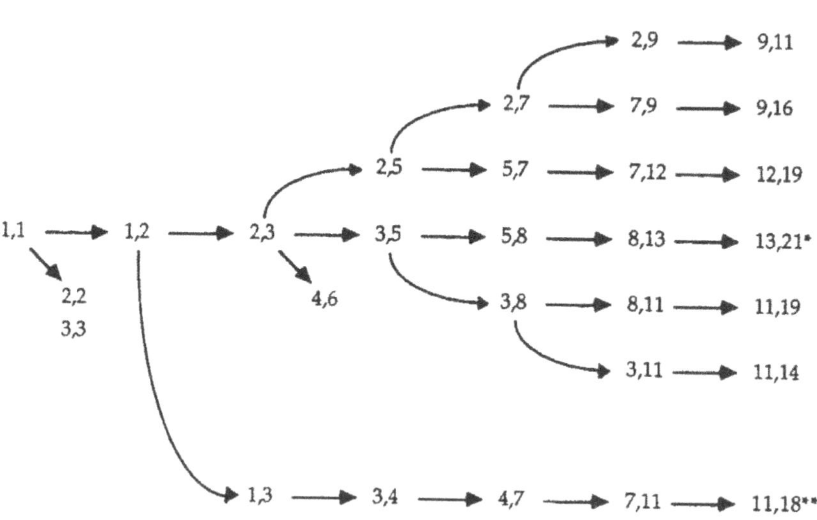

Figure 4.3 The transformation tree of possible phyllotaxis patterns.

have been invented over the years to account for different leaf arrangements in terms of "selective advantage", all of which have been proven irrelevant in one stroke. The power of dynamics — the syntax of form — is that it predicts the set of possible transformations, *excluding all others*. It also tells us how the possible forms are related by transformation.[67]

It is not known if all the possible forms in Fig. 4.3 actually exist in nature. The main Fibonacci sequence with a divergence angle of 137.5° is in the middle row (marked with *), where the successive number more and more approaches the golden ratio. At the bottom is an alternative Fibonacci sequence with a divergence angle of 99.5° (marked with **). Like the transformation tree in Fig. 4.2, it makes very definite predictions concerning neighbouring transformations. Thus, parastichies 8, 11 and 9, 11 (secondary spirals; the numbers indicate spirals to the right and left, respectively, starting from the centre of the flower or top of the cone), despite their apparent similarity, are quite far apart on the tree, whereas the neighbouring parastichies 8, 13 and 13, 21 appear superficially very different. As the tree is also an ontogenetic tree, it predicts that plants such as the Canadian pine (*Pinus resinosia*) with parastichies 8, 13 in the cone, goes through all of the main sequence in development. We do not know if that is true, but we did find that the leaf shoot bearing the cone has 3, 5 parastichies.

For the same reasons, we would predict that the decussate arrangement 2, 2 is the earliest divergence from the main Fibonacci sequence, followed by the alternative Fibonacci sequence beginning with 1, 3. Phylogenetic transformations are strictly predicted. For example, one would not expect an ancestor of a plant with parastichies 8, 13 to have had parastichies 7, 12, or even 2, 5, but most likely, 5, 8.

The dynamics of the processes are subject to contingent "complexification" (or simplification) in the course of evolution, by virtue of the lived experience of the organisms themselves. Nevertheless, it is highly constrained, when it comes to pattern formation.

It has become clear that directed genetic changes in given environments are just as non-random as morphological changes, and hence, possibly subject to comparable systemic constraints (see later).

Natural Selection and Molecular Evolution

Molecular evolution, the study of how proteins and nucleic acid sequences in different species evolve, has been dominated by the neutralist/selectionist controversy that continues to the present day.

Japanese geneticist Motoo Kimura (1924–1994) was best known for his neutral theory of molecular evolution,[68] which proposed that most of the amino acid and base changes in evolution resulted from random genetic drift of neutral mutations, i.e., mutations that did not influence the "fitness" of the organisms. In fact, he did not deny that natural selection could be operating, only that it was not reflected in the evolution of molecules. In effect, molecular evolution appears decoupled from the evolution of organisms, which, at least, is consistent with all other observations indicating the lack of simple translations between genes and phenotype, and is an independent corroboration of Waddington's concept of the epigenetic landscape.[38]

The neutral mutation theory was inspired by earlier discoveries that when the amino acid or DNA base sequence of genes in different organisms were compared, they diverged apparently linearly according to the time at which the organisms shared a common ancestor. This gave rise to the "molecular clock" hypothesis[69,70] according to which, the rate of amino acid or nucleotide substitution is approximately constant per year over evolutionary time and among different species.[40]

As more data became available, the molecular clock hypothesis ran into trouble. Although there is a correlation between genetic distance and time of divergence, such correlation is not universal, and is often violated.

Numerous studies on extant organisms show that mutation rates are far from constant.[71] For example, genetic differences between two subpopulations of medaka fish that had diverged for ~4 million years is three times that between two primate

Table 4.1 Genetic distance and estimated divergence time[71]

	% Identity				
	Prdm2	BTK	CytC	GCA1A	Div. time (MyBP)
H. sapiens vs D. rerio	39	61	80	66	450
H. sapiens vs X. laevis		55	85	75	360
H. sapiens vs G. gallus	71	85	87	81	310
H. sapiens vs M. musculus	91	98	91	91	91
F. rubripes vs D. rerio	45				420
	71				400
		89			200
			91		91

species, humans and chimpanzees, that are thought to have split 5–7 million years ago. Genetic distances measured on genealogical timescales of less than one million years are often an order of magnitude *larger* than those on geological timescales of more than a million years.

To illustrate the paradox, four randomly selected genes in different species are compared for their similarity (% identity). All four genes behave as good clocks in macroevolution from fish (*D. rerio*, zebrafish), to frog (*X. laevis*, African clawed toad), to bird (*G. gallus*, red jungle fowl), to mouse (*M. musculus*), and human (*H. sapiens*).

However, they give wildly contradictory timing at lower levels (see Table 4.1). When different species of fish are compared with each other, *F. rubripes* (puffer fish) versus *D. rerio*, divergence time ranged from 91 to 420 MyBP.

Epigenetic Complexity versus Genetic Diversity, Macroevolution versus Microevolution

Shi Huang at the Burnham Institute, La Jolla, California, proposed that an inverse relationship exists between genetic diversity and epigenetic complexity: multicellular organisms differentiated into tissues and cells are epigenetically complex and can tolerate less genetic variation (germline DNA mutation), whereas single-celled organisms, being epigenetically simple, can tolerate more.[71] Consequently, each level of epigenetic complexity will reach its maximum level of variations. This simple theory explains the major features of evolution, including the paradox of an overestimate of divergence times when some gene sequences in lower taxonomic levels are compared (see Table 4.1).

Humans are undoubtedly the most epigenetically complex species; but in terms of the number of genes, we have only roughly 1.6 times that of a fruit fly and about the same as a mouse or fish. However, the number of certain enzymes responsible for epigenetic gene organization, such as the PRDM subfamily of histone methyltransferases, increases during metazoan evolution from 0 in bacteria, yeasts and plants, to 2 in worms, 3 in insects, 7 in sea urchins, 15 in fishes, 16 in rodents and 17 in primates. The core histone genes H2A, H2B, H3 and H4 have been duplicated in humans but not chimpanzees, and the number of genes for microRNA (which play key regulatory functions) correlates well with organism complexity. Complex organisms also show complex gene expression patterns: 94% of human genes have alternative products or alternative splicing compared to only 10% in the nematode *C. elegans*.

According to Huang, organisms undergo epigenetic changes or genetic mutations in a certain range in accordance with their epigenetic complexity. More significantly, epigenetic complexity change is almost by definition, macroevolution, whereas genetic changes due to mutations causing minor variations in phenotypes that do not affect the epigenetic programmes are microevolution. Microevolution is a continuous process of accumulating mutations (see Box 4.1).

Macroevolution from simple to complex organisms is associated with a punctuational increase in epigenetic complexity and in turn a punctuational loss in genetic diversity. From a common ancestor, the genetic distance between two splitting descendants may gradually increase with time until a maximum is reached, remaining constant thereafter.

Box 4.1

Shi Huang's Theory of Molecular Evolution

- There is an inverse relationship between genetic diversity and epigenetic complexity.
- Multicellular organisms are epigenetically complex and tolerate less genetic diversity than simple single-celled organisms.
- Each level of epigenetic complexity has its maximum level of genetic diversity.
- From a common ancestor, the genetic distance between two splitting descendants gradually increases with time until a maximum is reached and remains constant thereafter; if time is long enough for genetic distance to reach the maximum, then genetic distance between two genera of the same family would be similar to that between two families or orders, or phyla.
- Macroevolution involves epigenetic changes and is decoupled from microevolution of genes.

The maximum genetic diversity hypothesis predicts that if time is long enough for genetic distance to reach the maximum, then the genetic distance between two genera of the same family should be similar to that between two families, or orders, or phyla. That was demonstrated to be the case for a very old group such as fungi; in contrast, the molecular clock hypothesis predicts that the genetic distance between two fungi genera of the same family should be *smaller* than that between families, and still smaller than that between orders, and so on.

His hypothesis, Huang claims, explains top-down evolution, which is also consistent with the epigenetic origin of evolutionary novelties (see earlier), and the decoupling of macroevolution from the microevolution of genetic distance.

Continuity between Epigenetic and Genetic Changes

Huang's theory does explain a lot and could, in principle, resolve nearly all the major paradoxes in molecular evolution, except perhaps the widely different rates of divergence between different genes within the same organism.

More importantly, I believe Huang's hypothesis that epigenetically complex organisms are less tolerant of genetic or germ line diversity is incomplete, because the level of germ line diversity is *actively maintained*.

A key feature of epigenetics in complex organisms is that they have become more efficient at generating the sequence diversity required at the precise local somatic level;[8] and incidentally, also more efficient at reducing it at the germ line level through mechanisms such as gene conversion and concerted evolution,[2,3,5,6] all part of the death of the central dogma of molecular biology that has been happening since the 1980s.

Epigenetic processes such as RNA editing, alternative splicing, trans-splicing, exonisation and somatic hypermutation, can generate huge sequence diversity wherever and whenever required.[8] Some of those processes, coupled with reverse-translation, are powerful mechanisms for generating sequence diversity that can be tested by function within the individual organism, and then used to overwrite the germ line sequence(s). I have reviewed these mechanisms in some detail elsewhere,[2] including a range of evidence indicating that mutations are far from random, with the organism choosing when and how to mutate, or not to mutate at all.[72]

DNA recoding — rewriting genome DNA — appears to be a central feature of both the immune and nervous systems. DNA recoding is involved at the level of establishing neuronal identity and neuronal connectivity during development, learning and brain regeneration. And it appears that the brain, like the immune system, also changes according to experience.

John Mattick at University of Queensland St. Lucia in Australia and Mark Mehler at Albert Einstein College of Medicine, New York in the United States suggested that the potential recoding of DNA in nerve cells (and similarly in immune cells) might be primarily a mechanism whereby productive or learned changes induced by RNA editing are *rewritten* back to DNA via RNA-directed DNA repair.[73] This effectively fixes the altered genetic message once a particular neural circuitry and epigenetic state have been established.[8] Australian geneticist Ted Steele has proposed a similar RNA-directed recoding of DNA for the immune system.[74]

Unlike Steele,[74] Mattick and Mehler[73] fall short of proposing that the RNA-templated recoding of the genome and the associated structural and functional adaptations could be transmitted to the next generation. This could be crucial for brain evolution in primates leading up to humans, so that the gains made by successive generations could be accumulated.[8]

If the analogy with the immune system holds, then as suggested by Steele and colleagues, edited RNA messages or their reverse transcribed DNA counterparts could become inherited via the sperm.[9,75,76] "Sperm-mediated gene transfer" is well documented as a process whereby new genetic traits are transmitted to the next generation by the uptake of DNA or RNA by spermatozoa and delivered to the oocytes at fertilization.

Macroevolution therefore involves epigenetic and epigenetically directed genetic changes, and is decoupled from the random microevolutionary accumulation of base sequence changes.

These processes (reviewed in greater detail elsewhere[2]) are part and parcel of the fluid genome,[77] a molecular "dance of life" that is necessary for survival. In summary, there is continuity of macroevolutionary epigenetic and genetic changes due to fluid genome processes (see Box 4.2).

Heredity and Evolution in the Light of the New Genetics and Epigenetics

How should we see heredity in the light of the new genetics and epigenetics? Where does heredity reside if the genome itself is dynamic and fluid? Clearly, heredity does not reside solely in the DNA of the genome.

Ten years after the announcement of the human genome sequence, it has brought little progress in understanding life, health

Box 4.2

Continuity of Macroevolutionary Epigenetic and Genetic Changes (Ho 2003, 2004, 2009a)

- Epigenetically complex organisms are more efficient at generating genetic diversity locally and as required, by directed mutation, exonization, somatic hypermutations, RNA editing, alternatively splicing, trans-splicing, etc.
- Locally diversity-generating mechanisms feedback to rewrite genomic and germline DNA with functionally tested gene sequences, especially in the brain and the immune system.
- Epigenetic complex organisms actively decrease genomic and germline diversity by gene conversion, concerted evolution and sperm-mediated gene transfer.
- Macroevolution involves epigenetic and epigenetically directed genetic changes and is decoupled from random microevolutionary accumulation of base sequence changes.

or disease. Herculean efforts to locate the genes responsible for common diseases yielded next to nothing,[78] not surprising at all, given the fluidity of the genome and associated complexity of epigenetic mechanisms.

It has been clear to some of us since before the Human Genome Project was conceived, and copiously corroborated by the findings since, that heredity resides in an epigenetic state, a dynamic equilibrium between genetic/epigenetic and other cellular processes. But heredity does not end at the boundary of cells or organisms. As organisms engage their environments in a web of mutual feedback interrelationships, they transform or maintain their environments, which are also passed on to subsequent generations as home ranges and other cultural artefacts.[16,17,41,79] Embedded between organisms and their environment are social habits and traditions, an inseparable part of the entire dynamical complex that give rise to the stability of the developmental process, and which we recognize as heredity.[16,17,79] Heredity is thus distributed over the whole web of organism-environment interrelationships, where changes and adjustments are constantly taking place, propagating through all space-time in the maintenance of the whole, and some of these changes may involve genomic DNA. Thus, the fluidity of the genome is a *necessary* part of the dynamic stability, for genes must also be able to change as appropriate to the system *as a whole* (see Fig. 4.4).

While the epigenetic approach fully reaffirms the fundamental holistic nature of life and discredits any theory ascribing putative group differences in human attributes to genes,[80] it also gives no justification to *simplistic mechanistic* ideas on arbitrary effects arising from use and disuse or the inheritance of acquired characters. It does not lead to any kind of determinism, environmental or genetic. Organisms are above all, complex, nonlinear dynamical systems,[43] and as such, they have regions of stability and instability

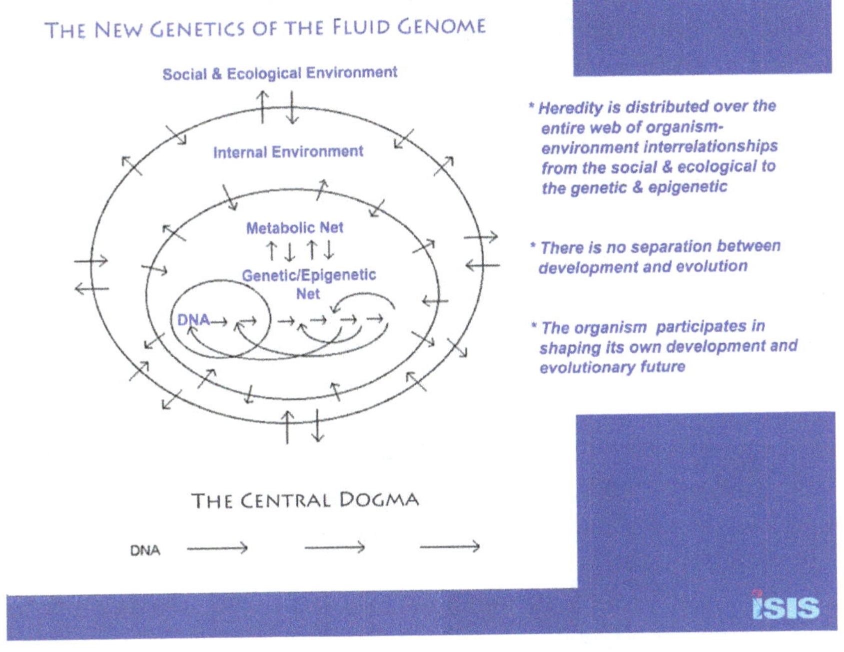

Figure 4.4 Heredity of the fluid genome versus the central dogma.

that enable them to maintain homeostasis, or to adapt to change, or not, as the case may be. The appearance of novelties and of mass extinctions alike in evolutionary history are but two sides of the same coin; we cannot be complacent about the capacity of organisms to adapt to any and all environmental insults that are perpetrated, the most pressing of which is anthropogenic global warming. The dynamics of the developmental process ultimately holds the key to heredity and evolution, in determining the sorts of changes that can occur, in its resilience to certain perturbations and susceptibility to others. And our knowledge in this crucial area is urgently required.

What implications are there for evolution? Just as interaction and selection cannot be separated, nor are variation (or mutation) and selection, for the "selective" regime may itself cause specific epigenetic variations or "directed" mutations. The organism experiences its environment in one continuous nested *process*, adjusting and changing, leaving imprints in its epigenetic system, its genome as well as on the environment, all of which are passed on to subsequent generations. Thus, *there is no separation between development and evolution*. In that way, the organism actively participates in shaping its own development as well as the evolution of its ecological and social community. We do hold the future in our hands, and we must be careful.

References and Notes

1. Ho, M.W. and Saunders, P.T. "Beyond Neo-Darwinism: An Epigenetic Approach to Evolution". *J Theor Biol* 78 (1979): 573–591.
2. Ho, M.W. "Development and Evolution Revisited". In *Handbook of Developmental Science, Behavior and Genetics* (K. Hood, C. Halpern, G. Greenberg and R. Lerner, eds.), pp. 61–109, Blackwell Publishing, New York, 2009a.
3. Ho, M.W. "Epigenetic Inheritance: What Genes Remember". *Sci Soc* 41 (2009b): 4–5.
4. Bird, A. "Introduction Perceptions of Epigenetics". *Nature* 447 (2007): 396–398.

5. Ho, M.W. "Caring Mothers Strike Fatal Blow Against Genetic Determinism". *Sci Soc* 41 (2009c): 6–9.
6. Ho, M.W. "From Genomics to Epigenomics". *Sci Soc* 41 (2009d): 10–12.
7. Ho, M.W. "Epigenetic Toxicology". *Sci Soc* 41 (2009e): 13–15.
8. Ho, M.W. "Rewriting the Genetic Text in Human Brain Development and Evolution". *Sci Soc* 41 (2009f): 16–19.
9. Ho, M.W. Epigenetic Inheritance through Sperm Cells, the Lamarckian Dimension in Evolution". *Sci Soc* 42 (2009g): 40–42.
10. Lamarck, J.B. *Philosophie Zoologique*, Paris, 1809.
11. Burkhardt, R. *The Spirit of Systems*, Harvard University Press, Cambridge, Massachusetts, 1977.
12. Barthelemy-Madaule, M. *Lamarck the Mythical Precursor*, translated by M.H. Shank, MIT Press, Cambridge Massachusetts, 1982.
13. Ho, M.W. "Lamarck the Mythical Precursor: A book Review". *Paleon Ass Circ* 14 (1983): 10–11.
14. Ho, M.W. "Where Does Biological Form Come From?" *Rivista di Biologia* 77 (1984a): 147–179.
15. Darwin, C. *On the Origin of Species by Means of Natural Selection, or the Preservation of Favoured Races in the Struggle for Life*, John Murray, London, 1859.
16. Ho, M.W. "Environment and Heredity in Development and Evolution". In *Beyond Neo-Darwinism: An Introduction to the New Evolutionary Paradigm* (M.W. Ho and P.T. Saunders, eds.), pp. 267–290, Academic Press, London, 1984b.
17. Ho, M.W. "Heredity As Process". *Rivista di Biologia* 79 (1986): 407–447.
18. Ho, M.W. (1998). *Genetic Engineering Dream or Nightmare: The Brave New World of Bad Science and Big Business*. Third World Network, Gateway Books, MacMillan, Continuum, Penang, Malaysia, Bath, UK, Dublin, Ireland, New York, 1998, 1999, 2007 (reprint with extended Introduction).
19. Ho, M.W. "Darwin's Pangenesis: The Hidden History of Genetics and the Dangers of GMOs". *Sci Soc* 42 (2009h): 42–45.
20. Ho, M.W. "Evolution by Process, Not by Consequence: Implications of the New Molecular Genetics on Development and Evolution". *Int J Comp Psychol* 1 (1987): 3–27.

21. Baldwin, J.M. "A New Factor in Evolution". *Am Nat* 30 (1896): 441–451, 536–553.
22. Goldschmidt, R.B. *The Material Basis of Heredity*. Yale University Press, New Haven, 1940.
23. Løvtrup, S. *Epigenetics*. John Wiley & Sons, London, 1974.
24. Gould, S.J. and Eldredge, N. "Punctuated Equilibria: An alternative to Phyletic Gradualism". In *Models in Paleobiology* (T.J.M. Schopf, ed.), pp. 82–115, Freeman, Cooper & Co., San Francisco, 1972.
25. Valentine, J.W. *On the Origin of Phyla*. University of Chicago Press, Chicago, 2004.
26. Ho, M.W. "Living with Oxygen". *Sci Soc* 43 (2009i): 9–12.
27. Carroll, S.B. *The New Science of Evo Devo, Endless Forms Most Beautiful*, W.W. Norton and C., New York, 2005.
28. Gilbert, S.F. "The Morphogenesis of Evolutionary Developmental Biology". *Int J Dev Biol* 47 (2003): 467–477.
29. Brakefield, P.M. "Evo-Devo and Constraints on Selection". *Trends Ecol Evol* 21 (2006): 362–367.
30. Blumberg, M.S. *Freaks of Nature: What Anomalies Tell Us About Development and Evolution*, Oxford University Press, Oxford, 2009.
31. Coyne, J.A. "Evolution's Challenges to Genetics". *Nature* 457 (2009): 382–383.
32. Saunders, P.T. "Development and Evolution". In *Beyond Neo-Darwinism: An Introdution to the New Evolutionary Paradigm* (M.W. Ho and P.T. Saunders, eds.), pp. 243–263, Academic Press, London, 1984.
33. Webster, G. and Goodwin, B.C. "The Origin of Species: A Structuralist Approach". *J Soc Biol Struct* 5 (1982): 15–47.
34. Gottlieb, G. "Normally Occurring Environmental and Behavioural Influences on Gene Activity: From Central Dogma to Probabilistic Epigenetics". *Psychol Rev* 105 (1998): 792–802.
35. Peel, A.D., Chipman, A.D. and Akam, M. "Arthropod Segmentation: Beyond the *Drosophila* Paradigm". *Nat Rev Genet* 6 (2005): 905–916.
36. Palmer, A.R. "Symmetry Breaking and the Evolution of Development". *Science* 306 (2004): 828–833.
37. Kondo, S. and Miura, T. "Reaction-Diffusion Model As a Framework for Understanding Biological Pattern Formation". *Science* 329 (2010): 1616–1620.

38. Waddington, C.H. *The Strategy of the Genes*, Allen & Unwin, London, 1957.
39. Saunders, P.T. "The Epigenetic Landscape and Evolution". *Biol J Linn Soc Lond* 39 (1990): 125–134.
40. Lowenstein, J.M. "Molecular Phylogenetics". *Ann Rev Earth Planet Sci* 14 (1986): 71–83.
41. Ho, M.W and Saunders, P.T. "Epigenetic Approach to the Evolution of Organisms: With Notes on Its Relevance to Social and Cultural Evolution. In *Learning, Development, and Culture* (H.C. Plotkin, ed.), pp. 343–361, Wiley, London, 1982.
42. Ho, M.W. and Saunders, P.T., eds. *Beyond Neo-Darwinism: An Introduction to the New Evolutionary Paradigm*, Academic Press, London, 1984.
43. Saunders, P.T. "The Organism As a Dynamical System". In *Thinking About Biology* (W. Stein and F.J. Varela, eds.), pp. 41–63, Addison-Wesley, Reading, Massachusetts, 1992a.
44. Kawai, M. (1962) "On the Newly-Acquired Behaviours of the Natural Troop of Japanese Monkeys on Koshima Island". Abstracts of the papers read in the seventh annual meeting of the Society for Priimate Researches, 22–24 November 1962, Japan Monkey Centre, Inuyama, http://www.springerlink.com/content/w85j34568088u82q/.
45. Ho, M.W., Tucker, C., Keeley, D. and Saunders, P.T. "Effects of Successive Generations of Ether Treatment on Penetrance and Expression of the Bithorax Phenocopy". *J Exp Zool* 225 (1983): 1–12.
46. Ho, M.W. "Nurturing Nature". In *Genetic Explanations: Sense and Nonsense* (S. Krimsky and J. Gruber, eds.), Harvard University Press, Harvard, 2011a.
47. Thompson, D'Arcy W. *On Growth and Form*, abridged edition (J.T. Bonner, ed.), Cambridge University Press, Cambridge, 1961.
48. Turing, A. "The Chemical Basis of Morphogenesis". *Phil Trans B* 237 (1952): 37–72.
49. Saunders, P.T., ed. *Alan Turing's Collected Works: Morphogenesis*. Elsevier, Amsterdam, 1992b.
50. Saunders, P.T. "Alan Turing and Biology". *IEEE Ann Hist Comput* 15 (1993): 33–36.

51. Saunders, P.T. "Mathematics, Structuralism and the Formal Cause in Biology". In *Dynamic Structures in Biology* (B.C. Goodwin, G.C. Webster and A. Sibatani, eds.), pp. 107–120, Edinburgh University Press, Edinburgh, 1989. (Japanese translation: Yoshioka Shoten, Kyoto, 1991.)
52. Ho, M.W. *The Rainbow and the Worm: The Physics of Organisms*, 3rd Edition. World Scientific, Singapore, 2008.
53. Ho, M.W., Stone, T.A., Jerman, I., Bolton, J., Bolton, H., Goodwin, B.C., Saunders, P.T. and Robertson, F. "Brief Exposure to Weak Static Magnetic Fields during Early Embryogenesis Cause Cuticular Pattern Abnormalities in *Drosophila* Larvae". *Phys Med Biol* 37 (1992): 1171–1179.
54. Ho, M.W. "Genes Don't Generate Body Patterns". *Sci Soc* 52 (2011b): 16–19.
55. Nusslein-Volhard, C. "Gradients That Organize Embryo Development". *Sci Am* 275 (2006): 54–55, 58–61.
56. Gregor, T., Wieschaus, E.F., McGregor, A.P., Bialek, W. and Tank, D.W. "Stability and Nuclear Dynamics of the Bicoid Morphogen Gradient". *Cell* 130 (2007): 141–152.
57. Gibson, M.C. "Bicoid by the Numbers: Quantifying a Morphogen Gradient". *Cell* 130 (2007): 14–16.
58. Goodwin, B.C., Webster, G. and Sibatani, A. eds. *Dynamic Structures in Biology*, Edinburgh University Press, Edinburgh, 1989.
59. Ho, M.W. "How Rational Can Rational Morphology Be? A Post-Darwinian Rational Taxonomy Based on the Structuralism of Process". *Rivista di Biologia* 81 (1988a): 11–55.
60. Ho, M.W., Matheson, A., Saunders, P.T., Goodwin, B.C., and Smallcombe, A. "Ether-Induced Segmentation Disturbances in *Drosophila melanogaster*". *Roux's Arch Dev Biol* 196 (1987): 511–524.
61. Ho, M.W. "An Exercise in Rational Taxonomy". *J Theor Biol* 147 (1990): 43–57.
62. Ho, M.W. and Saunders, P.T. "Rational Taxonomy and the Natural System with Particular Reference to Segmentation". *Acta Biotheoretica* 41 (1993): 289–304.
63. Saunders, P.T. and Ho, M.W. "Reliable Segmentation by Successive Bifurcation". *Bull Math Biol* 57 (1995): 539–556.

64. Gould, S.J. *Ontogeny and Phylogeny*. Belknap Press, Harvard University, Cambridge, Massachusetts, 1977.
65. Ho, M.W. and Saunders, P.T. "Rational Taxonomy and the Natural System: Segmentation and Phyllotaxis". In *Models in Phylogeny Reconstruction* (R.W. Scotland, D.J. Siebert and D.M. Williams, eds.), pp. 113–124, The Systematics Association Special Volume 52, Oxford Science, Oxford, 1994.
66. Douady S. and Couder Y. "Phyllotaxis As a Physical Self-Organized Growth Process". *Phys Rev Lett* 68 (1992): 2098–2101.
67. Ho, M.W. "In Search of the Sublime: Significant Form in Science and Art". *Sci Soc* 39 (2008): 4–11.
68. Kimura, M. "Evolutionary Rate at the Molecular Level". *Nature* 217 (1968): 624–626.
69. Zuckerkandl, E. and Pauling, L.B. "Molecular Disease, Evolution and Genetic Heterogeneity". In *Horizons in Biochemistry* (M. Kasha and B. Pullman, eds.), pp. 189–225, Academic Press, New York, 1962.
70. Margoliash, E. "Primary Structure and Evolution of Cytochrome C." *Proc Natl Acad Sci* 50 (1963): 672–679.
71. Huang, S. "Inverse Relationship between Genetic Diversity and Epigenetic Complexity". *Nature Precedings*: doi:10.1038/ripre.2009.1751.2: Posted 13 Jan 2009, http://precedings.nature.com/documents/1751/version/2.
72. Ho, M.W. "To Mutate or Not to Mutate". *Sci Soc* 24 (2004): 9–10.
73. Mattick, J.S. and Mehler, M.F. "RNA Editing, DNA Recoding and the Evolution of Human Cognition". *Trends Neurosci* 31 (2008): 227–233.
74. Steele, E.J. "Reflections on the State of Play in Somatic Hypermutation". *Molec Immunol* 45 (2008): 2723–2726.
75. Rothenfluh, H.S. and Steele, E.J. "Origin and Maintenance of Germ-Line V Genes". *Immunol Cell Biol* 71 (1993): 227–232.
76. Steele, E.J. *Somatic Selection and Adaptive Evolution*. Williams-Wallace International Inc., Toronto, 1979.
77. Ho, M.W. *Living with the Fluid Genome*. ISIS/TWN, London/Penang, 2003.

78. Ho, M.W. "Ten Years of the Human Genome: Reams of Data and No Progress in Sight". *Sci Soc* 48 (2010a): 22–25.
79. Ho, M.W. "Genetic Fitness and Natural Selection: Myth or Metaphor". In *Proc. 3rd Schneirla Conference,* Lawrence Erlbaum, New Jersey, 1988b.
80. Ho, M.W. "Celebrating the Uses of Human Genome Diversity and Dissecting the Controversies". *Sci Soc* 48 (2010b): 26–29.

5

Nurturing Nature

How parental care affects adolescent and adult offsprings' health, behaviour and genes, but neither the genes nor the environment actually determines who we are and what we must do to make the best of it.

Why a Leading Biologist Leads Social Critique against Genetic Determinism

I first met Ruth Hubbard, Emeritus Professor at Harvard University, and her husband George Wald (1906–1997), American Nobel laureate physiologist, at a conference on *Towards a Liberatory Biology* in Bressanone in the Italian Alps.[1,2] Ruth was already a leading light in the radical critique of genetic determinism — the idea that organisms are hardwired in their genetic makeup — from a broad socio-political perspective. As a research scientist who had worked on visual pigments together with her husband for many years, she was by no means unaware of hormones and enzymes encoded by genes that enable an organism to transform energy, grow and develop. But she insisted that there are social determinants for what people are, or perceived to be, much more powerful than biology and genes (and she has been proven right).

I suspect that she was getting rather impatient with the anodyne and frequently opaque rhetoric of many sociologists that fail to come to grips with the real issues, not to mention the obfuscation by "bio-ethicists", some of whom were a contradiction in terms. The unsuspecting public was left to the mercy of slick propaganda from vested interests intent on profiting by blaming people's ills on their genes and selling them both the diagnosis and appropriate remedies: abortion for the unborn, gene drugs and gene therapies for adults scared witless after having tested positive for genes that will give them incurable diseases. Ruth's book co-authored with her son Elijah Wald — *Exploding the Gene Myth: How Genetic Information Is Produced and Manipulated by Scientists, Physicians, Employers, Insurance Companies, Educators and Law Enforcers*[3] — is admirable for delivering its important message clearly, succinctly, and with punch and panache, true to how she is in real life.[4]

How Scientific and Social Critiques Converge

My own critique of genetic determinism began in the 1970s[5] and is much more based on science, though very mindful of how the science is manipulated and abused by vested interests without regard for ethics or safety, as in the promotion of genetic modification.[6] I take science to be *reliable knowledge of nature that enables us to live sustainably with her*.[7] This certainly does *not* understate the large influences that society and politics have on science and more to the point, what passes for science, which can be very much mistaken and unreliable as is the case of genetic determinism. Science as reliable knowledge of nature is what we live by, and hence has large implications on how we live, and choose to live.[8]

My critique converges with that of Ruth Hubbard, because social and environmental influences are indeed powerful determinants on how we grow and develop, precisely as Ruth has been saying; so much so that social and environmental influences can mark and change our genes for life. That is what the Human Genome Project to sequence the entire human and other genomes has ended up telling us, despite the fact that it was inspired and promoted by genetic determinism.

The new genetics of the "fluid genome" had already emerged by the early 1980s, long before the Human Genome Project was conceived.[9] It belongs in the organic paradigm of spontaneity and freedom[10] that defies any kind of determinism, biological or environmental.

A Decade of the Human Genome Yields Next to Nothing

In 2000, Clinton announced the first draft of the human genome sequence, and said it would "revolutionize the diagnosis, prevention and treatment of most, if not all human diseases." Francis Collins, then director of the genome agency at the US National Institutes of Health said that genetic diagnosis of diseases would be accomplished in ten years and that treatments would start to roll out perhaps five years after that.[11]

The anti-climax came just eight months later when the complete map was announced.[12] Chief gene-sequencer Craig Venter admitted: "We simply do not have enough genes for this idea of biological determinism to be right." The environment, he said, is critical.

Ten years on, and genomics research has yielded *no* cures, while the hope of identifying genes for common diseases is fast receding. Nina Paynter and her research team at Brigham and Women's Hospital in Boston looked at 101 genetic variants (single nucleotide polymorphisms) from whole genome scans that had been linked to heart disease. These, together, turned out to be of no value in predicting the disease among 19 000 women followed up for 12 years.[13] In contrast, family history was the most significant predictor, as it had been before genomics. As Harold Varmus, now director of the National Cancer Institute, said: "Genomics is a way to do science, not medicine."[12]

Demise of Genetic/Biological Determinism

The assumption that genes are stable and insulated from environmental influences is pivotal to neo-Darwinian theory, and the root and stem of genetic/biological determinism.[2,5,6,9] It was inspired by the theory of the separate and inviolable germplasm due to German evolution biologist August Weismann (1834–1914), which had been flawed from the start. Plants do not have separate germ cells at all; every somatic cell is potentially capable of becoming a germ cell, which is why plants can be propagated from cuttings (and genetically modified plants so readily created in cell cultures). Most animals also do not have germ cells that separate from the rest of the body early in development. Furthermore, there is no evidence that genes in germ cells are stable or immune from environmental influences. We now know that the environment can impact directly on the germ cells in the developing foetus and passed on to subsequent generations. Toxic environmental substances such as bisphenol A and other endocrine disruptors specifically affect germ cells in the developing foetus, giving rise to the grandmother effect.[14] Even more surprisingly, sperm cells are efficient vehicles for carrying foreign (altered) genes into egg cells at fertilization[15] and the foreign genes can be expressed in the embryos developed from the fertilized eggs.[16]

Evidence that genes are neither stable nor immune from direct environmental influence has been accumulating almost as soon as genetic engineering began in the mid-1970s and applied to unravelling the detailed molecular machinery of genetics. To their astonishment, molecular geneticists soon witnessed classical genetics being turned upside-down on their own lab benches. They found exceptions and violations to every tenet of classical genetics that had been accepted without question for decades. In direct contradiction to the concept of a relatively static genome with linear causal chains emanating from genes to the organism and the environment, they discovered constant cross talk between genome and environment. Feedback from the environment not only determines which genes are turned on where, when, by how much and for how long, but marks, moves and changes the genes themselves. By the early 1980s, molecular geneticists had already coined the term "fluid genome" to capture what I would later describe as a molecular "dance of life" necessary for survival.[9]

Just when we finally got used to thinking that a gene in molecular genetics was the coding sequences (eventually read out as amino acid sequence of proteins) equipped with various control regions for start and stop that would determine how actively the gene is expressed, we need to think again. New research by a large consortium of molecular geneticists is revealing that such "genes" are in bits dispersed throughout the genome, interweaving with bits of other genes. As genes are intertwined, so are the functions.[17] Multiple DNA sequences may serve the same function, and conversely the same DNA sequence can have different functions. It is futile to try and define a gene or a separable function for any piece of DNA. This is ultimately why genes for common diseases can never be found. And incidentally, it is also why genetic modification is both dangerous and futile: human genetic engineers do not know the steps of this incredibly complex molecular dance of life.[18] All they can do, even now, is to follow and marvel at some of the footprints of this dance, marks left on the DNA and the histone proteins bound to the DNA, a script that an individual will pass on to the next generation.

Myth of Genetic Determinism Still Perpetrated in Academia

Mainstream genetics research during the decades since the discovery of the DNA double helix in 1953 has focussed on identifying "genes" or "genetic predisposition" for every 'trait', real or imaginary, and continues on today.[9] Imaginary traits are rife in the hybrid discipline of "evolutionary psychology", long dedicated to inventing stories on "selective advantage" for each of the "traits" so that the corresponding gene could become "fixed" in the population by neo-Darwinian natural selection.

Another hybrid discipline, "behavioural genetics", formerly dedicated to studies based on identical twins, began identifying DNA (gene) markers for behaviour; and indeed claimed to have found one for increased tendency towards violent behaviour in boys who experienced maltreatment in childhood.[19] The gene encoding the enzyme monoamine oxidase A (MAOA) — involved in the metabolism of neurotransmitters — exists in two variants, one expressing high activity, the other, low activity. While all boys in the study showed increased "disposition towards violence" if they received maltreatment as children, those with low enzyme activity appeared to show a greater increase. The researchers claimed a weak residual effect due to the low-activity MAOA, while conceding the large effect of the environment. But even this weak genetic predisposition soon faded away as more data became available.[20]

Behavioural geneticists are not the only researchers wasting time and resources chasing "will o' the wisp" gene markers. The project to map genetic predisposition to diseases was the main rationale for the $3 billion Human Genome Project that decades later delivered next to nothing, basically because it is not genomic DNA but "epigenetic" environmental influences that overwhelmingly affect our health and well-being.[21]

Epigenetic Inheritance

The term "epigenetic" came from *epigenesis*, the process whereby an organism with differentiated organs, tissues and cells develop from a relatively featureless egg. Developmental geneticist and evolutionist Conrad Waddington (1905–1975) invented the concept of the "epigenetic landscape"[22] to represent the dynamical structure of the developmental system that defines the range of non-random changes for evolution, and that was the sense in which we had used "epigenetic" in 1979 in opposition to genetic determinism.[5] Nowadays, "epigenetic" usually refers to a heritable change that does not involve DNA sequence alteration,[23] but that is becoming rapidly obsolete, due to epigenetic mechanisms that actually change DNA sequences directly, or via an RNA intermediate that undergoes editing, alternative splicing, etc., coupled with reverse transcription of the RNA into DNA that may be slotted back into the genome. All that has large implications on how development can direct evolution.[24]

New research has abundantly confirmed the overriding importance of environmental influences across the disciplines: from nutrition to toxicology, and most dramatically in brain development.[25]

Epigenetic inheritance is effectively the inheritance of acquired characters[26] usually attributed to French evolutionist and polymath Jean-Baptiste de Lamarck (1744–1829), and it has come into its own in maternal effects. Before we go into molecular epigenetics, we look at how maternal/parental care in early childhood impacts subsequent development in adolescence and adulthood.

Early Maternal/Parental Effects on Subsequent Development

Maternal/parental effects on development are well known and demonstrated across many species. In mammals, the long period of gestation and postnatal mother-child relationship provide maternal influences that extend well into the adult life of the offspring.

Prenatal stress[27] and malnutrition[28] experienced by the mother affects her neuroendocrine system and in turn, the development of the nervous system in the foetus. The care received (usually from the mother, but can be substituted by surrogates) during early infancy can produce changes in the development of the nervous system that regulate its response to novelty and social behaviour.[29] Thus, the maternal environment experienced by a developing organism can play a critical role in shaping its adult behaviour.

Infant rhesus macaques socially isolated for periods of 3–12 months play much less, are highly aggressive with peers, perform poorly in learning and cognitive discrimination tasks, and are inhibited and fearful of novelty.[30] These behavioural patterns continue into adulthood and affect reproductive success, particularly in artificially reared females, who display high rates of infant abuse, neglect and infanticide. Maternally deprived macaques also have an elevated hypothalamic-pituitary-adrenal (HPA) response to stress (see later), impairments in learning and social behaviour, and altered serotonergic systems (which regulate anxiety), suggesting that it is the disruption of the mother-infant relationship rather than the general consequence of social isolation that contributes to these effects.

In humans, numerous studies have shown that environmental adversity early in life is associated with increased risks of both physical and psychiatric disorders in adolescence and adulthood. Experience of childhood adversities that form a maladaptive family function (MFF) cluster — parental mental illness; substance abuse disorder and criminality; family violence; physical abuse; sexual abuse; and neglect — is highly correlated with psychiatric disorders, with onset from childhood through adolescence to adults.[31] In a study involving 18 303 adult participants from ten countries, a history of three or more childhood adversities (physical abuse, sexual abuse, neglect, parental death, parent divorce, other parental loss, parental mental disorder, parental substance use, parental criminal behaviour, family violence and family economic adversity) was associated with the onset of all six physical conditions investigated: heart disease, asthma, diabetes, arthritis, chronic spinal pain and chronic headache (hazard ratios from 1.44 to 2.19), and early onset mental disorders were independently associated with the onset of five physical conditions (except diabetes, hazard ratios from 1.43 to 1.66).[32] The most recent large survey in the United States including ten states and the district of Columbia looked at nine adverse childhood experiences (ACEs) — physical, sexual and emotional abuse; household member mental illness; alcoholism; drug abuse; imprisonment; divorce and intimate partner violence — and their association with chronic physical illnesses. The results show that compared to those reporting no exposure to ACEs, the adjusted odds of myocardial infarction, asthma, fair/poor health, frequent mental distress and disability were higher for those reporting one to three, four to six, or seven to nine ACEs. The odds of coronary heart disease and stroke were higher for those who reported four to six and seven to nine ACEs; the odds of diabetes were higher for those reporting one to three and four to six ACEs.[33]

There is substantial evidence that lack of parental care or childhood abuse can contribute to subsequent criminal behaviour. It has been estimated that up to 70% of abusive parents were themselves abused[34] and 20–30% of abused infants are likely to become abusers. These findings in humans are replicated in experiments on primates.[35]

A meta-analysis compared rates of sexual and other forms of abuse reported in 17 studies involving 1,037 sex offenders and 1,762 non-sex offenders, and also examined the prevalence of different forms of abuse in 15 studies that compared adult sex offenders against adults (n = 962) and against children (n = 1 334) to see if the sexually abused-sexual abuser association is even more specific to individuals that sexually abuse children.[36] The results showed a higher prevalence of sexual abuse history among adult sex offenders than among non-sex offenders (odds ratio 3.36, 95% confidence interval, 2.23–4.82). The two groups did not differ significantly with regard to a history of physical abuse. There was a significantly lower prevalence of sexual abuse history among sex offenders against adults compared to sex offenders against children (odds ratio 0.51, 95% confidence interval, 0.35–0.74), whereas the opposite was found for physical abuse (odds ratio 1.43, 95% confidence interval, 1.02–2.02).

The effects of childhood adversities on untoward behaviour can appear much earlier during adolescence. A study on adolescent violent behaviour and ACEs was carried out on data from 136,549 students in the 6th, 9th and 12th grades who responded to the 2007 Minnesota Student Survey, an anonymous self-report survey.[37] The results showed that 28.9% reported at least one ACE, the most common being alcohol abuse by a household family member that caused problems. Every type of ACE on the survey was significantly associated with adolescent interpersonal violence, weapon-carrying on school grounds and self-directed violence (self-mutilation, suicidal ideation and suicide attempt). For each additional ACE reported, the risk of violence perpetration increased by 35–144 %.

There were significant gender differences in the results. For girls, the risk of violence perpetration was increased 1.7- to 5-fold by any ACE, the highest being suicide attempt associated with physical abuse by a household adult and sexual abuse by a non-family member. For boys, the risk of violence perpetration was increased 1.7- to 44-fold by any ACE, the highest being dating violence associated with sexual abuse by a family member. The likelihood of adolescent violence perpetration increased as the number of adverse events reported by the student increased. Compared with those who reported no ACE, the risk of violence perpetration adolescents with four or more ACEs was 2- to 7-fold for girls (bullying and suicide attempts, respectively) and 2.7- to 10-fold for boys (bullying and suicide attempts, respectively).

Clearly, the early environment has a large impact on the psychiatric and physical health of humans (and of model animals), but it does not *determine* whether children will inevitably grow up unhealthy or to be criminals, any more than their genetic makeup does. More importantly, changing the environment can do a lot to prevent adverse outcomes, and can often undo the harm that individuals or their parents have experienced in early life.

Thus, a public health approach to parenting — the Positive Parenting Program in the United States — has significantly reduced a number of child-maltreatment indicators for participating communities compared with control non-participating communities, such as rates of confirmed child abuse, out-of-home placements, emergency department visits, and hospitalization for child abuse injuries.[38]

At the same time, other programmes aimed at youth mental health intervention based on behavioural therapy have helped to identify negative or damaging thought patterns and generate alternative interpretations that are more hopeful, and may yet reverse the damage done to adolescents with histories of ACEs.[39]

It will be instructive to examine the epigenetics of maternal behaviour in an animal model that has been thoroughly investigated in the laboratory within the past decade. For as long as anyone can remember, people have been debating the following question: Is it our genetic makeup or the environment that determines who we are? The research findings on maternal care are telling us that the question is the wrong one to ask, for neither our genetic makeup nor the environment determines who we are. The question we should be asking is: How can we give everyone the best opportunity of being what they want to be?

Epigenetics of Maternal Behaviour

Researchers at McGill University in Montreal, Canada, and Columbia University in New York, United States, have been studying maternal behaviour in rats for many years. They found that mother rats who care adequately for their pups and others who do not, shape their offspring's responses to stress accordingly for the rest of their lives, and that is correlated with different states of expression in relevant genes.[40–45]

The mother rat licks and grooms her pups in the nest and while nursing them also arches her back. Some mothers (high-LG) do that more often than others (low-LG). The offspring of high-LG mothers grow up less fearful and able to cope with stress than those of low-LG mothers, and this involves the hypothalamus-pituitary-adrenal (HPA) pathway of response to stress. The magnitude of the HPA stress response is a function of the corticotrophin-releasing factor (CRF) secreted by the hypothalamus, which activates the pituitary-adrenal system. This is in turn modulated by glucocorticoid secreted in the hypothalamus, which feeds back to inhibit CRF synthesis and secretion, thus dampening the HPA response, and restoring homeostasis (dynamic balance).

The adult offspring of high-LG mothers show increased glucocorticoid expression in the hippocampus, and enhanced sensitivity to glucocorticoid feedback. This enhanced sensitivity was due to the increased expression of glucocorticoid receptor (GR), boosted in turn by the increased expression of transcription factor NGF-1-A that binds to the promoter of the GR gene. These differences in gene expression states are accompanied by significant differences in DNA-methylation (addition of methyl-group) of the GR promoter, with low methylation from offspring of high-LG mothers correlating with high expression, and high methylation from offspring of low-LG mothers correlating with low expression. The researchers also found significantly higher acetylation of histone in chromatin protein around the GR gene (as consistent with active gene expression) in the offspring of high-LG than in the offspring of low-LG mothers.

Interestingly, cross-fostering the offspring of low-LG to high-LG mothers and vice versa at day 1 after birth induced changes in the offspring in line with the *foster* mother, with correlated changes in the gene expression states. (So, foster parents can influence their children biologically!)

It turns out that the different gene expression states are acquired during the first week of life, and persist into adulthood. Pups of both high-LG and low-LG mothers start out practically the same. Just before birth, the entire region of the GR promoter is unmethylated in both groups. That is because most gene marks are erased in the germ cells. Changes develop according to the behaviour of the mother within the critical period of the first week of life, and remain stable thereafter.

Nevertheless, these changes in DNA methylation and histone acetylation could be reversed, even in adults, as demonstrated by the rather drastic method of infusing chemical activators or inhibitors into the brain, with concomitant changes in the adult's response to stress.[45] Thus, infusing the histone deacetylase inhibitor Trichostatin A (TSA) into the brains of offspring from low-LG mothers increased histone acetylation, and decreased methylation of the GR promoter, thus boosting GR expression to levels indistinguishable from the brains of offspring from high-performing mothers. And when tested for anxiety levels, they performed like offspring from high-LG mothers. On the other hand, injecting methionine, the precursor of S-adenosyl methionine (SAM), the co-factor of DNA methylase, into the brains of offspring from high-LG mothers increased methylation of the GR promoter to levels the same as those of offspring from low-performing mothers, thereby decreasing GR expression and causing them to switch their behaviour accordingly to resemble that of offspring from low-LG mothers.

Thus, epigenetic states are stable yet dynamic and plastic, giving no support to any kind of determinism, genetic or environmental.

Maternal Care and Sex Hormones

What predisposes mothers to be caring or otherwise?

Apparently, the female offspring inherit the characteristics of their mothers when it comes to maternal care, not genetically, but epigenetically.

The hippocampus is the "emotion centre" of the brain. It is vulnerable to stress and richly supplied with receptors for the sex and reproductive hormones, and maternal care is regulated by those hormones.

In the rat, the researchers found oxytocin receptors linked to the expression of maternal behaviour.[35] Oxytocin (OT) is a hormone secreted by the posterior pituitary gland, and it stimulates the contraction of the uterus and ejection of milk. Variations in OT receptor levels in critical brain regions, such as the medial preoptic area (MPOA) of the hypothalamus, are associated with differences in maternal care. OT receptor binding in the MPOA is increased in high-LG compared with low-LG mothers. Furthermore, differences in OT receptor binding in the MPOA between high-LG and low-LG females are dependent on oestrogen; it is eliminated by ovariectomy and reinstated with oestrogen replacement. However, whereas ovariectomized high-LG females respond to oestrogen with an increase in OT receptor binding, low-LG females show no such effect. Studies with mice suggest that oestrogen regulation of OT receptor binding in the MPOA requires the α-subtype of the oestrogen receptor (ERα).

ERα is a transcription factor that regulates gene transcription on binding oestrogen. The cellular response to oestrogen depends on the amount of ERα present.

The researchers found that by day 6 after birth, ERα expression in the MPOA of female offspring from high-LG mothers is significantly increased compared with that of female offspring from low-LG mothers, and this state continues into adulthood, which is correlated with the female offspring of high-LG and low-LG mothers becoming high-LG and low-LG mothers accordingly, and this epigenetic state perpetuates itself via the female line until and unless disrupted by environmental intervention.

Cross-Fostering Reverses the Damage

One effective environmental intervention is cross-fostering. The biological offspring of high-LG and low-LG mothers were reciprocally exchanged within 12 hours of birth and reared to adulthood. When these offspring were examined, ERα expression in the MPOA of the adult females born to low-LG mothers but cross-fostered to high-LG mothers became indistinguishable from that of the normal biological offspring of high-LG mothers; and conversely, ERα expression in the MPOA of adult females born to high-LG mothers but reared by low-LG mothers resembled that of normal biological offspring of low-LG mothers. Cross-fostering as such had no effect, so exchanging offspring between two low-LG mothers or two high-LG mothers did not alter the expression of ERα in the MPOA of the offspring.

Correlated with the high and low ERα expression in the MPOA were significant differences in the methylation of CpG sites across the entire ERα promoter. Overall, significantly elevated levels of methylation were found in the promoter of offspring with low ERα expression in the MPOA compared with high ERa expression in the MPOA.

Maternal Care Influences Brain Development and Many Gene Functions

Obviously, maternal care does not just influence a few genes. The McGill University team has previously found that in the rat, increased anxiety in response to stress in the offspring from low-LG mothers is associated with decreased neuronal development and density of synapses in the hippocampus. The offspring of high-LG mothers, on the other hand, show increased survival of neurons and synapses in the hippocampus, and improved cognitive performance under stressful conditions.[42] Researchers at University of Amsterdam and Leiden University in the Netherlands have also found that the pyramidal neurons in layers 2/3 of the brain cortex from high-LG and low-LG adult rats have different morphologies.[44] The high-LG rat neurons have more slender "dendritic trees" — the branching processes receiving inputs from other neurons — with fewer branches than those from low-LG rats. The density of dendritic spines (small projections from the surfaces of the dentritic trees) is also significantly lower in high-LG rats. These observations suggest a rather extensive influence of maternal care on brain development and gene expression.

In order to examine the effect on gene expression of high-LG and low-LG mothers and TSA or methionine infusion, the four different treatment groups were compared with their respective control groups using microarrays to monitor changes in 31 099 unique mRNA transcripts.[45] A total of 303 transcripts (0.97%) were altered in the offspring of high-LG mothers compared to offspring of low-LG mothers: 253 transcripts (0.81 %) up-regulated and 50 transcripts (0.15%) down-regulated. TSA treatment of offspring of low-LG mothers altered 543 transcripts (1.75%): 501 transcripts (1.61%) up-regulated and the rest, 42 transcripts (0.14%), down-regulated. Methionine treatment of offspring of high-LG mothers changed 337 transcripts (1.08%), with 120 (0.39%) up-regulated and 217 (0.7%) down-regulated.

The results suggest that maternal care during the first week of life determines the expression of hundreds of genes in the adult offspring, but gene expression is nevertheless reversible even in the adult. Caring mothers tend to activate more genes in their offspring than mothers that do not provide adequate care. TSA treatment results predominantly in gene activation as expected, and methionine treatment results predominantly in silencing genes.

Epigenetic Effects of Childhood

Epigenetics of Enriched Environments

Environmental effects on the epigenome are much more extensive than just parental care. Researchers at Tufts University School of Medicine, Boston, Massachusetts and Rush University Medical Center, Chicago, Illinois, in the United States demonstrated that exposure of 15-day old mice for two weeks to an enriched environment that includes novel objects, increased social interactions and voluntary exercise, enhances long-term potentiation, not just in the mice exposed, but also in the future offspring of female mice through early adolescence, even if the offspring never experienced the enriched environment.[46] Long-term potentiation (LTP) is a persistent increase in strength of synapses between neurons following high-frequency stimulation, and is a form of synaptic plasticity known to be important for learning and memory. The effect of the enriched environment lasts for about two months, and was not cancelled by cross-fostering, indicating that the trans-generational effect occurs before birth, during embryogenesis. The effect is age-dependent, as it cannot be induced in adult mice. In both generations of mice, LTP induction is accompanied by the new appearance of a whole new signalling pathway, the "cAMP/p38 MAP (mitogen activated protein) kinase-dependent signalling cascade". If the effect occurs in humans, it means that an adolescent's memory can be influenced by environmental stimulation experienced by the mother when she was young.

Epigenetic Footprints of Childhood Trauma

Are the epigenetic footprints of maternal care identified in detailed animal studies relevant to the human species? Michael Meaney and his colleagues at McGill University have extended their findings in rats to humans. They examined epigenetic differences in a neuron-specific promoter of the glucocorticoid receptor in post-mortem hippocampus (12 in each group) obtained from suicide victims with a history of childhood abuse, suicide victims with no history of childhood abuse, and non-suicide victims that died from other causes, none of whom had a history of childhood abuse.[47] They found decreased expression of glucocorticoid receptor and increased DNA methylation of the specific promoter that binds the transcription factor NGF1-A in suicide victims with a history of childhood abuse compared with suicide victims without childhood abuse, which were indistinguishable from the controls. These are the same epigenetic footprints that the team had previously discovered in rodents that did not provide adequate maternal care.

Psychiatric disorders such as major depression and post-traumatic stress disorder are commonly connected with disorders of the cardiovascular, metabolic and immune system. Recent studies suggest that accelerated aging of cells may be an explanation. Telomeres are DNA repeats that cap the ends of chromosomes and make them more stable, and they shorten with each cell division, making them a marker for biological age. Physiological stress such as radiation and toxins, oxidative stress and cigarette smoke can shorten telomeres. The body responds to stress by the coordinated activities of several systems, including the hypothalamic-pituitary-adrenal (HPA) axis, the sympathetic nervous system, and the immune system. These systems mobilize energy and prepare the individual to cope with the stress. Chronic stress, however, can damage the endocrine, immune and metabolic systems, and may result in shortening the telomeres. Individuals giving care to Alzheimer's patients experience chronic stress, and when their white blood cells were examined, the telomeres were found to be shortened. The same telomere shortening has been linked to pessimism in healthy postmenopausal women and in patients with unipolar and bipolar mood disorders.[48]

Researchers at Butler Hospital and Brown University Medical School, Providence, Rhode Island, in the United States have now found that stress in childhood due to maltreatment also leads to telomere shortening.[48] Telomere shortening is a major risk factor for a range of adverse conditions, including major depression, anxiety disorders and substance abuse. The researchers tested 31 adults (22 women and 9 men), aged 18 to 64 years recruited via advertisement in the community for a larger study of stress reactivity and psychiatric symptoms. Of these, 21 reported either no history of childhood maltreatment, and 10 reported a history of moderate or severe childhood maltreatment. None had acute or unstable medical illness, endocrine diseases or ongoing treatment with drugs that might influence HPA axis functions. The maltreatment group did not differ significantly from the control group with respect to age, sex, smoking status, BMI (body mass index, a measure of obesity), hormonal contraception use in female subjects, race, education, socioeconomic status or perceived stress. The maltreatment group had significantly shorter telomeres than the control group, and was associated with both physical neglect and emotional neglect. The sample size was small, so there was no association of telomere length with age in the sample, which made the association with childhood abuse/neglect all the more significant.

Implications for Health

Although the epigenetic effects of maternal (parental) care have been worked out in most detail in rodents, there is potential for similar effects in other species including primates and humans, as recent evidence indicates.

In humans, lack of parental care or childhood abuse can contribute to subsequent criminal behaviour (see earlier). Furthermore, lack of parental care and parental overprotection ("affectionless control") is also a risk factor for depression, adult antisocial personality traits, anxiety disorders, drug use, obsessive-compulsive disorder and attention-deficit disorders. Conversely, people who reported high levels of maternal care were found to have high self-esteem, low anxiety and less salivary cortisol in response to stress.[35] Longitudinal studies demonstrated that mother-child attachment is crucial in shaping the cognitive, emotional and social development of the child. Throughout childhood and adolescence, secure children are more self-reliant, self-confident and have more self-esteem. Secure infants also have better emotional regulation, express more positive emotion and respond better to stress. Infant disorganized attachment has been associated with the highest risk of developing later psychopathology, including dissociative disorders, aggressive behaviour, conduct disorder and self-abuse.

Nutrition, Environmental Enrichment and Mental Health

The dramatic effects of TSA and methionine infusion in altering gene expression patterns in rats also have obvious implications for drug intervention, or better yet, intervention/prevention through adequate nutrition.[49] Epigenetic drugs such as inhibitors of DNA methylation or histone deacetylation lack specificity and may well have unintended untoward side effects.

Intervention through adequate nutrition is supported by substantial evidence.[49] In rats, dietary L-methionine has been shown to be crucial for normal brain development, and its deficiency implicated in brain aging and neurodegenerative disorders. Synthesis of SAM (cofactor for DNA methyl transferase) is dependent on the availability of dietary folates, vitamin B12, methionine, betaine and choline. Developmental choline deficiency alters SAM levels and global and gene-specific methylation. And prenatal choline availability has been shown to impact on neural cell proliferation, learning and memory in adulthood. Several studies have shown that additional dietary factors, including zinc and alcohol, can affect the availability of methyl groups for SAM formation and thereby influence CpG methylation. Maternal methyl supplements positively affect the health and longevity of the offspring.

Other studies have shown that certain dietary components may act as histone deacetylase inhibitors (HDACIs), including diallyl disulfide, sulforaphane and butyrate. For example, broccoli, which contains high levels of sulforaphane, has been associated with H3 and H4 acetylation in peripheral blood mononuclear cells in mice three to six hours after consumption.

HDACIs are an active area of research as anti-inflammatory and neuroprotective agents in autoimmune diseases such as lupus and multiple sclerosis. Sodium butyrate has been shown to have antidepressant effects in mice.

The new findings on environmental intervention, such as environmental enrichment in reversing the damages of social isolation, and fostering to reverse the harm of parental neglect, are indicative of the huge potential in saving our children with the appropriate social policies.

All in all, these remarkable findings on the epigenetic effects of maternal care show how important it is for societies to look after the welfare of children and mothers-to-be, in order to ensure both mental and physical health of future generations.

References and Notes

1. See Hubbard, R. (1981). "The Theory and Practice of Genetic Reductionism — From Mendel's Laws to Genetic Engineering". In *Against Biological Determinism* (C.M. Barker, L. Birke, A.D. Muir and S.P.R. Rose, eds.), Allison and Busby, London, 1981.
2. Ho, M.W. and Saunders, P.T. "Adaptation and Natural Selection: Mechanism and Teleology". In *Towards a Liberatory Biology* (C.M. Barker, L. Birke, A.D. Muir and S.P.R. Rose, eds.), Allison and Busby, London, 1981.
3. Hubbard, R. and Wald, E. *Exploding the Gene Myth: How Genetic Information is Produced and Manipulated by Scientists, Physicians, Employers, Insurance Companies, Educators and Law Enforcers*, Beacon Press, Boston, 1997.
4. Exploding the Gene Myth, a conversation with Frank R. Aqueno, 1997, accessed 22 August 2015, http://gender.eserver.org/exploding-the-gene-myth.html.
5. See Ho, M.W. and Saunders, P.T. "Beyond Neo-Darwinism: An Epigenetic Approach to Evolution". *J Theor Biol* 78 (1979): 573–591.
6. Ho, M.W. *Genetic Engineering Dream or Nightmare? The Brave New World of Bad Science and Big Business*. Third World Network, Penang, and Gateway Books, Bath, 1998. 2nd edition, McMillan Books, Ireland, and Continuum, New York, 1999. Translated into many languages and reprinted with updated and extended introduction, TWN, 2006, http://www.i-sis.org.uk/genet.php.
7. Ho, M.W., Novotny, E., Webber, P. and Daniels, E.E. "Towards a Convention on Knowledge". Draft 7, ISIS-SGR-TWN Discussion Paper, 2002, http://www.i-sis.org.uk/conventiononknowledge.php.
8. For a thorough philosophical justification of science as reliable knowledge of nature, see Ho, M.W. "Toward an Indigenous Western Science: Causality in the Universe of Coherent Space-Time Structures". In *New Metaphysical Foundations of Modern Science* (W. Harman and J. Clark, eds.), pp. 179–213, Institute of Noetic Sciences, Sausalito, 1994a.

9. Ho, M.W. *Living with the Fluid Genome*. ISIS/TWN, London/Penang, 2003, http://www.i-sis.org.uk/fluidGenome.php.
10. See Riley, D., McCraty, R. and Schneider, S. "Quantum Jazz Biology: Interview with Mae-Wan Ho". *Sci Soc* 47 (2010): 4–9.
11. Wade, N. "A Decade Later, Genetic Map Yields Few New Cures". *New York Times*, 12 June 2010, http://www.nytimes.com/2010/06/13/health/research/13genome.html?_r=1&th&emc=th.
12. See Ho, M.W. (2001). "Human Genome Map Spells Death of Genetic Determinism". *ISIS News*, February 2001, http://www.i-sis.org.uk/isisnews/i-sisnews7.php.
13. Paynter, N.P., Chasman, D.I., Paré, G., Buring, J.E., Cook, N.R., Miletich, J.P. and Ridke, P.M. "Association Between a Literature-Based Genetic Risk Score and Cardiovascular Events in Women". *JAMA* 303 (2010): 631–637.
14. See Ho, M.W. "Epigenetic Toxicology". *Sci Soc* 41 (2009a): 13–15.
15. Ho, M.W. "Epigenetic Inheritance through Sperm Cells: The Lamarckian Dimension in Evolution". *Sci Soc* 42 (2009b): 40–42.
16. Sciamanna, I., Vitullo, P., Curatolo, A. and Spadafora, C. "Retrotransposons, Reverse Transcriptase and the Genesis of New Genetic Information". *Gene* 448 (2009): 180–186.
17. ENCODE Project Consortium. "Identification and Analysis of Functional Elements in 1% of the Human Genome by the ENCODE Pilot Project". *Nature* 447 (2007): 799–816.
18. Ho, M.W. "GM Is Dangerous and Futile". *Sci Soc* 40 (2008): 4–8.
19. Caspi, A., McClay, J., Moffitt, T.C., Mill, J., Martin, J., Craig, I.W., Taylor, A. and Poulton, R. "Role of the Genotype on the Cycle of Violence in Maltreated Children". *Science* 297 (2002): 851–854.
20. Balaban, E. and Lewontin, R. "Criminal Genes". *GeneWatch*, March–April 2007.
21. Ho, M.W. "From Genomics to Epigenomics". *Sci Soc* 41 (2009c): 10–12.
22. Waddington, C.H. *The Strategy of the Genes*, Allen & Unwin, London, 1957.
23. "Epigenetics", Wikipedia, accessed 4 November 2008, http://en.wikipedia.org/wiki/Epigenetics.
24. Ho, M.W. "How Development Directs Evolution: Lamarck versus Darwin". In *Handbook of Developmental Systems: Theory and*

Methodology (C.M. Molenaar, R.M. Lerner and K.M. Newel, eds.), pp. 131–153, Guildford Press, New York, 2014.
25. Ho, M.W. "Rewriting the Genetics Text in Brain Development". *Sci Soc* 41 (2009d): 16–19.
26. Ho, M.W. "Epigenetic Inheritance: 'What Genes Remember'". *Sci Soc* 41 (2009e): 4–5.
27. Lindqvist, C., Janczak, A.D., Nätt, D., Baranowska, I., Lindqvist, N., Wichman, A., Lundeberg, J., Lindberg, J., Torjesen, P.A. and Jensen, P. "Transmission of Stress-Induced Learning Impairment and Associated Brain Gene Expression from Parents to Offspring in Chickens". *PLoS One* 4 (2007): e364.
28. Weinstock, M. The Potential Influence of Maternal Stress Hormones on Development and Mental Health of the Offspring". *Brain Behav Immun* 19 (2005): 296–308.
29. Martin-Gronert, M.S. and Ozanne, S.E. "Maternal Nutrition during Pregnancy and Health of the Offspring". *Biochem Soc Trans* 34 (2006): 779–792.
30. Reviewed in Champagne, F.A. and Curley, J.P. "Epigenetic Mechanisms Mediating the Long-Term Effects of Maternal Care on Development". *Neurosci Biobehav Rev* 33 (2009): 593–600.
31. Green, J.G., McLaughllin, K.A., Berglund, P.A., Gruber, M.J., Sampson, N.A., Zaslavsky, A.M. and Kessle, R.C. "Childhood Adversities and Adult Psychiatric Disorders in the National Comorbidity Survey Replication: I. Associations with First Onset of DSM-IV Disorders". *Arch Gen Psychiatry* 7 (2010): 113–123.
32. Scott, K.M., Von Korff, M., Angermeyer, M.C., Benjet, C., Bruffaerts, R., de Girolamo, G., Haro, J.M., Lépine, J.P., Ormel, J., Posada-Villa, J., Tachimori, H. and Kessler, R.C. "The Association of Childhood Adversities and Early Onset Mental Disorders with Adult Onset Chronic Physical Conditions". *Arch Gen Psychiatry* 68 (2011): 838–844.
33. Gilbert, L.K., Breiding, M.J., Merrick, M.T., Thompson, W.W., Ford, D.C., Satvinder, S.D. and Parks, S.E. "Childhood Adversity and Adult Chronic Disease: An Update from Ten States and the District of Columbia, 2010". *Am J Prev Med* 48 (2015): 345–349.
34. Chapman, D. and Scott, K. "The Impact of Maternal Intergenerational Risk Factors on Adverse Development Outcomes". *Dev*

Rev 21 (2001): 305–325; Egeland, B., Jacobvitz, D. and Papatola, K. *Child Abuse and Neglect: Biosocial Dimensions*, Aldine, New York, 1987.

35. Reviewed in Champagne, F.A. "Epigenetic Mechanism and the Transgenerational Effects of Maternal Care". *Front Neuroendocrinol* 29 (2008): 386–397.
36. Jespersen, A.F., Lalumière, M.L. and Seto, M.C. "Sexual Abuse History among Adult Sex Offenders and Non-Sex Offenders: A Meta-Analysis". *Child Abuse Negl* 33 (2009): 179–192.
37. Duke, N.N., Pettingell, S.L., McMorris, B.J. and Borowsky, I.W. "Adolescent Violence Perpetration: Associations with Multiple Types of Adverse Childhood Experiences". *Pediatrics* 125 (2010): e778-e786.
38. Prinz, R.J., Sanders, M.R., Shapiro, C.J., Whitaker, D.J., Lutzker, J.R. "Population-Based Prevention of Child Maltreatment: The U.S. Triple P System Population Trial". *Prev Sci* 10 (2009): 1–12.
39. Gillham, J., Reivich, K. "Cultivating Optimism in Childhood and Adolescence". *Ann Am Acad Pol Soc Sci* 591 (2004): 146–163.
40. See Meaney, M.J. "Maternal Care Gene Expression and the Transmission of Individual Differences in Stress Reactivity Across Generations. *Ann Rev Neurosci* 24 (2001): 1161–1692.
41. Ho, M.W. "Caring Mothers Reduce Stress for Life". *Sci Soc* 24 (2004): 11–55.
42. Champagne, F.A., Weaver, I.C., Diorio, J., Dymov, S., Szyf, M. and Meaney, M.J. "Maternal Care Associated with Methylation of the Estrogen Receptor-Alpha1b Promoter and Estrogen Receptor-Alpha Expression in the Medial Preoptic Area of Female Offspring. *Endocrinology* 147 (2006): 2909–2915.
43. Ho, M.W. "Caring Mothers Strike Fatal Blow Against Genetic Determinism". *Sci Soc* 41 (2009f): 6–9.
44. Smit-Rigter, L.A., Champagne, D.L. and van Hooft, J.A. "Lifelong Impact of Variations in Maternal Care on Dentritic Structure and Function of Cortical Layer 2/3 Pyramidal Neurons in Rat Offspring". *PLoS ONE* 4 (2009): e5167, doi:10.1371/journal.pone.0005167.

45. Weaver, C.G., Meaney, M.J. and Szyf, M. "Maternal Care Effects on the Hippocampal Transcriptome and Anxiety-Mediated Behaviours in the Offspring That Are Reversible in Adulthood". *PNAS* 103 (2006): 3480–3485.
46. Arai, J.A., Li, S., Harley, D.M. and Feig, I.A. "Transgenerational Rescue of a Genetic Defect in Long-Term Potentiation and Memory Formation by Juvenile Enrichment". *J Neurosci* 29 (2009): 1496–1502.
47. McGowan, P.O., Sasaki, A., D'Alessio, A.C., Cymov, S., Labon, B., Szyl, M., Turecki, G. and Meaney, M.J. "Epigenetic Regulation of the Glucocorticoid Receptor in Human Brain Associates with Childhood Abuse". *Nat Neurosci* 12 (2009): 342–348.
48. Reviewed in Tyrka, A.R., Price, L.H., Kao, H.T., Porton, B., Marsella, S.A. and Carpenter, L.L. "Childhood Maltreatment and Telomere Shortening: Preliminary Support for an Effect of Early Stress on Cellular Aging". *Biol Psychiatry* 67 (2010): 531–534.
49. McGowan, P.O., Meaney, M.J. and Szyf, M. "Diet and the Epigenetic (Re)programming of Phenotypic Differences in Behaviour". *Brain Res* 1237 (2008): 12–24.

6

No Genes for Intelligence in the Fluid Genome

The "fluid genome" has been known since the early 1980s, which predictably makes identifying genes even for common disease well-nigh impossible. Genome-wide scans using state-of-the art technologies on extensive databases have failed to find a single gene for intelligence; instead, environment and maternal effects may account for most, if not all, correlation among relatives, while identical twins diverge genetically and epigenetically throughout life. Abundant evidence points to the enormous potential for improving intellectual abilities (and health) through simple environmental and social interventions.

The Bell Curve Illusion

The heritability of intelligence or IQ (intelligence quotient, see Box 6.1) has been hotly debated for decades. The most recent round of exchange was provoked by *The Bell Curve* by Richard Hernstein and Charles Murray, published in 1994. The book argued that IQ tests are an accurate measure of intelligence, that IQ is a strong predictor of academic and career achievement, that it is highly heritable and little influenced by the environment, and most controversially, racial differences in IQ are likely due to genes. Consequently, the authors were sceptical about the ability of public policy initiatives to have much impact on IQ or IQ-related outcomes.[1]

The Bell Curve sold 300,000 copies and attracted a great deal of uncritical media attention. The American Psychological Association (APA) commissioned a report from a panel of experts rebutting its

Box 6.1

What is IQ?

IQ, intelligence quotient, is a score resulting from one of several standardized tests designed to assess intelligence. Modern IQ tests were constructed to have a mean score of 100 and standard deviations of the mean, i.e., between 70 and 130.

Quite apart from the fierce debate over the heritability of IQ or intelligence, the claim that IQ assesses intelligence and the validity of any single measure of intelligence are both strongly contested.

main claims. This report was published in 1996. Now, almost 20 years later, a second report has been issued to take account of the many new findings, including the following:

- Almost no genetic variants have been discovered that are consistently associated with variation in IQ in the normal range.
- The heritability of IQ varies significantly by social class.
- The importance of the environment for IQ is established by the 12-point to 18-point increase in IQ when children are adopted from working-class into middle-class homes.
- Even when improvements in IQ produced by the most effective early childhood interventions fail to persist, there can be very marked effects on academic achievements and life outcomes.
- In most developed countries, studied gains on IQ tests have continued, and similar gains are beginning in the developing world.
- The IQ gap between Blacks and Whites has been reduced by 0.33 SD (standard deviation) in recent years.

Revolution in the Heartland of Genetics

The report does not quite capture the revolution breaking out in the heartland of genetics.

Simply put, there are no genes for intelligence in the human genome, however you choose to define intelligence (or a gene for that matter), and there has been much contention on that alone. For example, recent research shows that non-intellectual factors such as test motivation can increase IQ scores by an average of 0.64 SD, with larger effects for individuals with lower baseline IQ scores.[2]

More fundamentally, the heritability of IQ estimated in the conventional model is now widely seen as deeply flawed. Heritability — the component of population variation (variance) attributed to genes — has been inflated by gene interactions, gene-environment interactions and other non-linear effects, in the same way that the heritability for common diseases has been inflated.[3] Not only that, the heritability estimated from resemblance (correlation, covariance) between twins and siblings could be due to shared environments, especially maternal environments.

Even more seriously, the classical Mendelian inheritance on which all estimates of heritability depends has been severely compromised by pervasive epigenetic (and environmental cultural) inheritance. Epigenetic and cultural inheritances often go together, resulting in correlations between relatives that have been erroneously attributed to shared genes. On the other hand, epigenetic variations due to individual experiences, and somatic mutations from a host of DNA marking and changing processes, make even monozygotic twins diverge genetically from each other to substantial degrees. These observations, all part of the "fluid genome", strike at the very core of the conventional genetic determinist paradigm.

I shall start from the problems emerging in identifying genes for common diseases, which looks much more concrete than trying to identify genes for intelligence.

Where are All the Promised Genes?

When the human genome sequence was announced in 2000, President Clinton said it would "revolutionize the diagnosis, prevention and treatment of most, if not all human diseases". Ten years on, and *Fortune* magazine called it: "The great DNA letdown". A poll by science journal *Nature* returned the verdict: "The hoped-for revolution against human disease has not arrived."

Box 6.2

Genome-Wide Association Studies

Genome wide association studies (GWAS) involves rapidly scanning markers across the complete genomes of many people to find associations of genetic variants to particular diseases or traits. Typically, thousands or tens of thousands of individuals are scanned, simultaneously for up to 550 000 single nucleotide polymorphisms (SNPs) — common differences in single nucleotides at specific sites across the human genome with frequencies >5% — using DNA microarrays (chips).

That is as some of us had predicted in 2000 and before.[4,5] The Human Genome Project has generated reams and reams of data since its inception, but there is little progress even in the apparently simple task of finding the genes responsible for susceptibility to common diseases.[3,6]

Top geneticists recently admitted that human genetics has been haunted by the mystery of "missing heritability" of common traits. Genome-wide association studies (GWAS, see Box 6.2) — the gold standard for the most exhaustive gene hunt that can be performed — have identified ~2,000 genetic variants associated with 165 common diseases and traits, but these variants appear to explain only a tiny fraction of the heritability in most cases.[7,8]

Heritability is technically the proportion of the variability of the trait in a population that can be attributed to differences in genes. Variability is measured statistically as *variance*, the sum of the squared individual deviation from the population mean. Heritability is commonly referred to as the "genetic component" of the variance as opposed to the proportion due to the environment, the "environmental component". Note that heritability refers to the variation, and *not* to the trait itself. Heritability changes according to the environment. It is not uncommon for the heritability of traits such as milk yield or height of a plant from the same genetic strain to change substantially from one year to the next. However, there is a tendency for some scientists as well as the popular media to mistakenly assume that any trait with a large heritability means it is predominantly genetically determined, which is definitely not the case.

No Genes for Common Diseases?

Nevertheless, the hunt for genes determining susceptibility to common diseases has continued for decades, spurred on over the past five years by the availability of DNA chips that allow genome-wide scans for more than 500 000 SNPs simultaneously.

Eric Lander and his team at Broad Institute of Harvard and MIT in the United States are among those suggesting that much of the missing heritability never existed in the first place.[8] They base their argument on *biometrical genetics*, a mathematical discipline that deals with continuously varying traits, such as crop yields, height, body mass, IQ scores or disease states that fall on a continuum, as for example, blood glucose, blood pressure, or some measure of disease severity.

I should point out that one arrives at precisely the same conclusion given the pervasive epigenetic influences of the environment on development,[4–6] which have been abundantly confirmed and extended since the human genome was sequenced.[9–11]

This convergence of molecular and biometrical genetic analyses is the most conclusive refutation of the reductionist, genetic determinist paradigm of linear causation from genes to traits that had made the Human Genome Project seem such a compelling undertaking; only to thoroughly discredit it as a result, as argued in my book, *Living with the Fluid Genome*, published in 2003.[12]

We now know that much of the variation may come from *individual* experiences of the environment; furthermore, those experiences can mark and change genes, influencing the development of the individual and in many case, the individual's offspring. Genes and environment operate in enormously complex feed-forward and feedback networks that straddle generations. This fundamentally *circular causation* between genes and environment means that genetic and environmental contributions are inseparable, and *any attempt at assigning linear effects to single genes is doomed to failure.*

We shall see how genetic determinism is finally unravelling within the heart of the genetics establishment, beginning with the findings of Lander's team on common disease traits[3] and continuing with the intelligence and IQ debate.[13]

The Genetic Component has Been Greatly Overestimated

Specifically, Lander and colleagues show that the missing heritability arises from an overestimate of total heritability (the genetic component of the variation in the trait) which implicitly assumes that no gene interactions (or gene environment interactions) exist, an assumption clearly unjustified. Including gene interactions gives a much smaller total heritability. In short, "missing heritability need not directly correspond to missing variants, because current estimates of total heritability may be significantly inflated by genetic interactions".[8]

Actually, gene interactions do belong to the "genetic component" of heritability. In biometrical genetics, "broad sense heritability" H^2 includes additive genetic effects as well as effects due to gene interactions and any non-additive, non-linear effects due to genes. But broad sense heritability is very difficult to determine. In practice, only the "narrow sense heritability" h^2 (the additive, linear effects due to genes) can be estimated. *Narrow sense heritability applies strictly to "polygenic" traits due to many genes each with a small additive effect, and is implicitly assumed to apply to all polygenic traits,* beginning with the pioneers of biometrical genetics (see later).

Geneticists therefore define the proportion of heritability of a trait explained, $\pi_{explained}$, as a ratio of phenotypic variance explained by the additive effects of known genetic variants, h^2_{known}, to the phenotypic variance that can be attributed to the additive effects of all variants, including those not yet discovered, h^2_{all} (Equation 1).

$$\pi_{explained} = h^2_{known}/h^2_{all} \tag{1}$$

The nominator h^2_{known} can be calculated directly from the measured effects of the variants, but the denominator h^2_{all} must be inferred indirectly from population data.

The prevailing view among geneticists is that the missing heritability is due to additional variants yet to be discovered, either common alleles with moderate-to-small effects or rare alleles (frequency <1%) with large effects.[7,8]

The other possibility, favoured by Lander's team, is that the missing heritability does not actually exist, and is an artefact arising from the total heritability h^2_{all} being overestimated in the first place, by ignoring the impacts of gene interactions.

For example, Crohn's disease (inflammatory disease of the bowel) has so far 71 risk-associated loci identified. Under the usual assumption of additive effects, these loci explain 21.5% of the estimated total heritability. Genetic interactions could account for the remaining nearly 80% missing heritability. Why then, has genetic interaction never been detected in population analyses? Lander and colleagues point out that to detect gene interactions for Crohn's disease may require sample sizes in the range of 500 000 individuals, which is rarely attained.

Gene interaction, or *epistasis*, is well known and pervasive, even before the human genome was sequenced, as I have stressed in *Genetic Engineering Dream or Nightmare*,[4] which predicted why genetic modification is both dangerous and futile. Since the human genome was sequenced, gene interaction takes on a literal dimension, as epitomized in the findings of project ENCODE (Encyclopaedia of DNA elements) organized by the US National Human Genome Research Institute, in which a consortium of 35 research groups went through 1% of the human genome with a fine-tooth comb to find out exactly how genes work.[14] They discovered that "genes appear to operate in a complex network, and interact and overlap with one another and with other components in ways not fully understood."[15] Essentially, the "gene" as a well-defined, separate unit of structure or function no longer applies. Instead, genes exist in bits strewn across the genome, structurally *and* functionally intertwined with other genes. The same sequence of DNA can have very different functions, and very different sequences can have the same function.

How Phantom Heritability Arises

Lander and colleagues point out[8] that in calculating the explained heritability (Eq. 1), the numerator h^2_{known} is estimated based on the effects of the individual genetic variants. The problem comes in estimating the denominator h^2_{all}. Because not all the variants are known, their contribution must be inferred based on phenotypic correlations in a population. This gives an *apparent heritability*, h^2_{pop}. And the missing heritability is then estimated by assuming that $h^2_{all} = h^2_{pop}$.

However, there is no guarantee that $h^2_{all} = h^2_{pop}$, unless the trait is strictly additive, and neither gene-gene interaction nor gene-environment interaction exists. For traits with gene interaction, which would realistically apply to practically all common traits and diseases, h^2_{pop} may significantly exceed h^2_{all}. In that case, even when all the variants for the trait have been identified, the missing heritability $\pi_{missing}$ will not diminish to zero; instead, it converges to $1 - (h^2_{all}/h^2_{pop})$, which Lander and colleagues refer to as "phantom heritability", $\pi_{phantom}$.

Simple Model Shows how Genetic Interactions Create Phantom Heritability

To show how genetic interactions create phantom heritability, Lander and colleagues introduced a simple model in which a trait depends on input from more than one process. Phantom heritability — that which remains missing even when all genetic variants have been identified — grows quickly with the number of inputs, approaching 100% of the total variation. For Crohn's disease, for example, just three inputs are sufficient to account for 80% of the phantom heritability.

Similarly, gene-environment interactions can produce additional phantom heritability, (as indeed other unaccounted sources such as epigenetic effects).

Twin Studies Deeply Flawed

The typical framework for analysing human traits depends on a systematic denial of epistasis, assuming that genes act in a purely additive way, each gene contributing a small amount to the trait, which is summed up depending on how many of those genes are present.

One measure of apparent heritability h^2_{pop}(ACE) assumes additive genetic variance, as well as common environmental and unique environment variance components, and a usual definition for apparent heritability is h^2_{pop} (ACE) = $2(r_{MZ} - r_{DZ})$, where r_{MZ} and r_{DZ} are the phenotypic (measured trait) correlations between monozygotic twins (sharing 100% of their genes) and dizygotic twins (sharing 50% of their genes), while the environment they share is assumed to be common, including the maternal environment.

But realistically,

$$h^2_{pop}(\text{ACE}) = h^2_{all} + W \tag{2}$$

where W represents the sum of variances due to all possible higher-order additive and non-additive interactions between genes. The crucial point is that if there are any gene interactions, then $W > 0$, so h^2_{pop} (ACE) overestimates h^2_{all}.

Unfortunately, there has been no way to estimate W from population data. In most human studies, the solution is to assume there are no gene interactions, in which case $W = 0$. Thus, twin studies systematically overestimate the genetic contribution to disease and other traits, most notably and controversially IQ.[13]

Additive Assumption Fundamental to Biometrical Genetics

Lander and colleagues are not the first to expose the fundamentally flawed assumptions of classical biometrical genetics. Helen Wallace of UK-based GeneWatch had published a similar critique five years earlier:[16] gene-gene and gene-environment interactions could reduce the calculated heritability considerably below that predicted by the standard twin-studies method based on pioneering British geneticist Ronald Fisher's 1918 assumption that genes act additively.

Implications

The major implication is that the hunt for susceptibility genes is practically useless. Indeed, Lander and colleagues[8] and others[7] see the primary purpose of medical genetics as the identification of underlying pathways and processes analogous to the hunt for mutants in model organisms, and not in "explaining heritability" or "predicting personalized patient risk".

But there are much wider implications on health policies. Governments and companies have been keen to set up whole genome biobanks ever since the human genome sequence was announced.[17] The UK government is pushing to let companies gain access to public health records to drive discovery in disease genomics.[18] But if the genetic contribution to disease is largely a phantom, what is the point of integrating whole genome sequences with electronic medical records as most of this information is likely to be clinically useless for most people?[17,19]

There are vested interests that want to keep the genetic myth alive. As Wallace points out, the evidence she presented in 2006, and Lander and colleagues presented in 2011, has had no impact on gene testing companies such as Illumina and 23 and Me, which continue to claim that everyone will have their genome mapped or sequenced in future, at birth or as a routine part of health care. The Director of the National Institutes of Health Francis Collins has echoed those claims in his populist book *The Language of Life*.[20] Wallace is convinced, as I am, that[19] "whole genome sequencing of everyone, leading to the 'prediction and prevention' of disease, is a science fantasy and a massive waste of money".

A fraction of the resources divested into much-needed primary health care and disease prevention through nutritional and other environmental/social interventions will do infinitely more to improve the health (as well as brain power) of the nation, as we shall see.

The Elusive IQ Genes

The hunt for IQ genes has been inspired by the large heritability estimated in conventional biometrical models based on correlations between twins and other biological relatives.[3] But the results so far have been disappointing to say the least, even more so than the hunt for disease genes.

A genome-wide association study (GWAS) on 7 000 subjects published in 2008 found only six genetic markers (SNPs, single nucleotide polymorphisms) associated with cognitive ability, and only one of those remained statistically significant on further tests. Together, the six markers explained barely 1% of the variance in general cognitive ability.[21] Recently, the association between 12 specific SNPs and "general intelligence" factor g was put to test in an attempt to replicate the associations found in earlier studies, but only one SNP remained significant. The researchers conclude that "most reported genetic associations with general intelligence are probably false positives".[22]

As in the case of common disease traits,[3] IQ or intelligence is plagued by the problem of "missing heritability". Even the heritability of human height, estimated at ~90%, failed to turn up common variants contributing more than 0.5 cm; and the set of 180 height-associated SNPs identified by the most comprehensive meta-analysis (on pooled data from many studies) only explains about 10% of the population variance.

The usual explanation for the missing heritability is that it is difficult to detect genetic variants with a small effect. In the case of intelligence, much is made of the findings in a new study led by researchers at Edinburgh University in the United Kingdom, which claims to "establish that human intelligence is highly heritable and polygenic".[23] The group first used data from five different GWAS and failed to identify any individual marker associated with either "crystallized" or "fluid" intelligence. (Crystallized intelligence is the individual's store of knowledge about the nature of the world and learned operations such as arithmetic that can be drawn upon to solve problems; while fluid intelligence is the ability to solve novel problems that depend relatively little on stored knowledge, as well as the ability to learn). They then applied a new method that tests the cumulative effects of all the SNPs, essentially by calculating the overall genetic similarity between each pair of individuals in a sample, and correlating this genetic similarity with phenotypic similarity (in IQ) across all the pairs. The result is that *all* the ~550 000 SNPs together could jointly explain 40% of the variation in crystallized intelligence and 51% of the variation in fluid intelligence. This exercise sounds more like a counsel of despair than a solution to the problem, and the result certainly does not offer any useful predictive information. Nevertheless, this same method, now named genomewide complex-trait analysis (GCTA), is now widely used by genetic determinists to bolster their untenable position; see for example, psychologist Robert Plomin's group at King's College, London, in the United Kingdom.[24] In a sample of 3 154 pairs of 12-year-old twins, GCTA on 1.7 million DNA markers for cognitive abilities (language, verbal, nonverbal and general) finds that the genes account for 0.66 of the heritability. Obviously, this method, like all previous biometrical methods, still ignores gene-gene and gene-environment interactions.

Other researchers are tackling the problem at the more fundamental level of the heritability estimates.

Maternal Environment Accounts for Much of Heritability

One of the first rebuttals to *The Bell Curve* came from Bernard Devlin and colleagues at University of Pittsburgh Pennsylvania in the United States in a paper published in *Nature* in 1997.[25] They showed that covariance (correlation) between relatives may not be due only to genes, but also to shared environment, especially maternal environment, which is not taken into account in conventional models. In a meta-analysis of 212 previous studies supplemented with twin studies published after 1981, Devlin and colleagues showed that an alternative model with two maternal womb environments, one for twins — both monozygotic (MZ) and dizygotic (DZ) — and another for siblings, fit the data much better.

Maternal effects, often assumed to be negligible, account for 20% of the covariance between twins and 5% between siblings, thereby correspondingly reducing the effects of genes, so the two measures of heritability were both less than 50%: the broad and narrow sense heritability were 48% and 34%, respectively.

The shared maternal environment may explain the striking correlation between the IQs of twins, especially adult twins reared apart. It also accounts for age effects: an apparent increase in heritability with age. Devlin and colleagues pointed out that cultural inheritance and interaction between genes and environment may also be at work to boost the apparent heritability of intelligence.

There is substantial brain growth *in utero*, and the brain has 70% of its final mass within a year of birth. IQ is known to be affected by prenatal environment: it is positively correlated with birth weight. Twins usually weigh less than singletons, and score on average 4–7 points lower on IQ tests.

Devlin and colleagues rejected Hernstein and Murray's conclusion; instead, they believed that "Interventions aimed at improving the prenatal environment could lead to a significant increase in the population's IQ."

Devlin and colleagues may well have *under*estimated the shared maternal environment for MZ twins, which in addition to sharing the same womb as for DZ twins, usually share the same placenta, and more importantly, originate from the same egg with common cytoplasmic components, including mitochondrial DNA and transcripts and gene products that control early embryonic development.[26] Common cytoplasmic effects will be expected to further reduce heritability estimates.

Virtual Twin Studies and Rearing Environment

Nancy Segal and colleagues at California State University Fullerton in the United States have pioneered the study of behaviour in "virtual twins" (VTs): same-age, unrelated siblings reared together since infancy. VTs replicate the rearing environment of twins but without the genetic relatedness, thereby enabling direct assessment of shared environmental effects on behaviour. Virtual twins are created in adoption, in which infants were adopted before one year of age; the unrelated sibling differing by less than nine months in age and who attend the same school grade, the pair being free of adverse birth events, and at least four years old. The foster homes are predominantly upper middle class.

In an updated analysis of IQ data based on a sample of 142 VT pairs, the VTs mean IQ score was 105.83 (SD = 13.37) and correlation between VTs is 0.28 ($p < .001$), showing a substantial contribution of rearing environment during infancy.[27]

The mean IQ score of the biological siblings exceeded that of the adopted siblings and when the paired data for members of 49 adopted-biological pairs were examined, biological children scored 113.08 (SD = 14.64), whereas adopted children scored 105.67 (SD 12.53), a difference of 7.41 points.

Significantly, there was greater similarity in IQ scores between adopted-biological than adopted-adopted pairs, resulting in correlations of 0.47 versus 0.1. Similar results have been found by other research groups, suggesting that the environmental stimulation from a high-IQ biological child may also enhance adopted sibling's IQs. (Note that this is my own suggestion, not Segal's.[28])

The IQ correlation of the adopted-biological pairs (0.47, $p < .001$) approaches that of DZ twins (0.46) and full siblings (0.47) reported by others.

To me, the research of Segal's group and Devlin's group together makes it highly likely that common rearing environment during infancy and maternal effects could account for most, if not all, the heritability in IQ that has been attributed to genes.

Socioeconomic Status and IQ

The higher IQ scores of the biological children relative to the adopted children observed by Segal's team[27] are not surprising in view of the predominance of upper middle class parents in the study, whereas adopted children are predominantly from parents of lower socioeconomic status (SES). Studies dating to the 1990s have shown that adopted children typically score 12 points or higher than siblings left with birth parents of lower SES or children adopted by lower SES parents.[1] A meta-analysis published in 2005 found an average effect of adoption of 18 points when extremely deprived institutional settings were included in the comparison.[29]

What correlates SES with IQ? There are marked differences beginning in infancy, between the environment of higher SES families and lower SES families in factors that are likely to influence intellectual growth, including nutritional status.[30] A study published in 1995 showed that by the age of three, a child of professional parents would have heard 30 million words spoken, while a child of working-class parents would have heard 20 million words, and a child of unemployed African American mothers would have heard 10 million words.[31] The child of professional parents received six encouragements for every reprimand, the child of working-class parents received two encouragements per reprimand, and the child of unemployed African American mothers received two reprimands for every encouragement.[31] These findings were extended using the HOME (Home Observation for Measurement of the Environment) technique.[1] HOME researchers assess family environments for the amount of intellectual stimulation: how much parents talk to the child, how much access the child has to books, magazines, newspaper and computers, how much the parents read to the child, how many learning experiences the child has outside the home (trips to museums, visits to friends); degree of warmth of parents *versus* punitive behaviour towards the child, etc. Very substantial association was found between HOME scores and IQ scores: a one-SD difference in summed HOME scores is associated with a nine-point difference in IQ. These studies do not separate genetic and environmental contributions, but as the authors of the new report of the APA commented[1]: "It is almost surely the case, however, that a substantial fraction of the IQ advantage is due to the environments independent of the genes associated with them." That is because of the knowledge that adoption adds 12–18 points to the IQ of unrelated children, who are usually from lower SES backgrounds.

The shared environmental effect on IQ applies not just to children. According to a review of six well-conducted studies, the shared environment effect in adulthood is about 0.16 on average.

Consequently, most if not all twin studies, especially studies of adults, overestimate heritability of IQ, especially as lower SES individuals are difficult to recruit to laboratories and testing sites.

Epigenetic and Cultural Inheritance

It is epigenetic and cultural inheritance that really blurs the distinction between environmental and genetic effects. Epigenetic modifications arising from environmental interactions are often inherited, both good and bad, as I have reviewed elsewhere.[32] But even what is inherited in the DNA does not determine the amino acid sequences of proteins, as revealed by a widespread mismatch of expressed RNA to the DNA sequences. Somehow, the DNA sequences are profusely edited, possibly to make them fit for purpose and context at any one time and place.[33] And most surprisingly, geneticists have discovered that even the food we eat can affect the expression of our genes via small RNA sequences in the food,[34] giving a literal meaning to "we are what we eat", and incidentally, showing up yet another hazard from genetically modified food with unpredictable, uncontrollable changes in the repertoire of new RNAs produced.

As these epigenetic modifications usually do not involve DNA base sequence alterations, they are not detected in SNP scans, and are also independent of SNP variations. Instead, they involve chemical markings of DNA or histone proteins that bind to DNA, or other mechanisms that change the state of expression of certain genes that are often passed on to future generations.[32]

Cultural tradition is a form of environmental inheritance more tenacious than any classical Mendelian genetic inheritance, but is also ignored in heritability estimates.

Human population geneticist Marcus Feldman and his colleagues at Stanford University in the United States and University of Aarhus in Denmark have noted that current models are strictly based on Mendelian genetics, failing to consider non-Mendelian epigenetic modifications of genes in response to environmental states. They addressed the problem at the level of biometrical models by extending the models to include epigenetic and environmental effects.[35]

They found that variation in epigenetic state and environmental state can result in highly heritable phenotypes through a combination of epigenetic and environmental inheritance. These two inheritance processes together can produce familial covariances (correlations) significantly greater than those predicted by models of purely epigenetic inheritance, and similar to those expected from genetic effects. The results suggest that epigenetic variation, inherited both directly, and through shared environmental effects, may make a key contribution to the missing heritability.

In other words, epigenetic and environmental effects working together can account for practically all the variation now attributed to the genes.

Chief among environmental effects is cultural inheritance. Feldman and colleagues referred to the aggregation of the disease Kuru in families of the Fore tribe of Papua New Guinea due to the ingestion of a *prion* protein (infectious protein agent) during funeral rituals in which dead relatives or close acquaintances are consumed. This case of purely cultural inheritance was originally mistaken for a genetic disorder because of high disease correlations between relatives.

A well-studied case of environmental epigenetic inheritance is the mother's licking-grooming of offspring in mice, which induces epigenetic changes in the offspring, influencing its response to stress as adults, and perpetuates the maternal behaviour in her female offspring.[36] This results in highly correlated behaviour between mother and offspring in both maternal behaviour and response to stress as adults, even though the epigenetic modifications are erased during early embryogenesis. Consequently, cross-adoption between mothers with high and low licking-grooming behaviour will break the biological mother-offspring correlations in a single generation.

Identical Twins are *not* Genetically Identical

As noted above, biometrical genetics models are based on classical Mendelian inheritance, in which genes are immune to direct or predictable environmental influence and passed on unchanged to the next generation except for rare random mutations. The old paradigm has been discredited at least as far back as the late 1970s,[4,12] long before the Human Genome Project was conceived.

In the post-genomics era, an increasing number of geneticists have begun to take notice of non-Mendelian inheritance and its invalidation of the basic tenets of biometrical genetics.

In a paper published in *Behavioral and Brain Sciences*, Evan Charney at Duke University speaks of a "paradigm shift" in the science of genetics. He points to recent discoveries of numerous processes that create extensive mutations in genome sequences and structure, as well as epigenetic modifications, which are completely at odds with the Mendelian model of inheritance underpinning heritability estimates.[26] Individuals do not have genes that are immutable throughout life, nor do they have the same genes in every cell of the body. He highlights retrotransposons — jumping genes that replicate and integrate themselves into different sites in the genome — which alter the sequence and state of activity of many genes; copy number variation and chromosomal abnormalities (aneuploidy) similarly, occur frequently in somatic cells as well as germ cells, both as part of normal development and in response to noxious environmental stimuli. Different tissues show distinctly different propensity for change, brain cells being especially prone to such modifications. These add to the already large repertoire of epigenetic processes that modify genes in response to environmental stimuli,[4,12,32–34] and most notably in the brain.[37]

The fundamental assumption of twin studies — that monozygotic twins share 100% of their genes — is demonstrably false.

MZ twins differ, to begin with, in the mitochondrial DNA (mtDNA) complement allocated in cell division of the original oocyte that generated the twins. The oocyte may have had different sets of mtDNA, a condition referred to as heteroplasmy. MZ twins diverge substantially in epigenetic modifications as well as retrotransposition, copy number variations and aneuploidy throughout life. Although the numerous processes that alter genomes occur in normal development, perhaps as part of "natural genetic engineering",[4,12] the same processes are known to be involved in many behavioural, psychiatric and neurodegenerative diseases, leaving us in no doubt that they have phenotypic consequences.[26]

In addition, stochastic non-linear developmental changes account for substantial divergence in the activities of different brain regions between twins.[38]

To summarize, no genes for intelligence can be found in the human genome. Instead, common environments, including maternal and rearing environments, along with epigenetic and cultural inheritance create substantial correlations between genetically unrelated individuals, while even "identical twins" diverge genetically and epigenetically throughout life.

The fundamentally *circular causation* between genes and environment makes it futile to separate genetic from environmental contributions to development.[39] Consequently, we must redouble all efforts at appropriate interventions to improve the mental and physical well-being of the nation, for which there is already a great deal of evidence.[30]

Environmental Interventions can Improve IQ and Academic Achievement

The 2012 report on intelligence from the APA states:[1] "A large number of interventions have been shown to have substantial effects on IQ and academic achievement."

The collapse of the genetic paradigm should convince us to spare no effort at interventions that can improve the intellectual prowess of the nation and deliver substantial health bonuses. Let's look at some of the options.

Education and Enrichment Programmes

There is clear evidence that schooling affects intelligence, as reviewed in the APA report.[1] Children deprived of school for an extended period of time show IQ deficits as much as 2 SDs. A child entering fifth grade approximately a year earlier than one nearly the same age (who enters fourth grade) will have a verbal IQ more than 5 points higher at the end of the school year, and as much as 9% higher by the eighth grade.

Children lose IQ and academic skills over the summer, and the loss is much greater for children of lower SES. The knowledge and skills of children in the upper fifth of family SES, however, actually *increase* over the summer. This effect is so marked that by late elementary school, much of the difference in academic skills between lower and higher SES children may be due to the loss of skills over the summer for lower SES children as opposed to the gains for higher SES children. Intervention over the summer months targeted at low SES children should narrow this gap. The beneficial effects of schooling apparently continue at least through junior high school.

The best prekindergarten interventions for lower SES children have substantial effects on IQ, but this typically fades by late elementary school, perhaps because the environment of the children does not remain enriched. Two examples in which early gains from prekindergarten intervention remained both placed children in average or above-average elementary schools. Children in the Milwaukee Project had an average IQ 10 points above controls at adolescence, and children in the intensive Abecedarian programme had IQs 4.5 points higher than controls at 21 years of age. Regardless of whether high-quality interventions have sustained IQ effects, the effects on academic achievement and life outcomes can be very substantial. The gains are particularly marked for intensive interventions such as the Perry School Project and Abecedarian programme.

By adulthood, individuals who had participated were about half as likely to have repeated a grade in school or to have been assigned to special education classes, and were far more likely to have completed high school, attended college, and owned their own home. This suggests that some of the effects are produced by gains in attention, self-control and perseverance, rather than IQ.

Self-control and discipline, along with creativity and flexibility are considered the key qualities to success in life, and can be targeted by specific interventions as described in a recent review.[40] For example, martial arts such as *tae-kwon-do* that emphasize self-control, discipline and character development gave children substantial gains in those cognitive functions (referred to as "executive functions"), much more so than standard physical education. The children participating in *tae-kwon-do* also improved more when tested on mental mathematics. Other effective interventions include "mindfulness" practices that focus one's complete attention on present experience, and *Tools of the Mind* that develop social and socializing skills through play.

The quality of teaching in kindergarten has a measurable impact on academic success and life outcomes.[1] Data from Project STAR in Tennessee showed that students randomly assigned to small kindergarten classrooms were more likely to subsequently attend college, attend a high ranked college, and have better life outcomes in a number of respects. Students who had more experienced teachers had higher earnings as adults, as did students for whom the quality of teaching — as measured by test scores — was higher.

Memory Training for Fluid Intelligence

It is perhaps not surprising that training people in working memory skills can enhance fluid intelligence, while having no effect on crystallized intelligence.[1] This applies to both adults and children with attention deficit hyperactivity disorder (ADHD). Working memory training of low SES children using a variety of computer and non-computer games resulted in IQ gains of 10 points on a matrix reasoning task.

Similar memory training over an eight-month period was effective for elderly participants.[1] Training older adults in memory, speed in processing and particular narrow reasoning skills produces substantial improvements that remain over a period of years. A study in the United Kingdom showed that an extra year of work was associated with a delay in the onset of Alzheimer's disease on average by six weeks.

The Overriding Importance of Early Nutrition

The overriding importance of early nutrition for learning is highlighted in a comprehensive resource list for professionals provided by the Food and Nutrition Information Center of the US Department of Agriculture's National Agricultural Library.[41] It gives clear evidence that nutritional intervention in elementary school can improve both health and academic performance.

Retrospective analyses were conducted on school performance indicators associated with the implementation of the Healthy Kids, Smart Kids programme, a grassroots effort to enhance school food and physical activity environment in Browns Mill Elementary School, Georgia. Data from 1995 to 2006 showed that the number of nurse, counselling and disciplinary referrals per 100 students followed a downward trend, while standardized test scores followed an upward trend beginning in the year of programme implementation.

A second study demonstrated the effect of a two-year obesity prevention programme on body mass index (BMI) and academic performance in low-income elementary schoolchildren. There were four intervention schools and one control school totalling 4 588 schoolchildren, 48% Hispanic. The data were presented for the subset (1 197) of the children (68% Hispanic) who qualified for free or reduced-price school lunches. The results showed that significantly more intervention than control children stayed within normal BMI range for both years. Although not significantly so, more obese children in the intervention than in the control decreased their BMI. Overall, intervention children had significantly higher maths scores in both years, and Hispanic and White intervention children were significantly more likely to have higher maths scores. Although not significantly so, intervention children had higher reading scores in both years.

The association between intelligence and diet at 3.5 and 7 years of age was examined in more than 500 children of European descent in Auckland, New Zealand, approximately half of them with low birth weight (≤10th percentile).[42] The relationship between IQ and diet measured by food frequency was investigated using multiple regression analysis. There was no significant difference in IQ between children with low birth weight and normal birth weight at 3.5 and 7 years of age, and no differences in food frequencies.

Eating margarine at least daily was associated with significantly lower IQ scores at 3.5 years in the total sample, and at 7 years in children with low birth weight. After controlling for potential confounders, children who ate margarine daily scored 2.81 points lower than children who did not. In all children, eating the recommended daily number of breads and cereals — 4 or more times — was associated with significantly higher IQ scores at 3.5 years; the gain was 3.96 points after controlling for potential confounders. Children who ate fish at least weekly had significantly higher IQ scores at seven years than those who did not, a gain of 3.64 points after controlling for confounders.

Eating fish does make you smart, it appears, precisely as we have been told in the traditional folklore of many cultures. A large study was carried out in Sweden to evaluate the association between fish intake and academic grades of 9 488 adolescents using multiple linear regression models and adjusting for potential confounders such as parents' education.[43] The results showed that grades were higher by 14.5 points in adolescents who ate fish once a week compared with those eating fish less than once a week. Adolescents who ate fish more than once a week scored even higher by 19.9 points. In the model stratified for parents' education, there were still higher grades among children with frequent fish intake in all educational strata.

A review[44] published in 2008 summarized evidence indicating that food insecurity is a prevalent risk to the growth, health, cognitive ability and behaviour of poor children in the United States. Infants and toddlers, in particular, are at risk even at the lowest level of food insecurity. The data indicate an "Invisible epidemic" of a serious condition.

The effect of nutritional status on brain development and scholastic achievement was examined in 96 high school graduates selected from the public and private schools in the richest and poorest counties of Chile's Metropolitan region.[45] These graduates had no history of alcoholism, or symptoms of brain damage, epilepsy or heart disease, and their mothers had no history of smoking, alcoholism or drug intake before and during pregnancy (all known to affect foetal development). The object was to have a healthy balanced sample in terms of low and high IQ, sex, and SES. The results showed that independently of SES, high school graduates with similar IQ have similar nutritional, brain development and scholastic achievement. Multiple regression analysis revealed that maternal IQ ($p < 0.0001$), brain volume ($p < 0.0387$) and severe under-nutrition during the first year of life ($p < 0.0486$), were the independent variables with the greatest explanatory power for the IQ variance, without interaction with age, sex or SES. IQ ($p < 0.0001$) was the only independent variable that explained both scholastic achievement variance and academic aptitude test variance, without interaction with age, sex or SES.

Studies by the Institute of Nutrition of Central America and Panama (INCAP) showed that supplementary feeding of infants and young children — with drinks that provide energy only or with added protein, both containing micronutrients — resulted in significant increases in cognitive development and school performance through to adolescence.[46] The research also suggested that the link of malnutrition to later development is not only through the neurological system, but also through changes in behaviour that affect the kinds of care the individual child receives.

A longitudinal two-year study on school children in rural Kenya[47] found significant relationships on regression analyses between available Fe, available Zn, vitamin B12 and riboflavin with improved cognitive test scores, after controlling for confounders such as energy intake, school, SES and illness.

Interventions aimed at eliminating food insecurity and micronutrient deficiencies are easily within the means of all developed nations, and should be given top priority in both developed and developing nations.

Exercise Increases Brain Power by Making More Neurons

A sedentary lifestyle is associated with increased risk for cardiovascular and metabolic diseases as well as cancer, and it is well known that exercise can reduce the incidence of diabetes, cancers and heart disease. Less well known is the beneficial effects of exercise for the brain, described in a comprehensive review.[48] In humans and rodents, physical activity enhances cognitive functions and counteracts age-related decline of memory, delays the onset of neurodegenerative diseases and enhances recovery from brain injury and depression.

A meta-analysis of a large number of studies on older adults[49] has shown that aerobic exercise, at least for the elderly, is very important for maintaining IQ, especially for executive functions such as planning, inhibition and scheduling of mental procedures. The effect of aerobic exercise is more than 0.5 SD for the elderly, more for those past age 65 than those younger. It is possible to begin cardiovascular exercise as late as the seventh decade of life and substantially reduce the likelihood of Alzheimer's disease.

But exactly how does exercise work to increase brain power and help prevent degeneration? The most likely answer appears to be through neurogenesis, the ability of the brain to repair and renew itself by making new neurons.[50]

Not long ago, neurobiologists and the general public believed that we were born with the neurons we would have in life, and no new neurons would ever be generated in the brain. That dogma was overturned in the 1990s. New neurons are continually generated throughout adulthood, mainly in two regions of the brain: the dentate gyrus in the hippocampus, a paired brain structure involved in memory, learning and emotion, and the subventricular zone, a layer of cells found along the brain's lateral ventricles. The newly generated neurons form synapses and integrate into existing neuronal circuits.

Laboratory experiments have revealed that exercise not only significantly increases the number of new neurons in rats and mice, it also influences the morphology of individual newly generated cells and enhances their maturation, and is associated with increased plasticity in the hippocampus in forming synapses, thereby influencing learning and memory. In rodents, both voluntary wheel running and forced treadmill training have been shown to improve spatial learning with different types of mazes and training.

In rodents as well as non-human primates, aging is associated with decline in neurogenesis and cognitive functions. The age-dependent reduction in neurogenesis can be partially prevented when animals are housed with a running wheel over a six-month period. Furthermore, the decline in neurogenesis and cognitive functions associated with normal aging can be reversed in part by wheel-running. Mice that had been sedentary for 18 months were started on the running wheel for one month, after which they showed significant improvements in spatial memory in learning the water maze, and the survival of newly generated neurons was also increased to the level of young sedentary controls.

Correlation between neurogenesis and exercise was first established in mice through magnetic resonance imaging (MRI) measurements of angiogenesis (blood volume).[51] Among all hippocampus subregions, exercise was found to have a primary effect on the dentate gyrus cerebral blood volume (CBV); the dentate gyrus was the only subregion known to support adult neurogenesis. Moreover, exercise-induced increases in dentate gyrus CBV were found to correlate with postmortem measurements of neurogenesis. Using similar MRI technologies, CBV maps were generated over time in the hippocampus of exercising humans. As in mice, exercise was found to have a primary effect on dentate gyrus CBV, and the CBV changes were found to selectively correlate with cardiopulmonary and cognitive function.

Another significant effect of exercise is an increase of brain-derived neurotropic factor (BDNF) in the hippocampus, which supports the survival of existing neurons and encourages the growth and differentiation of new neurons and synapses.[26, 52] The levels of hippocampal BDNF are significantly higher in wheel-running as opposed to sedentary rodents after five days, and correlates with the level of activity. There is also 3.1-fold as many new neurons in the dentate gyrus of running mice compared to sedentary mice.

Most intriguingly, running also increases retrotransposon activity, reflected in the number of new insertions of long interspersed nucleotide elements (LINEs-1, of L1) in the hippocampus, and also activates silenced L1 insertions in other non-neurogenic brain regions.[52] Such regulated "natural genetic engineering" processes are now found to be particularly active in the brain, and are strongly associated with normal brain function.[26,37]

The correlations between BDNF, neurogenesis and L1 insertions are presented in Fig. 6.1.

To Conclude

There is now overwhelming evidence that perinatal nutrition, education and enrichment programmes, and physical exercise are all highly effective in improving brain function, as well as health and well-being, and for all age groups. For far too long, our policy-makers have been misled and misinformed into believing that intellectual ability and health are largely determined by the genes, and hence social and environmental interventions would have little or no effect. This pernicious genetic determinist ideology has now been definitively and thoroughly refuted by a convergence of findings in molecular genomics and biometrical genetics. It is our responsibility to take immediate action in all the appropriate remedial and proactive interventions to safeguard the physical and mental health and well-being of the nation for the present and future generations.

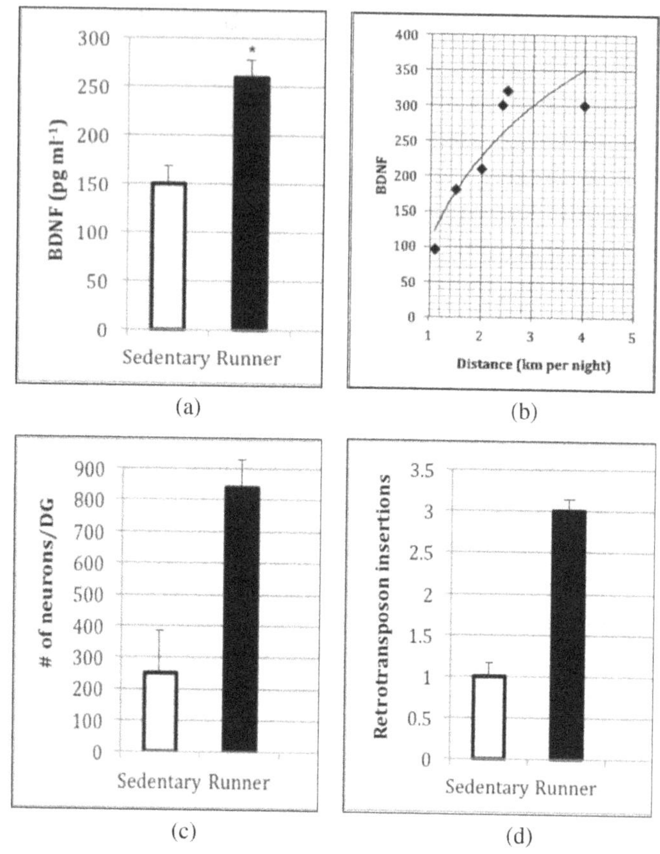

Figure 6.1 Correlations between physical exercise and increase in BDNF neuronal growth factor (a, b), number of new neurons generated (c) and retrotransposon insertions (d).

References and Notes

1. Nisbett, R.E., Aronson, J., Blair, C., Dickens, W., Glynn, H., Halpern, D.F. and Turkheimer, E. "Intelligence: New Findings and Theoretical Developments". *Am Psychol* 67 (2012): 130–159.
2. Duckworth, A.L., Quinn, P.D., Lynam, D.R., Loeber, R. and Stouthamer-Loeber, M. "Role of Test Motivation in Intelligence Testing". *Proc Natl Acad Sci* 108 (2011): 7716–7720.

3. Ho, M.W. "Mystery of Missing Heritability Solved?" *Science in Society* 53 (2012a): 28–31.
4. Ho, M.W. *Genetic Engineering Dream or Nightmare? The Brave New World of Bad Science and Big Business,* Third World Network, Penang, and Gateway Books, Bath, 1998 (reprinted in 2007 with extended Introduction), http://www.i-sis.org.uk/genet.php.
5. Ho, MW. (2000). "Human Genome: The Biggest Sellout in Human History". *ISIS TWN Report,* http://www.i-sis.org.uk/humangenome.php; a slightly different version in *ISIS News,* 6, September 2000, http://www.i-sis.org.uk/isisnews/i-sisnews6.php#huma.
6. Ho, M.W. "Ten Years of the Human Genome: Reams of Data and No Progress in Sight". *Science in Society* 48 (2010a): 22–25.
7. Visscher, P.M., Brown, M.A., McCarthy, M.I. and Yang, Y. "Five Years of GWAS Discovery". *Am J Hum Gen* 90 (2012): 7–24.
8. Zuk, O., Hechter, E., Sunyaev, S.R. and Lander, E.R. "The Mystery of Missing Heritability: Genetic Interactions Create Phantom Heritability". *Proc Natl Acad Sci* 109 (2012): 1193–1198.
9. Ho, M.W. "Death of the Central Dogma". *Science in Society* 24 (2004): 4.
10. Ho, M.W. "Epigenetic Inheritance: What Genes Remember". *Science in Society* 41 (2009): 4–5.
11. Ho, M.W. "Nurturing Nature: How Parental Care Changes Genes". In *Genetic Explanations: Sense and Nonsense* (S. Krimsky and J. Gruber, eds.), pp. 256–269, Harvard University Press, Cambridge, Massachusetts, 2013.
12. Ho, M.W. *Living with the Fluid Genome.* ISIS/TWN, London and Penang, 2003, http://www.i-sis.org.uk/fluidGenome.php.
13. Ho, M.W. "No Genes for Intelligence in the Human Genome". *Science in Society* 53 (2012b): 28–31.
14. ENCODE Project Consortium. "Identification and Analysis of Functional Elements in 1% of the Human Genome by the ENCODE Pilot Project". *Nature* 2, 447 (2007): 799–816.
15. Caruso, D. "Change to Gene Theory Raises New Challenges for Biotech", 3 July 2007, http://www.iht.com/articles/2007/07/03/business/biotech.php?page=1.

16. Wallace, H. A. "Model of Gene-Gene and Gene-Environment Interactions and Its Implications for Targeting Environmental Interventions by Genotype". *Theor Biol Med Model* 3 (2006): 35.
17. Ho, M.W. and Papadimitriou, N. "Human DNA 'Biobank' Worthless", 2002, http://www.i-sis.org.uk/DNAdatabaseproblems.php.
18. Kohane, I.S. "Using Electronic Health Records to Drive Discovery in Disease Genomics". *Nat Rev Genet* 12 (2011): 417–428.
19. Wallace, H. "'Phantom Heritability' Indicates Poor Predictive Value of Gene Tests", *Independent Science News*, 10 January 2012, http://independentsciencenews.org/health/phantom-heritability-indicates-poor-predictive-value-of-gene-tests/.
20. Collins, F.S. *The Language of Life*, Harper Perrenial, New York, 2010.
21. Butcher, L.M., Davis, O.S.P., Craig, I.W. and Plomin, R. "Genome-Wide Quantitative Trait Locus Association Scan of General Cognitive Ability Using Pooled DNA and 500K Single Nucleotide Polymorphism Microarrays". *Genes Brains Behav* 7 (2008): 435–436.
22. Chabris, C.F., Hebert, B.M., Beauchamp, J.P. *et al.* "Most Reported Genetic Associations with General Intelligence Are Probably False Positives". *Psychol Sci* 23 (2012): 1314–1323.
23. Davies, G., Tanesa, A. Payton, A. *et al.* "Genome-Wide Association Studies Establish That Human Intelligence Is Highly Heritable and Polygenic". *Mol Psychiatry* 16 (2011): 996–1005.
24. Plomin, R., Haworth, C.M.A., Meaburn, E.L. and Price, T.S., Wellcome Trust Case Control Consortium 2 and Davis, OS.P. "Common DNA Markers Can Account for More Than Half of the Genetic Influence on Cognitive Abilities". *Psychol Sc* 24 (2013): 4562–4568.
25. Devlin, V., Daniels, M. and Roeder, K. "The Heritability of IQ". *Nature* 388 (1997): 468–470.
26. Charney, E. "Behaviour Genetics and Post Genomics". *Behav Brain Sci* 35 (2012): 331–358.
27. Segal, N.L., McGuire, S.A. and Stohs, J.H. "What Virtual Twins Reveal About General Intelligence and Other Behaviors". *Pers Individ Dif* 53 (2012): 405–410.
28. Segal, N.L., e-mail to author, 29–30 January 2012.
29. Van Ijzendoorn M.H., Juffer, F. and Poelhuis, C.W.K. "Adoption and Cognitive Development: A Metaanalytic Comparison of Adopted

and Nonadopted Children's IQ and School Performance". *Psychol Bull* 131 (2005): 301–316.

30. Ho, M.W. "How to Improve the Brain Power and Health of a Nation". *Science in Society* 53 (2012c): 32–35.
31. Hart, B. and Risley, T. *Meaningful Differences in the Everyday Experiences of Young American Children*. Paul H. Brooks, Baltimore, Maryland, 1995.
32. Ho, M.W. "Epigenetic Inheritance: What Genes Remember". *Science in Society* 41 (2009a): 4–5.
33. Ho, M.W. "Mismatch of RNA to DNA Widespread". *Science in Society* 53 (2012d): 14–15.
34. Ho, M.W. "How Food Affects Genes". *Science in Society* 53 (2012e): 12–13.
35. Furrow, R.E., Christiansen, F.B. and Feldman, M.W. "Environment-Sensitive Epigenetic and the Heritability of Complex Diseases". *Genetics* 189 (2011): 1377–1387.
36. Ho, M.W. "Caring Mothers Strike Fatal Blow Against Genetic Determinism". *Science in Society* 41 (2009b): 6–9.
37. Ho, M.W. "Rewriting the Genetic Text in Human Brain Development and Evolution". *Science in Society* 41 (2009c): 16–19.
38. Molenaar, P.C.M., Boomsma, D.I., Smit, D. and Nesselroade, J.R. "Estimation of Subject-Specific Heritabilities from Intra-Individual Variation: iFACE" (preprint, courtesy of Peter Molenaar). *Twin Res Hum Genet* 15 (2012): 393–400.
39. Ho, M.W. "Development and Evolution Revisited". In *Handbook of Developmental Science, Behavior and Genetics* (K. Hood, C. Halpern, G. Greenberg and R. Lerner, eds.), pp. 61–108, Wiley-Blackwell, Chichester, 2010b.
40. Diamond, A. and Lee, K. "Interventions Shown to Aid Executive Function Development in Children 4 to 12 Years Old". *Science* 133 (2011): 959–964.
41. "Role of Nutrition in Learning and Behavior: A Resource List for Professionals". Food and Nutrition Information Center, National Agricultural Library, USDA, August 2011, http://www.nal.usda.gov/fnic/pubs/learning.pdf.
42. Theodore, R.F., Thompson, J.M.D., Waldie, K.E., Wall, C., Becroft, D.M.O., Robinson, E., Wild, C.J., Clark, P.M. and Mitchell, E.A.

"Dietary Patterns and Intelligence in Early and Middle Childhood". *Intelligence* 37 (2009): 506–513.

43. Kim, J.L., Winkvist, A., Åberg, M.A., Åberg, N., Sundberg, R., Torén, K. and Brisman, J. "Fish Consumption and School Grades in Swedish Adolescents: A Study of the Large General Population". *Acta Paediatrica* 99 (2010): 72–77.
44. Cook, J.T. and Frank, D.A. "Food Security, Poverty, and Human Development in the United States". *Ann N Y Acad Sci* 1136 (2008): 193–209.
45. Ivanovic, D.M., Leiva, B.P., Pérez, H.T., Almagià, A.F., Toro, T.D., Soledad, M., Urrutia, C., Inzunza, N.B. and Bosch, E.O. "Nutritional Status, Brain Development and Scholastic Achievement of Chilean High-School Graduates from High and Low Intellectual Quotient and Socio-Economic Status". *Br J Nutr* 87 (2002): 81–92.
46. Engle, P.L. and Fernández, P.D. "INCAP Studies of Malnutrition and Cognitive Behavior". *Food Nutr Bull* 31 (2010): 83–94.
47. Gewa, C.A., Weiss, R.E., Bwibo, N.O. *et al.* "Dietary Micronutrients Are Associated with Higher Cognitive Function Gains Among Primary School Children in Rural Kenya". *Br J Nutr* 101 (2009): 1378–1387.
48. Van Praag, H. "Neurogenesis and Exercise: Past and Future Directions". *Neuromol Med* 10 (2008): 128–140.
49. Colcombe, S. and Kramer, A.F. "Fitness Effects on the Cognitive Function of Older Adults: A Meta-Analytic Study". *Psychol Sci* 14 (2003): 125–130.
50. Pereira, A.C., Huddleston, D.E., Brickman, A.D., Sosunov, A.A., Hen, R., McKhann, G.M., Sloan, R., Gage, F.H., Brown, T.R. and Small, S.A. "An In Vivo Correlate of Exercise-Induced Neurogenesis in the Adult Dentate Gyrus". *Proc Natl Acad Sci* 104 (2007): 5638–5643.
51. Cotman, C.W. and Berchtold, N.C. "Exercise: A Behavioural Intervention to Enhance Brain Health and Plasticity". *Trends Neurosci* 25 (2002): 295–301.
52. Muotri, A.R., Marchetto, M.C.N., Zhao, C. and Gage, F.H. "Environmental Influence on L1 Retrotransposons in the Adult Hippocampus". *Hippocampus* 19 (2009): 1002–1007.

7

Natural versus Artificial Genetic Modification and Perils of GMOs

The rationale and impetus for artificial genetic modification was the central dogma of molecular biology that assumed DNA carries all the instructions for making an organism, which are transmitted via RNA to protein to biological function in linear causal chains. Since the mid-1970s, however, molecular geneticists have uncovered a remarkable circular causation network between organism and environment that not only determines which genes are expressed but marks and changes genes and genomes. In order to survive, the organism needs to engage in natural genetic modification in real time, an exquisitely precise molecular dance of life with RNA and DNA responding to and participating in "downstream" biological functions. Artificial genetic modification, in contrast, is crude and imprecise, and interferes with the natural processes, driving natural systems towards a state of maximum entropy as the perturbations are propagated and amplified down the generations with devastating consequences.

It has been almost 20 years since the first genetically modified organism (GMO) entered the market.[1] A GMO is simply any organism with synthetic genetic material inserted into its genome ("genome" in this context includes extrachromosomal plasmids and mitochondrial and chloroplast DNA); it is made in the laboratory with sterile techniques, which also means without the need for sexual reproduction between donor and recipient species of the genetic material. The basis for such genetic manipulation was the central dogma of molecular biology due to British biologist Francis Crick[2,3] (1916–2004) who shared the Nobel Prize with American biologist James Watson (and British biophysicist Maurice Wilkins (1916–2004)) for the DNA double-helix structure.[4] Though often acknowledged as an oversimplification, the central dogma has remained the basis for artificial genetic modification and the more sophisticated version, "synthetic biology".[5] It is supposed implicitly that individual "genetic messages" in DNA are faithfully copied or transcribed into RNA, which are then translated into proteins via a genetic code; each protein determining a particular trait, such as herbicide tolerance or insect resistance; one-gene-one-character. Hence inserting a new genetic message into an organism will give it the desired character to serve our every need. If it were really as simple as that, genetic modification would work perfectly every time. Unfortunately, things are vastly more complicated.

The Fluid Genome and Natural Genetic Engineering

In *Genetic Engineering Dream or Nightmare*[6] I described the new genetics of the "fluid genome" — a term coined by molecular geneticists that appeared on the back cover of a popular book published in the early 1980s[7] — and warned that the greatest danger of genetic modification is its being misguided by the ideology of genetic determinism (the central dogma of the old genetics). Genetic determinism made genetic modification seem compelling, but was totally contrary to the reality of the new genetics as documented in numerous reviews and books since,[8–15] though none of the publications except *Living with the Fluid Genome*[13] has dealt explicitly with the perspective presented here.

Genetics has been turned upside down beginning in the mid-1970s and especially since the human genome was announced in 2000. The tools of genetic manipulation have been advancing and improving in leaps and bounds. Today, geneticists can dissect and analyse the structure and function of genes and genomes in minute detail down to the base sequence of a nucleic acid in one single cell using next-generation sequencing (see Box 7.1).[16]

Empirical findings began to trickle in during the 1970s, then streamed and poured out of laboratories. The new genetics is telling

Box 7.1

Next-Generation Sequencing

Next-generation sequencing (NGS) extends sequencing across millions of reactions taking place in parallel. This enables rapid sequencing of large stretches of DNA base pairs spanning entire genomes, with instruments capable of producing hundreds of gigabase (Gb) data in a single sequencing run. To sequence a single genome, the genome is first fragmented into a library of small segments that can be uniformly and accurately sequenced in millions of parallel reactions. The newly identified strings of bases, called reads (of a defined length) are then reassembled using a known reference genome as a scaffold (re-sequencing), or in the absence of a reference genome (*de novo* sequencing), assembled by overlaps. The full set of aligned reads reveals the entire sequence of each chromosome in the genome.

NGS data output has been rising steeply since its invention in 2007, when a single sequencing run could produce a maximum of about 1 Gb data. By 2011, the rate had reached nearly a terabase (Tb, 10^{12}b), a thousand-fold increase. By 2012, researchers could sequence more than five human genomes in a single run, producing data in roughly one week at a cost of less than $5 000 per genome. The $1 000 genome is now within our grasp.

NGS high throughput capacity has enabled "deep sequencing" of genomes and transcriptomes to look for rare DNA variants or rare species of RNA transcripts. Deep sequencing means that the total number of reads is many times larger than the length of the sequence under study. "Depth" (coverage) is the average number of times a nucleotide is read.

us in no uncertain terms that the genome is fluid and dynamic. It is constantly conversing with the environment in a circular network that marks and changes genomic DNA in myriad ways, with both DNA and RNA taking part in transmitting genetic information and in executing and altering genetic information in real time. I use the term "natural genetic modification" for the totality in which an organism goes about its business of living, with effects reverberating and amplified down the generations.[12,13] This cross-talk not only involves different parts of the organisms, but extends to the microbiota within the organism (in the digestive tract, for example),[17,18] as well as species in the external environment[18] and the ecosystem as a whole.

The organism is doing its own natural genetic modification with great finesse, a molecular dance of life that is necessary for survival. Unfortunately, genetic engineers do not know the steps or the rhythm and music of the dance; they are only now tracing its footprints in the genome: how DNA and histone proteins are marked, or DNA and RNA are changed, for example. It is clearly impossible to modify one gene or one function at a time without affecting other functions, and ultimately the entire organism and its ecosystem. It is also this molecular dance of life that makes organisms and ecosystems vulnerable to the unintended effects of genetic modification, as will be made clear later on. Furthermore, the insults and injuries to organisms and ecosystems exposed to the GMOs can be passed on to future generations to influence the course of evolution.

In retrospect, natural genetic modification makes perfect sense, because the "instructions" written in the genomic DNA can never fully anticipate what the organism will experience throughout its life. Genomic DNA is not only a text whose meaning changes with experience;[19] it is subject to revision, a changing narrative that prepares us for life at the next stage, and also enables us to pass on our experiences to the next generation, for better or for worse. RNA is far from a simple transcript of the genetic message, but can change and corrupt the message through RNA editing,[20] destroy it by RNA interference[21] or rewrite the DNA text by reverse transcription and gene conversion.[19,22]

In reality, the organism shapes its own development and evolutionary future; that is why it is so important to take responsible action.[23] Indeed, new findings on the fluidity and responsiveness of the genome have made the hazards of genetically modified DNA and RNA even greater than anyone has envisaged.

Genetic Modification Inherently Hazardous

Reliable evidence obtained by scientists independent of the biotech industry going back to the 1990s and evidence obtained by farmers in the field both show that genetically modified (GM) feed invariably causes harm, regardless of the animal species or the food crops genetically modified or the genes and constructs inserted into the genome. A list is presented in Table 7.1, partly drawn from a review published in 2007[24] and updated with studies done and cases uncovered since, including an independent meta-analysis of data from 19 feeding trials carried out by the biotech industry.

Table 7.1 Accumulating evidence on the health hazards of GM crops

1. A study from Norway published in 2015 showed that *Daphnia magna* (water flea) chronically exposed to MON810 Bt-maize leaves (containing Cry 1Ab toxin) suffered reduced growth and fertility compared to controls exposed to leaves from a near-isogenic non-GM maize.[25]
2. A study from Egypt published in 2014 found that GM feed containing the cauliflower mosaic virus 35S (CaMV 35S) promoter damaged the liver, kidney and testis of male albino rats after 30 to 90 days and the CaMV 35S promoter was transferred into the tissues of the rats.[26–28]
3. A paper published in 2013 reported significantly higher rates of severe stomach inflammation in pigs on farms in the United States. The pigs were fed a mixed GM corn and soybean diet for 22.7 weeks compared with pigs fed an equivalent non-GM control diet: 3 2% compared to 12%. Female pigs on the GM diet also had uterus heavier by 25% on average.[29]
4. In 2012, researchers in Egypt reported that male rats fed Bt corn MON810 showed a wide range of organ and tissue abnormalities;[30,31] these effects were subsequently replicated in male and female rats and their offspring during a three-month feeding trial.[32]
5. A two-year lab feeding trial in France published in 2012 found rats of both sexes exposed to Roundup and/or Roundup-tolerant maize *not* sprayed with herbicide were two to three times as likely to die as controls and to develop large tumours, of mammary glands in females and of kidney and skin in males.[33–35] In other words, the GMO without the herbicide was also harmful in every respect. Pituitary disease was up more than two-fold in females and liver and kidney diseases up 1.5–2-fold in males on GM maize alone that was not sprayed with herbicide. These effects appeared *after* the 90 days period legally required for safety tests on GMOs, thereby exposing the inadequacy of the regulatory regime. (This published study was astonishingly unilaterally "retracted" by the journal a year after it was published, the journal editor acting on behalf of the biotech industry.[36] The paper was soon republished elsewhere.[37])

(*Continued*)

Table 7.1 *(Continued)*

6. A Danish farmer found excessive illnesses and deaths in his pigs fed GM soy meal including chronic diarrhoea, birth defects, reproductive problems, bloating, stomach ulcers, weak and smaller piglets, and reduced litter size. These were entirely reversed when he put them on a GM-free diet.[38, 39]
7. A meta-analysis pooling all available data on 19 feeding trials on GM soybean and maize, both glyphosate-tolerant and Bt crops representing 83% of commercialized GMOs carried out for 90 days — a period inadequate for detecting the most serious health impacts — nevertheless found significant disruption of liver and kidney functions.[40, 41]
8. Professor emeritus Don Huber at Purdue University in the United States warned of "pathogen new to science" associated with glyphosate-tolerant GM crops and livestock fed on them, causing unprecedented deaths and infertility.[42, 43]
9. Between 2005 and 2006, senior scientist Irina Ermakova at the Russian Academy of Sciences in Moscow reported that female rats fed glyphosate-tolerant GM soybeans produced excessive numbers of severely stunted pups and more than half of the litter dying within three weeks, while the surviving pups were completely sterile.[44, 45]
10. Between 2004 and 2005, hundreds of farm workers and cotton handlers in Madhya Pradesh, India reported allergy symptoms from exposure to Bt cotton containing Cry1Ac or both Cry1Ac and Cry1Ab proteins.[46]
11. Between 2005 and 2006, thousands of sheep died after grazing on Bt cotton crop residues in four villages in the Warangal district of Andhra Pradesh in India.[47]
12. In 2005, scientists at the Commonwealth Scientific and Industrial Research Organization in Canberra, Australia tested a transgenic pea containing a normally harmless protein in bean (α-amylase inhibitor 1), and found it caused inflammation in the lungs of mice and provoked sensitivities to other proteins in the diet.[48,49]
13. From 2002 to 2005, scientists at the Universities of Urbino, Perugia and Pavia in Italy published reports indicating that GM-soya fed to young mice affected cells in the pancreas, liver and testes.[50–54]
14. In 2003, villagers in the south of the Philippines suffered mysterious illnesses when a Monsanto Bt maize hybrid containing Cry1Ab protein came into flower; antibodies to the Cry1Ab protein were found in the villagers; there have been at least five unexplained deaths and some remain ill to this day.[55]
15. In 2004, Monsanto's research dossier, kept confidential for commercial reasons, showed that rats fed MON863 GM maize containing Cry3Bb protein developed serious kidney and blood abnormalities.[56]

(Continued)

Table 7.1 (*Continued*)

16.	Between 2001 and 2002, a dozen cows died in Hesse, Germany after eating Syngenta GM maize Bt176 containing Cry1Ab/Cry1Ac plus glufosinate-tolerance; and more in the herd had to be slaughtered due to illnesses.[57]
17.	In 2012, biotech giant Syngenta was criminally charged with denying knowledge it had since 1996 that its GM maize kills livestock during a civil court case brought by the farmer that ended in 2007.[58]
18.	In 1998, senior scientist Arpad Pusztai and colleagues formerly of the Rowett Institute in Scotland reported damage in every organ system of young rats fed GM potatoes containing snowdrop lectin, including a stomach lining twice as thick as that of controls.[59]
19.	Also in 1998, scientists in Egypt found similar effects in the gut of mice fed Bt potato containing a Cry1A protein.[60]
20.	In 2002, Aventis company (later Bayer Cropscience) submitted data to UK regulators showing that chickens fed glufosinate-tolerant GM maize Chardon LL were twice as likely to die compared with controls.[61]

A consistent picture of GM-linked deaths and illnesses has emerged, with scientists confirming what farmers have been experiencing and witnessing for years. This is all the more significant, as independent scientific studies are hard to come by because of lack of support; and scientists find it very difficult to obtain the GM material from the biotech companies for their research or to publish their results.[62] The inevitable conclusion one comes to is that GM is *inherently* hazardous.

Sources of Hazards from GMOs

There are many possible sources of hazards of GMOs associated with the fluid and responsive genome; I have put them into five categories in Table 7.2.[6,63,64] Apart from the impacts to health (Table 7.1), there is also evidence that GM crops yield less, poison crops and soil, and cause the emergence and re-emergence of many crop diseases.[65]

The weight of evidence against the safety of GMOs is substantial, and hundreds of scientists have openly declared that there is no consensus on the safety of GMOs[66] in order to counteract the disinformation campaign that has taken over the mainstream media

Table 7.2 Categories of hazards from GMOs

1. **Uncontrollable, unpredictable impacts on safety due to the genetic modification process***
   Scrambling the host genome*
   Widespread mutations*
   Inactivating genes*
   Activating genes*
2. **Creating new transcripts (RNAs) including those with regulatory functions***
   Creating new proteins*
   Creating new metabolites or increasing metabolite to toxic levels*
   Activating dormant viruses*
   Creating new viruses by recombination of viral genes in GM insert with those in the host genome*
3. **Toxicity of transgene protein(s) introduced (intentionally or otherwise)**
   Transgene protein toxic*
   Transgene protein allergenic or immunogenic*
   Transgenic protein becoming allergenic or immunogenic due to processing*
   Unintended protein created by sequence inserted may be toxic or immunogenic
4. **Effects due to the GM insert and its instability***
   Genetic rearrangement with further unpredictable effects*
   Horizontal gene transfer and recombination*
   Spreading antibiotic and drug resistance*
   Creating new viruses and bacteria that cause diseases
   Creating mutations in genomes of cells to which the GM insert integrates including those associated with cancer*
5. **Toxicity of herbicides used with herbicide-tolerant GM crops***

*Documented in scientific literature, as reviewed in the main text.

including top scientific journals.[67] But we are still largely in the dark as to the precise nature of the hazard(s) associated with *individual* GMOs. Toxicity has been found for transgene products such as the Bt proteins from different strains of the soil bacteria *Bacillus thuringiensis* expressed in many GM crops,[65] while the multiple toxicities and carcinogenicity of glyphosate herbicides, heavily used with glyphosate-tolerant GM crops, and increasingly also as desiccant for non-GM crops and weed control in parks and residential areas, are no longer in doubt. In March 2015, the World Health Organization (WHO) expert panel on cancer re-assessed publicly available evidence on glyphosate and concluded it is "probably carcinogenic".[68] One recent review blames glyphosate for most, if not all, increases in chronic diseases in recent years.[69] The evidence has come from all over the world. Glyphosate herbicides have brought rising tides of birth defects, cancers, fatal kidney disease, sterility and dozens of other illnesses.[70] An Independent Scientists Manifesto on Glyphosate calling on governments at all levels to ban the spraying of the herbicide based on overwhelming evidence of harm from scientific studies and witness testimonies compiled has been signed by 547 scientists and 919 non-scientists from 70 countries as of 11 December 2015.[71]

There remains a range of hazards associated with GMOs that are not so easily identified, even though evidence exists for most if not all of them in the scientific literature. These are due to the unpredictable and uncontrollable nature of the genetic modification process (Table 7.2, category 1), which can activate or inactivate genes; scramble genomes; create new proteins, new nucleic acids, new metabolites and others due to the transgenic DNA and its instability (Table 7.2, category 3). There is also the danger of horizontal gene transfer — the direct transfer of DNA into the genomes of cells — from the GMO to all other species interacting with the GMO. *These are all consequences of interference with the natural genetic modification processes.* To identify the actual sources of hazards require genome sequencing and detailed comparisons of transcriptome, epigenome, proteome and metabolome, techniques all easily available today, and should be mandatory in the current regulatory regimes for risk assessment of GMOs, but are not.

Natural Genetic Modification and the New Genetics

Natural genetic modification employs the same copy, cut and splice tools as artificial genetic modification. (Artificial genetic modification is possible only by usurping the tools of the natural processes.) It enables organisms to express genes in different parts of the genome at the appropriate duration and levels, or mark and modify them, as and when required in specific cells and tissues.

To produce even one protein in our body — originally thought to be a single continuous message — requires elaborate cut-and-splice operations. The international research consortium project ENCODE (Encyclopedia of DNA Elements) data have revealed that the vast majority of genomic DNA in higher organisms (above the level of bacteria) include "non-coding" segments.[72,73] The "gene", a theoretical construct that has never been possible to define rigorously, is now known to be scattered in bits across the genome, overlapping with bits of multiple genes that have to be spliced together before translating into a protein. The term used for the bits is "coding sequences" or *exons*, which are typically separated by non-coding bits called *introns*.

The expression of each gene already requires the assembly of a small army of special molecular engineers. The human genome contains about 20 000 protein-coding genes, most of which would be active in one cell or other of the body at any one time.

That's not all. Humans contain practically the same number of protein-coding genes as nematodes that have only 1 000 cells compared to humans' 10^{14} cells. In contrast, non-protein-coding DNA, largely absent from bacteria, increases with increasing complexity of organisms, reaching 98.8% of the human genome.[74,75] Much of that previously considered "junk DNA" is dynamically and differentially transcribed (latest estimate >85%[76]) into many families of short and long non-coding (nc)RNAs. These ncRNAs regulate gene expression and genome architecture by interacting with DNA, RNAs, proteins and other cofactors.

Cells and tissues also respond to their environments by recruiting different contingents of molecular engineers for marking and modifying, cutting and splicing specific RNA or DNA, or remodelling chromatin (complex of DNA and histone proteins) at specific genome locations. We know only a small fraction of the vast amount of details involved. But it is already so remarkable that leading molecular geneticist James Shapiro at University of Chicago and a pioneer of the fluid genome is saying that practically nothing happens at random.[77–79] Cells and their genomes are not "passive victims of replication errors or DNA damage."[77,80] Instead, just about everything, including so-called random mutations, happens by "natural genetic engineering" (almost the same as what I call "natural genetic modification").

Indeed, cells have special proofreading and error-correcting functions to eliminate and repair damaged/mutated bases in the genetic material, getting errors down to below 1 in a billion bases under normal conditions. But during starvation, bacteria can also target precise mutations to specific sites in the genome to generate new metabolic functions.[77,80] Such "directed" or "adaptive" mutations are now well documented in bacteria as well as human cells.[81–83] The human immune system executes accurate cut-and-splice genome rearrangements to create a large variety of immunoglobulin chains and also targets hypermutations to specific immunoglobulin variable sites to generate huge diversities of antibodies for defence against invading pathogens.

I have only given a tiny sampling of organisms' remarkable feats of natural genetic modification, which are precisely targeted, context-dependent and complex, and negotiated by the organism as a whole. It is a well-known paradox that both plant and animal cells maintained in culture undergo uncontrollable mutations and chromosomal rearrangements (somaclonal variations),[84] in contrast to cells within the healthy organism, which show extremely low levels of "random" mutations.

The new genetics of the fluid genome belongs in the organic paradigm of quantum biology.[85] The organism is actually the living incarnation of the elusive quantum computer[86] that has been the holy grail of quantum physicists for more than 30 years, and is said to have powers of rapidly and precisely solving problems beyond the ordinary computer, which is effectively what natural genetic modification achieves. Quantum coherent organisms intercommunicate via coherent electromagnetic signals, and there is indeed evidence that molecules do just that. Molecules that react together share common frequencies to which they resonate, thereby attracting one another;[87] and that may well be how specific genes are expressed at the appropriate time and place, or directed to mutate during metabolic stress, the genes being just those coding for enzymes that can metabolize a new substrate presented to the starved cells.[83]

Natural Versus Artificial Genetic Modification

The targeted precision and complexity of natural genetic engineering/modification makes clear that genetically modified organisms (GMOs) created by the crude methods generally used until very recently can only be highly unsafe.[88,89] Much current effort is dedicated to "genome editing" using guided or otherwise specific DNA-cutting enzymes to alter DNA sequence at target locations in the genome. But off-target, cytotoxic effects continue to dog the latest attempts.[90–92] Artificial genetic modification invariably interferes with natural genetic modification, and it is well-nigh impossible to avoid doing so. It depends on disrupting and overriding the organism's own minutely choreographed process; the result is uncontrollable and unpredictable off-target effects.

To override the natural system, the synthetic GM DNA molecules are forced into the cells in large numbers with stressful methods such as gene gun[93] or electric shock[94] carried by vectors (such as the *Agrobacterium* binary vector) designed to invade genomes. Further, the transgenes are equipped with aggressive promoters such as the cauliflower mosaic virus (CaMV) 35s and similar viral promoters in order to force the cells to express the foreign genes.[88] These and other stresses (as Shapiro points out) are well known to mobilize endogenous transposons (jumping genes) that scramble and destabilize genomes. Consequently, transgenic lines are unstable,[63–65] both from silencing and loss of transgenes, which make horizontal transfer of transgenic DNA more likely than non-transgenic DNA. For the same reasons, transgenic lines often suffer yield drag,[95–97] while complete crop failures have been reported in India[98] and during the recent drought in the United States while non-GM crops survived.[99] The same transgene instability may have been responsible for the persistent failure in the pilot commercial planting of Bt brinjal in Bangladesh.[100]

Table 7.3 Artificial versus natural genetic modification

Artificial	Natural
Context-inappropriate, hence uncontrollable and unpredictable hazards: scrambled genomes, new nucleic acids, proteins and metabolites	***Context-dependent precise targeting and accurately choreographed with little or no off-target effects***
Depends on disrupting the natural process, hence adverse and hazardous interference Stressful methodology — gene guns, electric shocks, invasive vectors (*Agrobacterium*), aggressive virus promoters (CaMV 35S) — multiply hazards and destabilize genomes, resulting in transgene instability, yield drag and horizontal transfer of transgenes	***Always appropriate to context and hence no adverse interference***
Reductionist aims, without regard to the whole organism (uncontrollable somaclonal variations of cells in culture)	***Negotiated by the organism as a whole*** (very low random mutation rates in cells unless targeted)

Natural genetic modification is not something that happens only occasionally; it is constant and all-pervasive in the life of the organism, interconnecting millions of molecular players in a cell at any one time. That is why artificial genetic modification has failed to produce any ecologically beneficial or complex traits, while even the "single gene" traits are unstable. Artificial and natural genetic modification are contrasted in Table 7.3.

Artificial Genetic Modification Imperils the Biosphere by Hijacking the Natural Processes

Despite its failures and inefficacy, artificial genetic modification can nevertheless endanger organisms exposed to its sphere of influence, which is why GMOs are invariably found to be harmful (see Table 7.1). Specific hazards involving the CaMV 35S promoter and the *Agrobacterium* binary vector, both widely used in genetic modification, are fully reviewed elsewhere.[65,88] Here, I highlight aspects that are increasingly important in "new generation" GMOs and more sophisticated genetic modification techniques. These are associated with new nucleic acids intentionally introduced or arising from unintended artificial modifications. They harm organisms and ecosystems by hijacking the natural processes; especially via nucleic acids in food, horizontal gene transfer and trans-generational inheritance.

New Nucleic Acids Enter the Human Food Chain to Alter Gene Expression and Worse

A research team from China first reported in 2011 that short regulatory micro (mi)RNAs (~22 nt) originating from plants eaten can resist digestion, enter the bloodstream, and get into cells to change the expression of specific genes.[101] It raised serious concerns over the safety of GMOs, for they introduce entirely new nucleic acids into the human food chain,[102] both intentionally created and unintended within the GMOs. Monsanto orchestrated an attempt to discredit this finding[103] but it has been abundantly confirmed and extended since.

A survey of human plasma for miRNAs using NGS (Box 7.1) carried out by Kai Wang and David Galas at the Institute for Systems Biology and Paul Wiles at University of Luxembourg found extensive and widespread presence of miRNAs originating from grains and other food items including soybean, tomato and grape. Some of the miRNAs or miRNA-like molecules were synthesized and transfected into mouse fibroblasts, and found to alter the expression profiles of a number of genes.[1,104]

Researchers at Moringga Milk Industry Zama Kanagawa, Japan, using more conventional microarray and quantitative PCR analyses, identified 102 miRNA in cow's milk, 100 in colostrum and 53 in mature milk, with 51 common to both.[105] In addition, some messenger(m)RNAs were found in the milk. These miRNAs and mRNAs were wrapped inside lipoprotein vesicles rather like the exosomes identified in the bloodstream of animals (as well as in cell culture medium) that are believed to be part of the nucleic acid intercommunication system of the body (see below). Both miRNAs and mRNAs were also present in infant formulas bought from Japanese supermarkets.

A team at University of Louisville, Kentucky in the United States isolated exosome-like nanoparticles from the edible plants ginger root, grape, grapefruit and carrot, which contain proteins, lipids and miRNA. These nanoparticles were taken up by intestinal macrophages and stem cells of mice and preferentially induced the expression of antioxidant genes and genes involved in the maintenance of intestinal homeostasis[106] that protect against all kinds of chronic diseases including cancer. This serves to remind us that epigenetic effects can be beneficial or harmful, and why it is important to eat good wholesome food and avoid GMOs.

Not only RNA, but also DNA from meals eaten could be identified. A study led by Sándor Spisák at the Hungarian Academy of Sciences in Budapest and Harvard Medical School in Boston, Massachusetts analysed over 1 000 human adult samples from four independent studies using NGS and NGS databases, and found DNA fragments derived from food in all plasma samples, some large enough to code for complete genes.[107] The team found DNA from dozens of plant species differing between individuals, mostly likely reflecting their diet, including grains, beans and vegetables. There was also meat DNA, but because animal DNA is more similar to human DNA, it is more difficult to ascertain.

There is increasing evidence that cells in the body intercommunicate via circulating nucleic acids actively secreted into the bloodstream.[88,108] These circulating nucleic acids are able to influence gene expression in other cells and to transform other cells by horizontal gene transfer (direct uptake of the nucleic acids into the cells and incorporation into the cell's genome). Cancer cells use the system to spread cancer around the body. Thus, nucleic acids from meals eaten, including those containing GMOs, may also enter the bloodstream to influence gene expression and to transfer horizontally into the cells' genome with potentially harmful consequences associated with insertion mutagenesis, including cancer development and genome instability. The CaMV 35S promoter, used to drive the expression of transgenes in almost all commercially grown GM crops, known to contain a recombination hotspot (hence prone to horizontal gene transfer), is promiscuously active in all kingdoms of organisms including human cells, and specifically induces transcription factors required for CaMV and HIV replication.[109] And after it had been widely used in commercially grown GM crops for 20 years, regulators "discovered" it overlaps with another dangerous virus gene involved in RNA silencing,[110] most likely involved in host defence against virus attacks.

The new GM crops exploiting RNA interference (RNAi) are obviously hazardous in this regard, as RNAi is based on sequence-specific interactions between regulatory RNA and target(s),[111] but are known to tolerate numerous mismatches, changing in different cells at different times, and certainly beyond control.[112,113] The potential off-target effects are huge.

Horizontal Transfer of GM Nucleic Acids

Horizontal gene transfer is part and parcel of natural genetic modification. In its simplest form, horizontal gene transfer involves uptake of foreign nucleic acids into cells and incorporation into the cell's genome. For this very reason, GMOs carrying bacterial and viral genes and other synthetic genetic elements can readily exploit this natural avenue to spread antibiotic resistance and to create new pathogens, as some of us have been warning against since the late 1990s.[114] All the more so, as GM constructs are designed to overcome natural barriers and to invade genomes.[88,89] There is already evidence that widespread unintended horizontal transfer of GM DNA has probably occurred. The most decisive evidence was provided in 2012 by Li Jun Wen, Jin Min and colleagues at Sichuan University in China.[115] (It appears that scientists in China are taking the lead in biosafety research.) The team set out to look for horizontal transfer of the ampicillin antibiotic resistance marker (arm) gene *blá*, which has been extensively deployed in artificial genetic modification. By using the appropriate molecular probes (primers), sufficiently sensitive polymerase chain reaction (PCR) for detection, and constructing a metagenomics plasmid library, they detected the GM arm gene in *all* of China's rivers, despite the fact that the country has not been growing any GM crops commercially, but field trials of GM crops containing the arm gene have been carried out.[116] This is the first study of its kind in the world. The researchers concluded that horizontal transfer of GM antibiotic resistance gene may be linked to the rise in antibiotic resistance in livestock and humans in China. The possibility that genetic engineering biotechnology may have contributed to the increase in antibiotic resistance and the emergence of new viral and bacterial pathogens has been raised by some of us since the 1990s,[114] but it has never been admitted by the World Health Organization or any other agency monitoring the spread of antibiotic resistance and infectious diseases.

New findings suggest that even very short (~20 bp) and damaged pieces of DNA can be taken up and incorporated in the bacterial genome,[117] making it clear that GM nucleic acids can indeed spread antibiotic resistance and create new viruses and bacteria that cause diseases by horizontal gene transfer and recombination.[118, 119] But regulatory agencies in the United States, Europe and elsewhere are still denying that horizontal transfer of GM DNA has taken place, based on unfounded assumptions and the failure to use sufficiently sensitive up-to-date detection methods with the correct molecular probes. It is a case of "don't look, don't find."

Transgenerational Inheritance

Finally, the effects of GMOs are perpetrated and amplified across generations, because they can be inherited. As mentioned earlier, the scope of genetic information passed onto the next generation of cells and organisms has greatly expanded to include, besides genomic DNA, DNA marks (such as methylation), histone marks, chromatin structure (whether inactive heterochromatin or active) plus a host of small RNA regulators of gene expression. It appears that different RNAs not only register so-called epigenetic change as the organism responds to the environment, but they also transmit acquired genetic information to subsequent generations independently of

DNA.[120] Once again, this highlights the potential perils of using RNA interference in GMOs (see above). The exposure of organisms to regulatory RNA molecules (without transgenesis) could already result in the transmission of effects to subsequent generations.

Certain small regulatory RNAs can be independently replicated by RNA-dependent RNA polymerase, an enzyme present in RNA viruses that do not go through a DNA intermediate, while another form of this enzyme is present in all eukaryote genomes,[121] and is suspected to be involved in the maintenance of transcriptional silencing.

Regulatory RNAs are passed on via germ cells from one generation to the next, and they may be stabilized by RNA methylation to survive the maternal-to-zygote transition during early embryogenesis to influence gene expression in the development of the offspring.[120] RNA also operates in a RNA-memory system to distinguish "self" and "non-self" via viral and other sequences integrated into the genome that can defend the host from viral infections and animal predators. This memory system is centrally involved in maintaining active as well as silenced genes across generations.

Female germ cells carry maternal RNAs, and maternal effects are well known and generally accepted. Much less known is that male germ cells are particularly adept at picking up somatic RNA and DNA and carrying the cargo into the egg at fertilization in a process that has come to be known as "sperm-mediated gene transfer".[122,123] While most of the extraneous nucleic acids added to mature sperm *in vitro* are taken in and transmitted as extra-chromosomal DNA in mosaic fashion (present in some cells), integration into the genome can also occur. The inheritance of acquired characters via the male germ cells has been demonstrated in all species examined.

The first hint that fathers can pass on acquired characters was the discovery that the experience of young boys could affect not just their health in later life, but also the health of their sons and grandsons. That was the beginning of the epigenetic revolution.[124] All kinds of life experiences, good and bad, from caring mothers to environmental toxins, leave epigenetic imprints that are passed on for generations afterwards.[125,126] In the case of environmental toxins, Michael Skinner's reproductive biology lab at Washington State University in the United States first reported in 2005 that injecting pregnant rats with endocrine disruptor fungicide vinclozolin caused sperm abnormalities that persisted in the male progeny for at least four generations.[128] The effects on reproduction correlate with altered DNA methylation pattern in the germ line (though the methylation differences vary widely among the animals, and failed to satisfy his critics.[129] Subsequently, they found that the insecticides DDT and permethrin, jet fuel, plastic additives phthalates and bisphenol A, and dioxin can all trigger trans-generational health effects in rats such as obesity and ovarian disease, and each resulted in a different pattern of methylation in sperm DNA.

In the context of epigenetic toxicology, we should also highlight the abundant evidence on the toxicity of glyphosate, the top herbicide used worldwide. It is an endocrine disruptor at very low concentrations, implicated in male infertility, birth defects and cancers (see above), but its epigenetic effects have not been investigated at all.

The processes and agents responsible for transmitting trans-generational effects are summarized in Box 7.2.[120,123,124]

The importance of natural genetic modification and the numerous molecular mechanisms for the inheritance of acquired characters have large implications for social policy.[80] A pioneer of molecular

Box 7.2

Processes and Agents for Transmitting Transgenerational Effects

- DNA marks
- Histone marks
- miRNA and other small RNAs
- Chromatin remodelling
- mRNA and other RNAs (via RNA methylation to stabilize through maternal to zygote transition)
- RNA memory (via integrated sequences)
- Sperm-mediated gene transfer
- Integration of reverse transcribed RNA

genetics, Joshua Lederberg (1925–2008), invented the term *euphenics*,[129] practices intended to improve phenotypes, as opposed to *eugenics*, practices intended to improve genotypes. He was remarkably prescient. In the light of the fluid genome, optimizing the environment for euphenics will automatically guarantee the good genes desired in eugenics, on account of circular causation in the fluid genome. For the same reasons, no amount of eugenics or good genes will protect you from a hostile adverse environment, gene therapy and genetic modification notwithstanding.

Is euphenics so idealistic that it is just a fantasy? Not at all! They are the things most, if not all, people have always wanted: social equality, the benefits of which are backed up by a lot of serious data,[130–132] non-stressful work places, creative collaborative atmosphere at schools and universities as well as in society, good wholesome non-GM food produced ecologically[133] while safeguarding natural biodiversity, renewable energies and a circular non-polluting green economy[134] just around the corner.

To Conclude

The rationale and impetus for genetic engineering and genetic modification was the central dogma of molecular biology that assumed DNA carries all the instructions for making an organism. The mechanistic fallacy is inherent in the very term "genetic *engineering*", for it goes against the grain of the concept of the fluid and responsive genome that had already emerged since the early 1980s.

In order to survive, the organism needs to engage in a kind of natural genetic modification in real time, an exquisitely precise and intricate molecular dance of life with RNA and DNA responding to the environment and participating fully in "downstream" biological functions. Artificial genetic modification interferes fundamentally with the natural processes and it is well-nigh impossible to avoid doing so. In addition, the artificial GM constructs are invasive and can take over the natural mechanisms for transfer into cells. GM nucleic acids have the ability to scramble the intricate molecular dance of life of a coherent, healthy organism, the effects of which can be passed down to future generations, amplified and propagated to the wider ecosystem with devastating consequences.

References and Notes

1. Bruening, G., Lyons, J.M. The Case of the FLAVR SAVR Tomato. *Calif Agric* 54 (2000): 6–7.
2. Crick, F.H.C. "On Protein Synthesis". *Symp Soc Exp Biol* 12 (1958): 139–163.
3. Crick, F. "Central Dogma of Molecular Biology". *Nature* 227 (1970): 561–563.
4. Watson, J.D. and Crick, F. "A Structure for Deoxyribose Nucleic Acid". *Nature* 171 (1953): 73–738.
5. Synthetic biology has been recently defined as the artificial design and engineering of biological systems and living organisms for purposes of improving applications for industry or biological research. See Osbourn, A.E., O'Maille, P.E., Rosser, S.J. and Lindsey, K. Synthetic biology. *New Phytol* 196 (2012): 671–677.
6. Ho, M.W. (1998). *Genetic Engineering Dream or Nightmare? The Brave New World of Bad Science and Big Business.* Gateway Books, Bath, 1998. 2nd edition, Continuum, New York, 1999. Reprinted with extended introduction and update, Third World Network, Penang, 2007.
7. Dover, G. and Flavell, D. *Genome Evolution*, Oxford University Press, Oxford, 1982.
8. Ho, M.W. "Evolution by Process, Not by Consequence: Implications of the New Molecular Genetics for Development and Evolution". *Int J Comp Psychol* 1 (1987): 3–27.
9. Pollard, J.W. "The Fluid Genome and Evolution". In *Evolutionary Processes and Metaphors* (M.W. Ho and S. Fox, eds.), pp. 63–84, Wiley, London, 1988.
10. Shapiro, J. "Genome Organization, Natural Genetic Engineering and Adaptive Mutation". *Trends Genet*, 13 (1997): 98–104.
11. Jablonka, E., Lamb, M. *Epigenetic Inheritance and Evolution: The Lamarckian Dimension.* Oxford University Press, Oxford, 1995.
12. Ho, M.W. "Death of the Central Dogma". *Sci Soc* 24 (2004a): 4.

13. Ho, M.W. *Living with the Fluid Genome*. ISIS, London, 2003.
14. Mattick, J.S. and Mehler, M.F. "RNA Editing, DNA Recoding and the Evolution of Human Cognition". *Trends Neurosci* 31 (2008): 227–233.
15. Cheng, C.C., Johnson, T.L. and Hoffmann, A. "Epigenetic Control: Slow and Global, Nimble and Local". *Gene Dev* 22 (2008): 1110–1114.
16. Ho, M.W. "Nucleic Acid Invaders from Food Confirmed". *Sci Soc* 63 (2014a): 24–25.
17. Dietert, R. and Dietert, J. "The Completed Self: An Immunological View of the Human-Microbiome Superorganism and Risk of Chronic Diseases". *Entropy* 14 (2012): 2036–2065.
18. Kosoy, M. "Deepening the Conception of Functional Information in the Description of Zoonotic Infectious Disease". *Entropy* 15 (2013): 1929–1962.
19. Ho, M.W. "Subverting the Genetic Text". *Sci Soc* 24 (2004b): 6–8.
20. Tang, W., Ge, Y. and Page, M. "Biological Significance of RNA Editing in Cells". *Mol Biotechnol* 52 (2012): 91–100.
21. Agrawal, N., Dasaradhi, P.V.N., Mohmmed, A., Malhotra, P., Bhatnager, R.K and Mukherhee, S.K. "RNA Interference: Biology, Mechanism and Applications". *Microbiol Mol Biol Rev* 67 (2005): 657–85.
22. Derr, L.K. and Strathern, J.N. "A Role for Reverse Transcripts in Gene Conversion". *Nature* 361 (1995): 170–173.
23. Ho, M.W. "Development and Evolution Revisited'. In *Handbook of Developmental Science, Behavior and Genetics* (D. Hood, C. Halpern, G. Greenberg and R. Lerner, eds.), pp. 61–108, Wiley, Chichester, 2010.
24. Ho, M.W., Cummins, J. and Saunders, P.T. "GM Food Nightmare Unfolding in the Regulatory Sham". *Microb Ecol Health Dis* 19 (2007): 66–77.
25. Ferreira Holderbaum, D.F., Cuhra, M., Wickson, F., Orth, A., Nodari, R.O. and Bøhn, T. "Chronic Responses of *Daphnia magna* under Dietary Exposure to Leaves of a Transgenic (Event MON810) Bt-Maize Hybrid and Its Conventional Near-Isoline". *J Toxicol Environ Health A* 78 (2015): 993–1007.
26. Oraby, H.A., Kandil, M.M., Hassan, A.A.M. and Al-Sharawi, H.A. "Addressing the Issue of Horizontal Gene Transfer from a Diet

Containing Genetically Modified Components into Rat Tissues". *Afr J Biotechnol* 13 (2014): 4410–4418.
27. Oraby, H.A., Kandil, M.M., Shaffie, N. and Ghal, I. "Biological Impact of Feeding Rats with Genetically Modified-Based Diet". *Turkish Journal of Biology* 38 (2014): 1–11.
28. Ho, M.W. "CaMV 35S Promoter in GM Feed that Sickened Rats Transferred into Rat Blood, Liver and Brain Cells". *Sci Soc* 65 (2015a): 32–34+37.
29. Carman, J.A., Vlieger, H.R., Ver Steeg, L.J., Sneller, V.E., Robinson, G.W., Clinch-Jones, C.A., Haye, J.I. and Edwards, J.W. "A Long-Term Toxicology Study on Pigs Fed a Combined Genetically Modified (GM) Soy and GM Maize Diet". *J Org Sys* 8 (2013): 39–54.
30. Gab-Alla, A.A., El-Shame, Z.S., Shatta, A.A., Moussa, E.A. and Rayan, A.M. "Morphological and Biochemical Changes in Male Rats Fed on Genetically Modified Corn (Ajeeb YG)". *J Am Sci* 8 (2012): 1117–1123.
31. El-Shamei, Z.S., Gab-Alla, A.A., Shatta, A.A., Moussa, E.A. and Rayan, A.M. "Histopathological Changes in Some Organs of Male Rats Fed on Genetically Modified Corn (Ajeeb YG)". *J Am Sci* 10 (2012): 984–996.
32. Abdo, E.M., Barbary, O.M. and Shaltout, O.E.S. "Feeding Study with Bt Corn (MON810: Ajeeb YG) on Rats: Biochemical Analysis and Liver Histopathology". *Food Nutr Sci* (2014): 5, 185–195, http://dx.doi.org/10.4236/fns.2014.52024.
33. Séralini, G.E., Clair, E., Mesnage, R., Gress, S., Defarge, N., Malatesta, M., Hennequin, D. and de Vendômois, J.S. "Long Term Toxicity of a Roundup Herbicide and a Roundup-Tolerant Genetically Modified Maize". *Food Chem Toxicol* 50 (2012): 4221–4231.
34. Saunders, P.T. and Ho, M.W. "GM Cancer Warning Can No Longer Be Ignored". *Sci Soc* 56 (2012): 2–4.
35. Saunders, P.T. "Excess Cancers and Deaths with GM Feed: The Stats Stand Up". *Sci Soc* 56 (2012): 4–5.
36. Ho, M.W. and Saunders, P.T. "Retracting Séralini Study Violates Science and Ethics". *Sci Soc* 61 (2014): 20–21.
37. Séralini, G.E., Clair, E., Mesnage, R., Gress, S., Defarge, N., Malatesta, M., Hennequin, D. and de Vendômois, J.S. "Republished

Study: Long Term Toxicity of a Roundup Herbicide and a Roundup-Tolerant Genetically Modified Maize". *Environ Sci Eur* 26 (2014): 14.
38. Sirinathsinghji, E. "GM Soy Linked to Illnesses in Farm Pigs". *Sci Soc* 55 (2012a): 8–9.
39. Pedersen, I. "Changing from GMO to Non-GMO Natural Soy: Experiences from Denmark". *Sci Soc* 64 (2014): 8–12.
40. Séralini, G.E., Mesnage, R., Clair, E., Gress, S., Spiroux de Vendôme, J. and Cellier, D. "Genetically Modified Crops Safety Assessment: Present Limits and Possible Improvements". *Environ Sci Eur* 23, doi:10.1186/2190-4715-23-10.
41. Sirinathsinghji, E. "GM Feed Toxic, Meta-Analysis Reveals". *Sci Soc* 52 (2011): 30–32.
42. Ho, M.W. "Emergency! Pathogen New to Science Found in Roundup Ready GM crops". *Sci Soc* 50 (2011a): 10–11.
43. Ho, M.W. "Scientist Defends Claim of New Pathogen Linked to GM Crops". *Sci Soc* 50 (2011b): 12–13.
44. Ermakova, I.V. "Genetically Modified Soy Leads to the Decrease of Weight and High Mortality of Rat Pups of the First Generation: Preliminary Studies". *EcosInform* 1 (2006): 4–9.
45. Ho, M.W. "GM Soya-Fed Rats: Stunted, Dead, or Sterile". *Sci Soc* 33 (2007): 4–6.
46. Ho, M.W. "More Illnesses Linked to Bt Crops". *Sci Soc* 30(2006a): 8–10.
47. Ho, M.W. "Mass Deaths in Sheep Grazing on Bt Cotton". *Sci Soc* (2006b): 30: 12–3.
48. Prescott, V.E., Campbell, P.M., Moore, A., Mattes, J., Rothenberg, M.E., Foster, P.S., Higgins, T.J.V., Hogan, S.P. "Transgenic Expression of Bean A-Amylase Inhibitor in Peas Results in Altered Structure and Immunogenicity. *J Agric Food Chem* 53 (2005): 9023–9030.
49. Ho, M.W. "Transgenic Pea That Makes Mice Ill". *Sci Soc* 29 (2006c): 28–29.
50. Malatesta, M., Caporaloni, C., Rossi, L.,Battistelli, S., Rocchi, M.B.L., Tonucci, F. and Gazzanelli, G. "Ultrastructural Analysis of Pancreatic Acinar Cells from Mice Fed on Genetically Modified Soybean". *J Anat* 201 (2002): 409–415.

51. Malatesta, M., Biggiogera, M., Manuali, E., Rochhi, M.B.L., Baldelli, B. and Gazzanelli, G. "Fine Structural Analyses of Pancreatic Acinar Cell Nuclei from Mice Fed on Genetically Modified Soybean". *Eur J Histochem* 47 (2003): 385–388.
52. Malatesta, M., Caporaloni, C., Gavaudan, S., Rocchi, M.B.L., Serafini, S.,Tiberi, C. and Gazzanelli, G. "Ultrastructural Morphometrical and Immunocytochemical Analysis of Hepatocyte Nuclei from Mice Fed on Genetically Modified Soybean". *Cell Struct Funct* 27 (2002): 175–180.
53. Malatesta, M., Tiberi, C., Baldelli, B., Battistelli, S., Manuali, E. and Biggiogera, M. "Reversibility of Hepatocyte Nuclear Modifications in Mice Fed on Genetically Modified Soybean". *Eur J Histochem* 49 (2005): 237–242.
54. Vecchio, L., Cisterna, B., Malatesta, M., Martin, T.E. and Biggiogera, M. "Ultrastructural Analysis of Testes from Mice Fed on Genetically Modified Soybean". *Eur J Histochem* 48 (2006): 449–454.
55. Ho, M.W. "GM Ban Long Overdue". *Sci Soc* 29 (2006d): 26–27.
56. Kempf, H. "French Experts Very Disturbed by Health Effects of Monsanto GM Corn" (translated from article in *LeMonde*). Lobbywatch.org, 22 April 2004. Accessed on 10 January 2016, http://www.lobbywatch.org/archive2.asp?arcid=3308.
57. Ho, M.W. and Burcher, S. "Cows Ate GM Maize and Died". *Sci Soc* 21 (2004): 4–6.
58. Sirinathsinghji, E. "Syngenta Charged for Covering Up Livestock Deaths from GM Corn". *Sci Soc* 55 (2012b): 4–5.
59. Pusztai, A., Bardocz, S. and Ewen, S.W.B. "Genetically Modified Foods: Potential Human Health Effects". In *Food Safety: Contaminants and Toxins* (J.P.F. D'Mello, ed.), pp. 347–371, Scottish Agricultural College, Edinburgh, 2003.
60. Fares, N.H. and El-Sayed, A.K. "Fine Structural Changes in the Ileum of Mice Fed on δ-Endotoxin-Treated Potatoes and Transgenic Potatoes". *Nat Toxins* 6 (1998): 219–233.
61. Novotny, E. "Animals Avoid GM food, for Good Reasons". *Sci Soc* 21 (2004): 9–11.

62. The Editors. "Do Seed Companies Control GM Crop Research?" *Scientific American*, 2009. http://www.scientificamerican.com/article/do-seed-companies-control-gm-crop-research/.
63. Ho, M.W. and Lim, L.C. "The Case for a GM-Free Sustainable World". Independent Science Panel Report, Institute of Science in Society and Third World Network, London and Penang, 2003.
64. Ho, M.W. and Lim, L.C. *GM-Free: Exposing the Hazards of Biotechnology to Ensure the Integrity of Our Food Supply*, Vitalhealth Publishing, Ridgefield, 2004.
65. Ho, M.W. and Sirinathsinghji, E. "Ban GMOs Now", ISIS special report, ISIS, London, 2013.
66. "Scientists Declare No Consensus on GMO Safety", *Sci Soc* 60 (2013): 46–49.
67. Ho, M.W., Sirinathsinghji, E. and Saunders, P.T. "Scientific American Disinformation on GMOs". *Sci Soc* 60 (2013): 2–3.
68. Ho, M.W. and Swanson, N. "Glyphosate 'probably carcinogenic to humans': Latest WHO Assessment". *Sci Soc* 66 (2015): 16–17+23.
69. Samsel, A., Seneff, S. "Glyphosate's Suppression of Cytochrome P450 Enzymes and Amino Acid Biosynthesis by the Gut Microbiome: Pathways to Modern Diseases". *Entropy* 15 (2013): 1416–1463.
70. Sirinathsinghji, E. and Ho, M.W. "Banishing Glyphosate", ISIS special report, London, 2015.
71. Swanson,N., Ho, M.W. and Saunders, P.T. "Independent Scientists Manifesto on Glyphosate". *Sci Soc* 67 (2015): 11.
72. ENCODE Project Consortium. "Identification and Analysis of Functional Elements in 1% of the Human Genome by the ENCODE Pilot Project". *Nature* 447 (2007): 799–816.
73. The ENCODE Project Consortium. "An Integrated Encyclopedia of DNA Elements in the Human Genome". *Nature* 489 (2012): 57–74.
74. Mattick, J.S. "The Central Role of RNA in Human Development and Cognition". *FEBS Lett* 585 (2011): 1600–1616.
75. Ho, M.W. "Non-Coding RNA and the Evolution of Complexity". *Sci Soc* 63 (2014b): 30–33.
76. Hangauer, M.J., Vaughn, I.W. and McManus, M.T. Pervasive transcript of the human genome produces thousands of previously

unidentified long intergenic noncoding RNAs. *PLoS Genet* 9 (2013): e1003569.
77. Shapiro, J.A. "How Life Changes Itself: The Read-Write (RW) Genome". *Phys Life Rev* 10 (2013): 287–323.
78. Shapiro, J.A. "Physiology of the Read-Write Genome". *J Physiol* 592 (2014a): 2319–2341.
79. Shapiro, J.A. "Epigenetic Control of Mobile DNA As an Interface Between Experience and Genome Change". *Front Genet* 5 (2014b): 87.
80. Ho, M.W. "Evolution by Natural Genetic Engineering". *Sci Soc* 63 (2014c): 18–23.
81. Selman, M., Dankar, S.K., Forbes, N.E., Jia, J.J. and Brown, E.G. "Adaptive Mutation in Influenza A Virus Non-Structural Gene Is Linked to Host Switching and Induces a Novel Protein by Alternative Splicing". *Emerg Microbe Infections* 1 (2012): e53.
82. Ho, M.W. "To Mutate or Not to Mutate". *Sci Soc* 24 (2004c): 9–10.
83. Ho, M.W. "Non-Random Directed Mutations Confirmed". *Sci Soc* 60 (2013a): 30–32.
84. Rubin, H. "Cancer Development: The Rise of Epigenetics". *Eur J Cancer* 28 (1992): 1–2.
85. Ho, M.W. *The Rainbow and the Worm: The Physics of Organisms*, 3rd edition. World Scientific, Singapore, 2008.
86. Ho, M.W. (2001). "Quantum Computer? Is It Alive"? ISIS review, 15 October 2001, http://www.i-sis.org.uk/QuantumComputing.php.
87. Ho, M.W. "The Real Bioinformatics Revolution: Proteins and Nucleic Acids Singing to One Another?" *Sci Soc* 33 (2007): 42–45.
88. Ho, M.W. "The New Genetics and Natural versus Artificial Genetic Modification". *Entropy* 15 (2013b): 4748–4781.
89. Ho, M.W. "Why GMOs Can Never Be Safe". *Sci Soc* 59 (2013c): 14–17.
90. Fu, Y., Foden, J.A., Khayter, C., Maeder, M.L., Reyon, D., Young, J.K. and Sander, J.D. "High Frequency Off-Target Mutagenesis Induced by CRISPR-Cas Nucleases in Human Cells". *Nat Biotech* 3 (2013): 822–826, http://www.ncbi.nlm.nih.gov/pmc/articles/PMC3773023/.
91. Kuscu, C., Arslan, S., Singh, R., Thorpe, J. and Adli, M. "Genome-Wide Analysis Reveals Characteristics of Off-Target Sites Bound by the Cas9 Endonuclease". *Nat Biotech* 32 (2014): 677–683.

92. Turkki, V., Schenkwein, D., Timonen, O., Husso, T., Lesch, H.P. and Ylä-Herttuaia, S. "Lentiviral Protein Transduction with Genome-Modifying HIV-1 Integrase-I-PpoI Fusion Proteins: Studies on Specificity and Cytotoxicity". *BioMed Research Internat*, article ID 370340, 2014, http://www.hindawi.com/journals/bmri/2014/379340/.
93. "Gene Gun", Wikipedia, 6 June 2014, http://en.wikipedia.org/wiki/Gene_gun.
94. "Electroporation", Wikipedia, 9 July 2014, http://en.wikipedia.org/wiki/Electroporation.
95. Benbrook, C. "Evidence of the Magnitude and Consequences of the Roundup Ready Soybean Yield Drag from University-Based Varietal Trials in 1998, 13 July 1999, http://stopogm.net/sites/stopogm.net/files/EvidenceBenbrook.pdf.
96. Jost, P., Shurley, D., Culpepper, S., Roberts, P., Nichols, R., Reeves, J. and Anthony, S. Economic "Comparison of Transgenic and Montransgenic Cotton Production Systems in Georgia". *Agron J* 100(2008): 42–51.
97. Ho, M.W. and Saunders, P.T. "Transgenic Cotton Offers No Advantage". *Sci Soc* 38 (2008): 30.
98. Lim, L.C. and Matthews, J. "GM Crops Failed on Every Count". *Sci Soc* 13/14 (2002): 31–33.
99. Sirinathsinghji, E. "GM Crops Destroyed by US Drought but Non-GM Varieties Flourish". *Sci Soc* 56 (2012c): 6–7.
100. Ho, M.W. "Bt Brinjal Fails Two Years Running Risks Spreading Disease". *Sci Soc* 65 (2015b): 28.
101. Zhang, L., Hou, D., Chen, X. *et al.* "Exogenous Plant MIR168a Specifically Targets Mammalian LDLRAP1: Evidence of Cross-Kingdom Regulation by MicroRNA. *Cell Res* advance online publication, 20 September 2011, doi:10.1038/cr 2011.158.
102. Ho, M.W. "How Food Affects Genes". *Sci Soc* 53 (2012a): 12–13.
103. Sirinathsinghji, E. "The Paradigm Shift from Genetic to Epigenetics and Its Implications for GM Crops Utilising RNAi Technologies". Presentation at 1st Forum of Development and Environmental Safety, Beijing, China, 25–26 July 2014.
104. Wang, K., Li, H., Yan, Y., Etheridge, A., Zhou, Y., Huang, D., Wilmes, P. and Galas, D. "The Complex Exogenous RNA Spectra in

Human Plasma: An Interface with Human Gut Biota". *PLoS One* 7 (2012): e51009.

105. Izumi, H., Kosaka, N., Shimizu, T., Sekine, K., Ochiya, T. and Takase, M. "Bovine Milk Contains MicroRNA and Messenger RNA That Are Stable Under Degradative Conditions". *J Dairy Sci* 95 (2012): 4831–4841.
106. Mu, J., Zhuang, X., Wang, Z., Jiang, H., Deng, Z.B., Wang, B., Zhang, L., Kakar, S., Yun, Y., Miller, D. and Zhaing, H.G. "Interspecies Communication Between Plant and Mouse Gut Host Cells Through Edible Plant Derived Exosome-Like Nanoparticles". *Mol Nutr Food Res* 58 (2014): 1561–1573.
107. Spisák, S., Solymos, N., Ittzés, P., Bodor, A. et al. "Complete Genes May Pass from Food to Human Blood". *PLoS One* 8 (2013): e69805.
108. Ho, M.W. "Intercommunication via Circulating Nucleic Acids". *Sci Soc* 42 (2009a): 46–48.
109. Ho, M.W. and Cummins, J. "New Evidence Links CaMV 35S Promoter to HIV Transcription". *Microb Ecol Health Dis* 21 (2009): 172–4; *Sci Soc* 43 (2009): 26–27.
110. Ho, M.W. "Hazardous Virus Hene Fiscovered in GM Vrops after 20 Years". *Sci Soc* 57 (2013d): 2–3.
111. Ho, M.W. "New GM Nightmares with RNA". *Sci Soc* 58 (2013e): 6–7.
112. Helwak, A., Kudla, G., Dudnakova, T. and Tollervey, D. "Mapping the Human miRNA Interactome by CLASH Reveals Frequent Noncanonical Binding". *Cell* 153 (2013): 654–665.
113. Ho, M.W. "RNA Interference 'complex and flexible' and Beyond Control". *Sci Soc* 59 (2013f): 18–19.
114. Ho, M.W., Traavik, T., Olsvik, O., Tappeser, B., Howard, V., von Weizsacker, C. and McGaein, G.C. "Gene technology and gene ecology of infectious diseases. *Microb Ecol Health Dis* 10 (1998): 33–39.
115. Chen, J., Jin, M., Ziu, Z.G., Guo, C., Chen, Z.L., Shen, Z.Q., Wang, X.W. and Li, J.W. "A Survey of Drug Resistance *Blá* Genes Originating from Synthetic Plasmid Vectors in Six Chinese Rivers". *Environ Sci Technol* 45 (2012): 13448–13454.

116. Sirinathsinghji, E. "GM Antibiotic Resistance in All China's Rivers". *Sci Soc* 57 (2013): 6–7.
117. Overballe-Petersen, S., Harms, K., Oriando, L.A.A., Mayar, J.V.M., Rasmussen, S., Dahl, T.W., Tosing, M.T., Poole, A.M., Sicheritz-Ponten, T., Brunak, S., Inselmann, S., de Vries, J., Wackernagel, W., Pybus, O.G., Nielsen, R., Johnsen, P.J., Nielsen, K.M. and Willerslev, E. "Bacterial Natural Transformation by Highly Fragmented and Damaged DNA". *Proc Natl Acad Sci early edition* 2014, www.pnas.org/cgi/doi/10.1073/pnas.1315278110.
118. Ho, M.W. "Horizontal Spread of GM DNA Widespread, But No One is Looking, Almost". *Sci Soc* 63 (2014d): 26–29.
119. Ho, M.W. "Horizontal Transfer of GM DNA: Why Is Almost No One Looking?" Open letter to Kaare Nielsen in his capacity as a member of the European Food Safety Authority GMO panel. *Microb Ecol Health Dis* 25 (2014e): 25919.
120. Ho, M.W. "RNA Inheritance of Acquired Characters". *Sci Soc* 63 (2014f): 34–37.
121. Bircher, J.A. "Ubiquitous RNA-Dependent RNA Polymerase and Gene Silencing". *Genome Biol* 10 2009: 243, 3 pp.
122. Spadafora, C. "Sperm-Mediated 'reverse' Gene Transfer: A Role of Reverse Transcriptase in the Generation of New Genetic Information". *Hum Reprod* (2008): 735–740.
123. Ho, M.W. "Sperm-Mediated Inheritance of Acquired Characters". *Sci Soc* 63 (2014g): 38–41.
124. Ho, M.W. "Epigenetic Inheritance: 'What Genes Remember'". *Sci Soc* 41 (2009b): 4–5.
125. Ho, M.W. "Caring Mothers Strike Fatal Blow Against Genetic Determinism". *Sci Soc* 41 (2009c): 6–9.
126. Ho, M.W. "Epigenetic Toxicology". *Sci Soc* 41 (2009d): 13–15.
127. Anyway, M.D, Cupp, A.S, Uzumcu, M. and Skinner, M.K. "Epigenetic Transgenerational Actions of Endocrine Disruptors and Male Fertility". *Science* 308 (2005): 1466–1468.
128. Kaiser, J. "The Epigenetics Heretic", News Focus, *Science* 343 2014: 3610–3613.
129. Lederberg, J. "Molecular Biology, Eugenics and Euphenics". *Nature* 198 (1963): 428–429.

130. Saunders, P.T. "Global Inequality and Its Ills. *Sci Soc* 63 (2014a): 2–3.
131. Saunders, P.T. "Capitalism and the Inexorable Rise of Inequality". *Sci Soc* 63 (2014b): 14–15.
132. Saunders, P.T. "Equality Is Good for You". *Sci Soc* 63 (2014c): 16–17.
133. Ho, M.W., Burcher, S., Lim, L.C. *et al.* (2008). "Food Futures Now, Organic*Sustainable*Fossil Fuel Free", ISIS/TWN, London/Penang, http://www.i-sis.org.uk/foodFutures.php.
134. Ho, M.W. "Living, Green and Circular". *Sci Soc* 53 (2012b): 20–23.

8

Non-Random Directed Mutations and Quantum Electrodynamics

The neo-Darwinian theory of evolution by natural selection of random mutations should be consigned to history where it belongs; electromagnetic intercommunication and resonance may be involved in activating and mutating just the right genes.

An Obsolete Theory Challenged by Directed Mutations

Conventional neo-Darwinian theory of evolution is firmly based on the natural selection of random mutations plus the central dogma assumption that environmental influences cannot change nucleic acids or become inherited. The central dogma has been invalidated at least since the early 1980s concomitantly with the emergence of the new genetics of the fluid genome.[1,2] Similarly, the randomness of mutations has been called into question since the 1970s in experiments demonstrating that cells subject to non-lethal selection come up repeatedly with just the right "adaptive" or "directed" mutations in specific genes that enable the cells to grow and multiply.[3,4]

Note that mutations adaptive for cells in enabling them to grow and multiply are not so for the organism as a whole, as they give rise to tumours and cancers; hence it is more appropriate to refer to them as "directed" mutations. In fact, it was the idea that cancers may involve directed mutations that prompted British physician and molecular geneticist John Cairns, then at

Harvard School of Public Health in Boston, Massachusetts, to study the phenomenon in bacteria.[5]

In one of the first experiments demonstrating directed mutations, a strain of *E. coli* bacterium — with a mutation in the *lacZ* gene that terminates the polypeptide prematurely and hence renders it unable to use lactose — only mutated back to the wild-type when lactose is present in the medium, not when lactose was absent.[3] Even more remarkably, another strain with the *lacZ* completely deleted was able to mutate a cryptic gene *ebgA* coding for an enzyme that could hydrolyze lactose and the regulator *ebgR* that normally represses *ebgA*. During growth, each of these point mutations occurs at a frequency of less than 10^{-8}; neither on its own would allow the *lacZ*-deleted strain to use lactose.[6]

Cairns and colleagues concluded their 1988 review on the origin of mutants as follows:[3]

> "Curiously, when we come to consider what mechanism might be the basis for the forms of mutation described in this paper, we find that molecular biology has, in the interim, deserted the reductionists. Now, almost anything seems possible. In certain systems, information freely flows back from RNA to DNA, genomic instability can be switched on under conditions of stress, and switched off when the stress is over, and instances exist where cells are able to generate extreme variability in localized regions of their genome. The only major category of informational transfer that has not been described is between proteins and the messenger RNA (mRNA) molecules that made them. If a cell discovered how to make that connection, it might be able to exercise some choice over which mutations to accept and which to reject."

As it happens, reverse translation as suggested by Cairns is not needed, and the same mechanisms for directed mutations could apply from *E. coli* bacteria to humans, as recent findings indicate.

The Unity of Biochemical Reactions

Mutagenesis mechanisms *in vivo* are essentially the same in all living cells, which is yet another instance of the unity of biochemistry.

The unity of biochemistry has been a guiding principle in the study of chemical processes in living organisms; as for example, the core energy metabolism in living organisms is reduction and oxidation of carbohydrates and water.[7] But studying biochemical reactions in the test-tube can be very misleading as the biochemical reactions *in vivo* typically happen within an environment crowded with enzymes where substrates are limiting. In an attempt to represent biochemical reactions *in vivo*, Henrik Kacser (1918–1995) and colleagues pioneered *metabolic control analysis*,[8,9] a kinetic framework for analysing rates and fluxes due to changes in enzymes and metabolites at steady state.

Barbara Wright and colleagues at the University of Montana, Missouri, in the United States are among the few biologists who see the need for such kinetic models as biochemical knowledge grows and the relevance of *in vitro* data to metabolism *in vivo* becomes increasingly questionable.[10] The usual hierarchical schemes of control based on any single cellular event of "gene activation" or "enzyme-induction" is simply not adequate to account for the biological effects that necessarily involve complex relationships among numerous molecular entities and biochemical pathways. However, models are only valid if they have predictive value. Using their computer programme *mfg*, the team has indeed created kinetic models that successfully predict directed mutations in three very different systems: *E. coli* under phosphate starvation, oxidative stress and osmotic stress; and in humans, the p53 tumour suppressor gene activated by genotoxic stress, and somatic hypermutation during the immune response to foreign antigens, where transcription frequency increases 10 000-fold and mutation frequency a million-fold.

The research team started from the observation that metabolic reactions unique to a particular environmental stress apparently target specific genes for increased rates of transcription and mutation, "resulting in higher mutation rates for those genes most likely to solve the problem". Stressors typically lead to activating ~1% of the genome to transcription and mutation, thereby directing and selecting those mutations that correct the problem (overcome the stress). However, they are silent on how this might occur; I shall return to this important point later.

For example, in starving (stressed) *E. coli* bacteria presented with lactose, the obvious limiting enzyme is β-galactosidase, encoded by the *lacZ* gene, which can split lactose into glucose and galactose, simple sugars that can be metabolized by downstream enzymes to provide energy and material for growth. Hence bacteria with defective *lacZ* genes are directed to mutate the gene until the normal functioning enzyme is restored; or in the case of bacteria with *lacZ* completely deleted, a new cryptic gene and its regulator are mutated until the functioning new enzyme that can break down lactose has been created and become expressed.

Similarly, in the systems analysed by Wright and colleagues,[10] *E. coli* starved for inorganic phosphate directs mutations to de-repress the *pho* regulon (phosphate regulated suite of genes) resulting in a new high-affinity phosphate transport system that gets phosphate into the cell at much lower concentrations, and also activates a hydrolytic enzyme able to get phosphate from new sources. In humans, genotoxic stress activates transcription of the p53 gene, resulting in mutations that inactivate the gene. And foreign antigen stress activates the transcription and mutation of hypervariable regions of immunoglobulins in B cells to produce the antibodies that can bind to and neutralize the foreign antigens.

Predicted Mutagenesis Hotspots Based on Secondary Stem Loop Structures of ssDNA

In all cases, transcription provides the single-stranded DNA (ssDNA being transcribed) that exposes unpaired, intrinsically mutable bases to mutation; and the mutable bases are guanines (Gs) and cytosines (Cs). Why they are mutable depends on the secondary stem loop structures (SLSs) the transcribed ssDNA adopts: sequences complementary to each other pair up to form stems leaving the unpaired bases as loops, and it is the unpaired Gs and Cs in loops that are vulnerable to mutations.

To simulate *in vivo* conditions in response to increased rates of transcription, the computer algorithm *mfg* was developed. During transcription, the vulnerability of a base to mutation depends on the stability of the secondary SLSs adopted by the ssDNA, and on the extent to which the base is unpaired. The *mfg* programme interfaces with the *mfold* programme that folds single-stranded segments of a specified length and sequence and reports all possible secondary structures that can form from each folded segment in descending order of stability. *Mfg* reports the stability of the most stable secondary structure in which a mutable base is unpaired, and also the percent of total folds in which it is unpaired. The mutability index of each unpaired base is the product of the two variables.

The *mfg*-identified highly mutable Gs and Cs in p53 for example, are actually located in ssDNA loops of predicted SLSs, as confirmed by analysis for codon 175 in exon 5 using S1 endonuclease, which cuts at single-stranded DNA and RNA. Similar analysis demonstrated that the hypermutable codons 245, 248, 273 and 282 of p53 are also located in single-stranded loops.

The unpaired Gs and Cs are intrinsically mutable; unpaired Gs primarily mutate to As (adenines) and unpaired Cs mutate to Ts (thymines). Compelling evidence for the underlying instability of unpaired Gs and Cs is shown by examples of silent mutations that do not change amino acid sequence, and hence are not subject to any selection. In all cases examined, the mutable bases are located in loops of identified secondary structures.

The somatic hypermutation (SHM) in pre-B cell involves a 10 000-fold increase in transcription, which is linked to a million-fold increase in mutation frequency, especially during phase 1 of the SHM. Activation-induced deaminase (AID), an RNA-editing enzyme, is implicated in the mutational mechanism in SHM as well as in all the other systems. In liver cancers induced by genotoxins and involving p53, the availability of unpaired intrinsically mutable Gs in ssDNA is rate-limiting for mutation frequency. Circumstances *in vivo* at low endogenous levels of transcription show that the majority of intrinsic G mutations are to A, and that the availability of Gs in ssDNA is rate-limiting for mutation frequency. The dual effects of oxyradicals (reactive oxygen species arising from incomplete oxidation[11]) which both activate transcription (about four-fold) and increase G to T mutations (to 85.8%), are accompanied by corresponding decreases in G to A mutations. Thus oxyradicals compete for the fate of rate-limiting directed mutations.

Other Evidence of Non-Random Mutations

In humans, SHM and class switch recombination (CSR, a genetic rearrangement) result in distinct genetic alterations at different regions of the immunoglobulin genes in B lymphocytes in generating antibody diversity: point mutations in variable regions and large deletions in S (switch) regions, respectively;[12] yet both depend on AID. B cell stimulation that induces CSR but not SHM leads to AID-dependent accumulation of SHM-like point mutations in the switch mu region independently of CSR. These findings strongly suggest that AID itself or some single molecule generated by RNA editing function of AID may mediate a common step of SHM and CSR, which is likely to be involved in DNA cleavage.

Another study suggests that DNA double-strand breaks (DSBs) are responsible for mutation hotspots in stress-induced mutation in *E. coli* by means of two mechanisms.[13] The first involves mutations occurring maximally within the first 2 kb and decrease logarithmically to ~60 kb. The second involves a weak mutation tail extending to 1 Mb from the double-strand break. Hotspots occur independently upstream and downstream in the replication path. The enzyme Rec D which allows DSB-exonuclease activity is required for strong local but not long-distance hotspot mutations, indicating that double-strand resection (cutback) and gap-filling synthesis underlie local hotspot mutations. Hotspots near DSBs open the possibility that specific genomic regions could be targeted for mutagenesis, and could also promote concerted (simultaneous) evolution within genes/gene clusters.

Finally, a study combining phylogenetic and population genetic techniques to compare 34 *E. coli* genomes carried out by researchers at the European Bioinformatics Institute, Welcome Trust Genome

Campus, Cambridge in the United Kingdom found that the rate of neutral mutations — neither advantageous nor deleterious — varies by more than an order of magnitude across 2 659 genes, with mutational hot and cold spots spanning several kilobases (entire operons).[14] The variation is not random; a lower rate in highly expressed genes and in genes undergoing stronger "purifying selection" (which implies they are preferentially protected from mutation or by repair mechanisms). According to the researchers, the findings suggest that mutation rate has been "evolutionarily optimized to reduce the risk of deleterious mutations".

However, current knowledge of factors influencing the mutation rate — including transcription-coupled repair and context-dependent mutagenesis — do not explain these observations, indicating additional mechanisms must be involved.

More than 12 000 single-nucleotide polymorphisms have been examined. Given that transcription is mutagenic (as found in the other studies described earlier), the negative association between expression and mutation rate is unexpected, indicating that there are indeed repair pathways coupled to transcription. But this mechanism alone is insufficient, as the non-transcribed strand also displays dependence between expression and mutation rate, and molecular experiments have reported that transcription-induced mutagenesis occurs in the presence of transcription-coupled repair. Therefore additional mechanisms that generally target highly expressed genes but are not directly coupled with the transcriptional machinery must exist.

How Does the Cell Know Which Genes to Mutate?

The findings indicate that there are numerous mechanisms for the cell to direct mutations to specific genes and specific sites in these genes, but they give no indication as to how the cell is able to do that. This same problem besets the "ordinary" processes of gene expression, in which armies of molecular genetic engineers have to be assembled and precisely choreographed at the right time and place to carry out the "natural genetic engineering" that we now know to be necessary for survival (see the previous "Chapter 7 — Natural versus Artificial Genetic Modification and Perils of GMOs" in this book). I suggest that electromagnetic signals are involved in natural genetic engineering as in directed mutations, which is part and parcel of natural genetic engineering. There are already good reasons to suspect that molecules intercommunicate by electromagnetic signals, and molecules that interact share common frequencies so they can attract one another through resonance.[15] This is increasingly relevant in the context of quantum electrodynamics theory, which predicts that interaction between light and water is essential for life (see "16 — Illuminating Water and Life: Essay in Honour of Emilio Del Giudice" in this book).

Light interacting with water creates coherent domains (CDs) that oscillate between the coherent ground state and an excited state close to the ionizing potential of water. Hence, a plasma of almost free electrons are created at the surface of the CDs, making them an ideal "redox pile" to drive redox reactions, the main energy metabolism of living organisms. The CDs, predicted to exist at ordinary temperatures and pressures, are also able to capture electromagnetic energy at different frequencies from its surroundings and to produce coherent excitations in the frequencies of the external fields. These coherent frequencies will attract reactants that share the same frequencies to the surfaces of CDs, enabling them to carry out the requisite reactions.

Thus, lactose supplied during starvation, for example, will send strong electromagnetic signals *via* the CDs to its normal metabolic enzyme, β-galactosidase, as well as to its gene, *lacZ*, causing it in turn to respond by transcription and to attract the requisite mutagenic machinery until the gene can be transcribed and translated into the enzyme that breaks down lactose, thereby restoring normal metabolic flux. The same applies to other situations of stress and stress relief. Resonance to electromagnetic signals is very precise, and will have all the appearance of being directed, particularly if the cell and organism is quantum coherent, as argued in detail in my books, *Living Rainbow H2O*[7] and *The Rainbow and the Worm: The Physics of Organisms*.[16] This is a testable hypothesis, as the signals could be revealed by appropriately sensitive detectors and analysers.

To Conclude

Mutations are highly non-random and directed; numerous mechanisms for generating mutations are involved that appear to be under the control of the cell or organism as a whole in different environmental contexts, leading to repeatable mutations in specific genes. These results are contrary to the fundamental neo-Darwinian tenet that evolution depends on the natural selection of random genetic mutations. I suggest that specific electromagnetic signals emitted by key molecules that can relieve the stress are communicated directly to activate the transcription and mutation of the requisite gene(s) in accordance with the quantum electrodynamics theory of life.

References and Notes

1. Dover, G. and Flavell, D. *Genome Evolution*. Oxford University Press, Oxford, 1982.
2. Ho, M.W. *Living with the Fluid Genome*, ISIS/TWN, London/Penang, 2003, http://www.i-sis.org.uk/fluidGenome.php.
3. Cairns, J., Overbaugh, J. and Miller, S. "The Origin of Mutants". *Nature* 335 (1988): 142–145.
4. Ho, M.W. "To Mutate or Not to Mutate". *Sci Soc* 24 (2004): 9–10.
5. Cairns, J. "Mutation and Cancer: The Antecedents to Our Studies of Adaptive Mutation". *Genetics* 148 (1998): 1433–1440.
6. Hall, B.G. and Hartl, D.L. "Regulation of Newly Evolved Enzymes: 1. Selection of a Novel Lactase Regulated by Lactose in *Escherichia coli*". *Genetics* 76 (1975): 391–400.
7. Ho, M.W. (2012). Living Rainbow H2O, World Scientific/Imperial College Press, Singapore/London, http://www.i-sis.org.uk/Living_Rainbow_H2O.php.
8. Kacser, H. and Burns, J.A. "The Control of Flux". *Symp Soc Exp Biol* 27 (1973): 65–104.

9. Ho, M.W., ed. *Living Processes: Book 2, Bioenergetics*, pp. 121–137. Open University Press, Milton Keynes, 1995.
10. Wright, B.E., Schmidt, K.H. and Minnick, M.F. "Kinetic Models Reveal the *In Vivo* Mechanisms of Mutagenesis in Microbes and Man". *Mutat Res* 752 (2013): 129–137.
11. Ho, M.W. "The Body Does Burn Water". *Sci Soc* 43 (2009): 14–16.
12. Naqaoka, H., Muramatsu, M., Yamamura, N., Kinoshita, K. and Honjo, T. "Activation-Induced Deaminase (AID)-Directed Hypermutation in the Immunoglobulin Smu Region: Implication of AID Involvement in a Common Step of Class Switch Recombination and Somatic Hypermutation". *J Exp Med* 195 (2002): 529–534.
13. Shee, C., Gibson, J.L. and Rosenberg, S.M. "DNA Double-Strand Breaks Provoke Mutation Hotpots Via Stress-Induced Mutation in *E. coli*: Two Mechanisms Produce Mutation Hotspots at DNA Breaks in *E. coli*". *Cell Rep* 2 (2012): 714–721.
14. Martincorena, H., Seshasayee, A.S.N. and Luscombe, N.M. "Evidence of Non-Random Mutation Rates Suggests an Evolutionary Risk Management Strategy". *Nature* 485 (2013): 95–98.
15. Ho, M.W. "The Real Bioinformatics Revolution: Proteins and Nucleic Acids Singing to One Another?" *Sci Soc* 33 (2007): 42–45.
16. Ho, M.W. (2008). *The Rainbow and the Worm: The Physics of Organisms*, 3rd edition. World Scientific, Singapore and Imperial College Press, London, 2008, http://www.i-sis.org.uk/rnbwwrm.php.

9

Consciousness and Neuroscience

9

The Biology of Free Will

How the new biology transcends the laws of mechanistic physics and frees itself from mechanical determination and control, presenting the organism as a sentient, coherent being that is free, from moment to moment to explore and create its possible futures.

Brain Science in Crisis

Distinguished neurophysiologist Walter Freeman at University of California Berkeley, begins his latest book by declaring brain science "in crisis": his personal quest to define constant psychological states arising from given stimuli has ended in failure after 33 years.[1] Patterns of brain activity are simply unrepeatable; every perception is influenced by all that has gone before. The *impasse*, he adds, is conceptual, not experimental or logical. This acknowledged breakdown of mechanical determinism in brain science is really long overdue, but it should not be misconstrued as the triumph of vitalism. As Freeman goes on to show, recent developments in nonlinear mathematics can contribute to some understanding of these non-repeatable brain activities.

The traditional opposition between mechanists — who believed that life can in principle be explained in terms of mechanical physics and chemistry — and their opponents the vitalists — who held that living things contain a vital principle irreducible to chemistry and physics — had already began to dissolve at the turn of the present century. Newtonian physics had given way to quantum theory at the very small scales of elementary particles and to general relativity at the large scales of planetary motion and beyond. The static, deterministic universe of absolute space and time is replaced by a multitude of contingent, observer-dependent space-time frames. Instead of mechanical objects with simple locations in space and time, one finds delocalized, mutually entangled quantum entities that carry their histories with them, like evolving organisms.[2] These significant developments gave birth to an *organicist* philosophy based firmly on contemporary physics.

French philosopher Henri Bergson (1859–1941) was a key figure in organicist philosophy. In his book *Time and Free Will*[3] he showed how Newtonian concepts — which dominate biological sciences then and now — negate psychology's claims to understand our inner experience at the very outset. In the science of mechanistic psychology, words that express our feelings — love and hate, joy and pain — emptied of their experiential content, are taken for the feelings themselves. They are then defined as individual psychological states each uniform for every occasion across all individuals, differing only in magnitude or intensity. (It is significant that Freeman's 33 years of trying to define just such constant psychological states had ended in failure, as mentioned above.) In particular, Bergson drew attention to the inseparability of space and time, both tied to real processes that have characteristic *durations*. The other major figure in organicist philosophy was English mathematician-philosopher Alfred North Whitehead (1861–1941) who saw physics itself and all of nature as unintelligible without a thoroughgoing theory of the organism that *participates* in knowing.[4]

Organicist philosophy was taken very seriously by a remarkable group of scientists, all Fellows of the Royal Society in Britain, who formed the multidisciplinary Theoretical Biology Club.[5] Its membership included Joseph Needham (1900–1995), eminent embryologist and biochemist later to be renowned for his work on the history of Chinese science; muscle physiologist and biochemist Dorothy Needham (1896–1987); geneticist Conrad H. Waddington (1905–1975); crystallographer J.D. Bernal (1901–1971); mathematician Dorothy Wrinch (1894–1976); philosopher J.H. Woodger (1894–1981) and physicist Nevill Mott (1905–1996). They acknowledged the full complexity of living organization, not as axiomatic, but as something to be explained and understood with the help of philosophy as well as physics, chemistry, biology and mathematics, as those sciences advance, and in the spirit of free enquiry, leaving open whether new concepts or laws may be discovered in the process.

A lot has happened since the project of the Theoretical Biology Club was brought to a premature end when they failed to obtain funding from the Rockefeller Foundation. Organicist philosophy has not survived as such, but its invisible ripples have spread and touched the hearts and minds, and the imagination of many who remain drawn to the central enigma that Austrian quantum physicist Erwin Schrödinger (1887–1961) later posed in his book *What is Life?*[6]

In the intervening years, the transistor radio, the computer and lasers have been invented. Whole new disciplines have been created: non-equilibrium thermodynamics, solid-state physics and quantum optics to name but a few. In mathematics, nonlinear dynamics and chaos theory took off in the 1960s and 1970s. Perhaps partly on account of that, many nonlinear physical and physicochemical phenomena have been actively investigated only within the past ten years, as physics became more and more organic in its outlook.

In a way, the whole of science is now tinged with organicist philosophy, as even "consciousness" and "free will" are on the scientific agenda. Bergson[3] had made a persuasive case that the traditional problem of free will is simply misconceived and arises from a mismatch between the quality of authentic, subjective experience and its description in language, in particular, the language of the mechanistic science of psychology. In my book *The Rainbow and the Worm: The Physics of Organisms*[2] I have shown how *contemporary* Western scientific concepts of the organism are leading us beyond conventional thermodynamics as well as quantum theory, and offering rigorous insights which reaffirm and extend our intuitive, poetic and even romantic notions of spontaneity and free will.

The New Organicist Science

I am making a case for a new organicist science. It is not yet a conscious movement but a Zeitgeist I personally embrace, so I really mean to persuade you to do likewise by giving it a more tangible shape. The new organicist science, like the old, is dedicated to the knowledge of the organic whole; hence, it does not recognize any discipline boundaries. It is to be found in and between *all* disciplines. Ultimately, it is an entire knowledge system by which one lives. There is no escape clause allowing one to plead knowledge "pure" or "objective", and hence having nothing to do with life. Most of all, the knowing being participates in knowing as much as in living. Participation implies responsibility, which is consistent with the truism that there can be no freedom without responsibility, and conversely, no responsibility without freedom. There is no placing mind outside nature as French mathematician philosopher René Descartes (1596–1650) had famously done for mechanistic science; the knowing being is wholeheartedly within nature: heart and mind, intellect and feeling. It is non-dualist and holistic. In all those respects, its affinities are with the participatory knowledge systems of traditional indigenous cultures all over the world.[7]

From a thoroughgoing organicist perspective, one does not ask, "What is life?" but, "What is it to be alive?" Indeed, the best way to know life is to live it fully. It must be said that we do not yet have a full-fledged organicist science. But I shall present some new snapshots of the organism, starting from the more familiar and working up, perhaps to the most sublime, from which a live portrait of the organism as a free, spontaneous being will begin to emerge. I shall show how the organism succeeds in freeing itself from the "laws" of mechanistic physics, from mechanical determinism and control, thereby becoming a *sentient*, *coherent* being that from moment to moment freely explores and creates its possible futures.

The Organism Frees Itself from the "Laws" of Physics

I put "laws" in quotation marks in order to emphasize that they are not laid down once and for all, and especially not to dictate what we can or cannot think. They are tools for helping us think; and most of all, to be transcended if necessary.

Many physicists have marvelled at how organisms seem able to defy the second law of thermodynamics, starting from British mathematical physicist William Thompson, also known as Lord Kelvin (1824–1907) and co-inventor of the second law, who nevertheless excluded organisms from its dominion:

> "The animal body does not act as a thermodynamic engine ... consciousness teaches every individual that they are, to some extent, subject to the direction of his will. It appears therefore that animated creatures have the power of immediately applying to certain moving particles of matter within their bodies, forces by which the motions of these particles are directed to produce derived mechanical effects.[8]

What impresses Lord Kelvin is how organisms seem to have energy *at will*, whenever and wherever required, and in a perfectly coordinated way. Another equally puzzling feature is that, contrary to the second law — which says all systems should decay into equilibrium and disorder — organisms develop and evolve towards ever increasing organization.

Of course, there is no contradiction, as the second law applies to isolated systems, whereas organisms are open systems. But how do organisms manage to maintain themselves far away from thermodynamic equilibrium and to produce increasing organization? Schrödinger wrote (pp. 70–71):[6]

> "It is by avoiding the rapid decay into the inert state of 'equilibrium' that an organism appears so enigmatic ... What an organism feeds upon is negative entropy, or, to put it less paradoxically, the essential thing in metabolism is that the organism succeeds in freeing itself from all the entropy it cannot help producing while alive.[11]

Schrödinger was severely reprimanded[9] by Linus Pauling and others for using the term "negative entropy", which does not correspond to any rigorous thermodynamic entity.

However, the idea that open systems can "self-organize" under energy flow became more concrete in the discovery of "dissipative structures" by Nobel laureate physical chemist Ilya Prigogine (1917–2003).[10] An example is the Bénard convection cells that arise in a pan of water heated uniformly from below. At a critical temperature difference between the top and the bottom, a phase transition occurs: bulk flow begins as the lighter, warm water rises from the bottom and the denser, cool water sinks. The whole pan eventually settles down to a regular honeycomb array of flow cells. Before phase transition, all the molecules move randomly with respect to one another. However, at a critical rate of energy supply, the system self-organizes into global dynamic order in which all the astronomical numbers of molecules are moving in formation as though choreographed to do so.

A still more illuminating physical metaphor for the living system is the laser[11] in which energy is pumped into a cavity containing atoms capable of emitting light. At low levels of pumping, the atoms emit randomly as in an ordinary lamp. As the pumping rate is increased, a threshold is reached when all the atoms oscillate together in phase, and send out a giant light track that is a million times as long as that emitted by individual atoms.

Both examples illustrate how energy input or energy pumping and dynamic order are intimately linked. These and other considerations led me to identify Schrödinger's "negative entropy" as "stored mobilizable energy in a space-time structured system".[12, 13] The key to understanding the thermodynamics of living systems turns out not so much to be energy flow but energy *storage* under energy flow (Fig. 9.1). Energy flow is of no consequence unless the energy can be trapped and stored within the system where it circulates to do work before dissipating. A reproducing life cycle, i.e., an organism, arises when the loop of circulating energy is closed. At that point, we have a life cycle, within which stored energy is mobilized, remaining largely stored as it is mobilized.

The life cycle is a highly differentiated space-time structure; the predominant modes of activity are themselves cycles spanning an entire gamut of space-times from the local and fast (or slow) to the

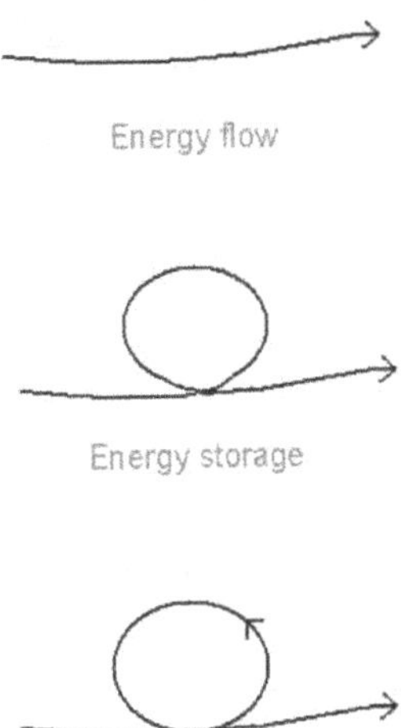

Figure 9.1 Energy storage under energy flow to form a reproducing life cycle.

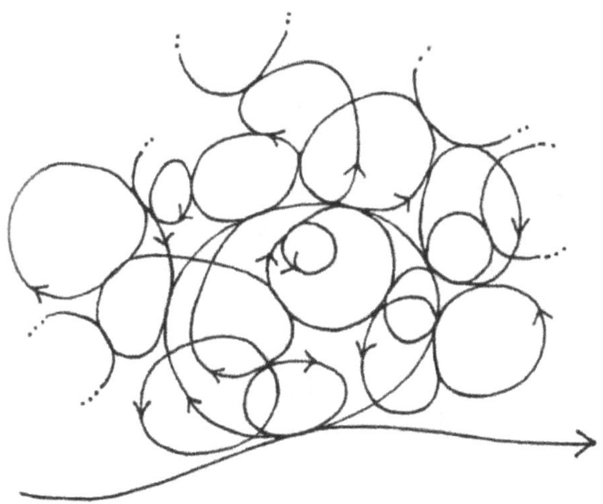

Figure 9.2 The many-fold cycles of life coupled to energy flow.

global and slow (or fast), all of which are coupled together. These cycles are most familiar to us in the form of biological rhythms extending over 20 orders of magnitude of time, from electrical activities of neurons and other cells to circadian and circa-annual rhythms and beyond. An intuitive picture is given in Fig. 9.2, where coupled cycles of different sizes are fed by the one-way energy flow. This complex, entangled space-time structure is strongly reminiscent of Bergson's "durations" of organic processes, which necessitates a different way of conceptualizing space-time as heterogeneous, nonlinear, multidimensional and nonlocal.[2, 14]

On account of the complete spectrum of coupled cycles, energy is stored and mobilized over all space-times according to the relaxation times (and volumes) of the processes involved. So organisms can take advantage of two different ways of mobilizing energy with maximum efficiency — non-equilibrium transfer in which stored energy is transferred before it is thermalized, and quasi-equilibrium transfer, for which the free energy change approaches zero according to conventional thermodynamic considerations.[2] Energy input into any mode can be readily delocalized over all modes, and conversely, energy from all modes can become concentrated into any mode. In other words, energy coupling in the living system is *symmetrical,* which is why we can have energy at will, whenever and wherever required.[2,12,13,15] The organism is in effect a closed self-sufficient energetic domain of cyclic non-dissipative processes coupled to the dissipative processes. In the formalism of conventional thermodynamics, the life cycle can be considered to first approximation, as consisting of all those cyclic processes — for which the net entropy change balances out to zero — coupled to those dissipative processes necessary for keeping it going, for which the net entropy change is greater than zero (see Fig. 9.3). This representation is derived from the thermodynamics of the steady state.[16,17] (Note that a later more significant version of this circular thermodynamics theory presents the entropy exported to the environment as $\Sigma\Delta S > 0$, validating Prigogine's hypothesis of minimum entropy production.[18])

Consequently, the organism has freed itself from the immediate constraints of energy conservation — the first law — as well as the second law of thermodynamics. *There is always energy available within the system, which is mobilized at close to maximum efficiency and over all space-time modes.*

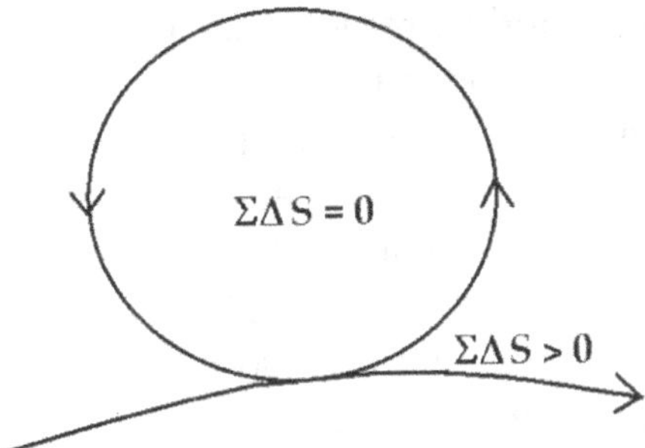

Figure 9.3 The organism frees itself from the constraints of energy conservation and the second law of thermodynamics.

The Organism Is Free from Mechanical Determinism

It was Waddington who first introduced nonlinear dynamical ideas into developmental biology in the form of the "epigenetic landscape" — a general metaphor for the dynamics of the developmental process.[19] The developmental paths of tissues and cells are seen to be constrained or canalized to "flow" along certain valleys and not others due to the "force" exerted on the landscape by the various gene products which define the fluid topography of the landscape.[20] This fluid topography contains multiple potential developmental pathways that may be realized as the result of "fluctuations", or if the environmental conditions, the genes or gene products change. This metaphor has been made much more explicit recently by Peter Saunders, a mathematician at King's College, London. He showed that the properties of the epigenetic landscape are "common not just to developing organisms but to most nonlinear dynamical systems."[21]

The Polychromatic Organism

A particular kind of nonlinearity that has made headlines recently is "deterministic chaos", a complex dynamical behaviour that is locally unpredictable and irregular while globally bounded by a "strange attractor" and extremely sensitive to initial conditions. It has been used to describe many living functions including the collective behaviour of ant colonies.[26] The unrepeatable patterns of brain activities that persuaded Freeman to declare brain science in crisis (see earlier) are typical of systems exhibiting deterministic chaos. Another putative example is the heartbeat, which is found to be much more irregular in healthy people than in cardiac patients. American physiologist Ary L. Goldberger at Harvard Medical School came to the conclusion that healthy heartbeat has "a type of variability called chaos", and that loss of this "complex variability" is associated with pathology and with aging.[23] Similarly, the electrical activities of the functioning brain, apart from being unrepeatable from moment to moment, also contain many frequencies. But during epileptic fits, the spectrum is greatly impoverished.[24] There is much current debate as to whether the complex variabilities associated with the healthy, functional state constitute chaos in the technical sense, so the question is by no means settled.[25]

A different understanding of the complex activity spectrum of the healthy state is that it is polychromatic, approaching "white" in the ideal, in which all the modes of energy storage are equally represented. It corresponds to the so-called $f(\lambda)$ = const. rule that German quantum biophysicist Fritz-Albert Popp has generalized from the spectrum of light or "biophotons" emitted from all living systems.[26] I have proposed that this polychromatic ideal distribution of stored energy is the state towards which all open systems capable of energy storage naturally evolve.[12] It is a state of both maximum and minimum in entropy content: maximum because energy becomes equally distributed over all the space-time modes (hence the "white" ideal), and minimum because the modes are all coupled or linked together to give a coherent whole; in other words, to a single degree of freedom.[2, 26] In a system where there is no impedance to energy mobilization, all the modes are intercommunicating and hence all the frequencies will be represented. Instead, when coupling is imperfect, or when the subsystem, say, the heart, or the brain, is not communicating properly, it falls back on its own modes, leading to impoverishment of its activity spectrum. Living systems are necessarily a polychromatic whole; they are full of colour and variegated complexity that nevertheless cohere into a singular being.

The Organism is a Free Sentient Being and Hence Able to Decide its Own Fate

One distinguishing feature of the living system is its exquisite sensitivity to weak signals. For example, the eye can detect single photons falling on the retina, and the presence of several molecules of pheromones in the air is sufficient to attract male insects to their appropriate mates. That extreme sensitivity of the organism applies to all levels and is the direct consequence of its energy self-sufficiency.[2] No part of the system has to be pushed or pulled into action, nor be subjected to mechanical regulation and control. Instead, coordinated action of all the parts depends on rapid *intercommunication* throughout the system. The organism is a system of "excitable media"[27] or excitable cells and tissues poised to respond specifically and disproportionately (i.e., nonlinearly) to weak signals because of the large amount of energy stored, which can thus amplify the weak signal into macroscopic action. It is by virtue of its energy self-sufficiency therefore that an organism is a *sentient* being, a system of sensitive parts all set to intercommunicate, to respond and to act appropriately as a whole to any contingency.

The organism is indeed free from mechanical determinism, but it does *not* thereby fall prey to indeterminacy. Far from surrendering its fate to the indeterminacy of nonlinear dynamics (or quantum theory, for that matter), the organism maximizes its opportunities inherent in the multiplicity of futures available to it. I have argued that indeterminacy is really the problem of the ignorance of the external observer, and not experienced by the quantum coherent being itself, which has full knowledge of its own state, and can readily adjust, respond and act in the most appropriate manner.[2] In a very real sense, *the organism is free to decide its own fate because it is a sentient being who has moment-to-moment, up-to-date knowledge of its own internal milieu as well as the external environment.*

The Organism Frees Itself from Mechanistic Control as an Interconnected, Intercommunicating Whole

The organism is not controlled mechanistically by "control" genes or molecules that tell the rest what to do via "messenger" molecules. Instead, it works as an interconnected, intercommunicating whole.

A Molecular Democracy of Distributed Control

There are thousands of enzymes catalyzing thousands of energy transactions and metabolic transformations in our body. The product of one enzyme is acted on by one or more other enzymes, resulting in a highly interconnected metabolic network. British biochemist and geneticist Henrik Kacser (1918–1995) was among the first to realize that once we have a network, especially one as complicated as the metabolic network, it is unrealistic to think that there could be special enzymes controlling the flow of metabolites under all circumstances.[28] He pioneered *metabolic control analysis* to discover how the network is actually regulated under different conditions.

After more than 20 years of investigation by many biochemists and cell biologists, it is now generally recognized that so-called "control" is invariably distributed over many enzymes (and metabolites) in the network, and moreover, the distribution of control differs under different conditions. The metabolic network turns out to be a "molecular democracy" of distributed control. As we shall see later, this applies all the way to the genes.

Long-Range Energy Continua in Cells and Tissues

Recent studies have also revealed that energy mobilization in living systems is achieved by protein or enzyme molecules acting as "flexible molecular energy machines",[29] which transfer energy directly from the point of release to the point of utilization, without thermalization or dissipation. These direct energy transfers are carried out in collective modes extending from the molecular to the macroscopic domain. The flow of metabolites is channelled coherently at the molecular level, from one enzyme to the next in sequence, in multienzyme complexes. At the same time, high-voltage electron microscopy and other physical measurement techniques reveal that the cell is more like a "solid state" than the "bag of dissolved enzymes" that generations of biochemists had previously supposed, as pointed out by American cell biologist James Clegg at University of California Davis.[30] Not only are almost all enzymes bound to an intricate "microtrabecular lattice" in the cell, but a large proportion of metabolites as well as water molecules are also structured on the enormous surfaces available. Aqueous channels are now thought to be involved in the active transport of solutes within the cell in the same way that the blood stream transport metabolites and chemical messengers within the organism.[31] British biochemist and famed historian of Chinese science Joseph Needham and his colleagues were already aware of all that in the 1930s.[32]

As American theoretical biophysical chemist Rick Welch and his colleagues proposed, the whole cell is linked up by "long-range energy continua" of mechanical interactions, electric and eletro-chemical fluxes and in particular, proton currents that form a "protoneural network", whereby metabolism is regulated instantly and down to minute detail.[33] In addition, the possibility that cells and tissues are also linked by electromagnetic phonons and photons is increasingly entertained.[2,34,35] As I shall show later, the cell (as well as organism) is not so much a solid state as liquid crystalline. Living systems, therefore, possess just the conditions that favour the rapid propagation of influences in all directions, so that local and global can no longer be easily distinguished. Global phase transitions may often take place, which can be initiated at any point within the system or subsystem. Abrupt, phase-transition like changes that typically occur in the electroencephalogram (EEG) of whole areas of the brain have been recorded with a large array of electrodes in Freeman's laboratory,[36] for which no definite centre(s) of origin can be identified. I have suggested that nonlocal intercommunication based on quantum coherence may be involved in these simultaneous changes in brain activity.[14]

Organism and Environment: A Mutual Partnership

Biology today remains dominated by the genetic determinist paradigm. Genes are seen to be the repository of information that controls the development of the organism, but are otherwise insulated from environmental influences, and passed on unchanged to the next generation except for rare random mutations. The much publicized Human Genome Project is being promoted on that very basis.[37] The genetic paradigm has already been fatally undermined at least since 1980, when a plethora of "fluid genome" processes were first discovered, and many more have come to light since. These processes destabilize and alter genes and genomes in the course of development; some of the genetic changes are so well correlated with the environment that they are referred to as "directed mutations". Many of the genetic changes are passed on to the next generation. I pointed out at the time that heredity can no longer be seen to reside solely in the DNA passed on from one generation to the next. Instead, the stability and repeatability of development — which we recognize as heredity — is distributed in the whole gamut of dynamic feedback interrelationships between organism and environment, from the socioecological to the genetic. All these may leave imprints that are passed on to subsequent generations, in the form of cultural traditions or artefacts, maternal or cytoplasmic effects, gene expression states, as well as genetic (DNA sequence) changes.[38]

The organism is highly interconnected and intercommunicating at all levels, extending from within the cell to the socioecological environment. It is on that account that the organism has freed itself from mechanistic controls of any kind. It is not a passive object at the mercy of random variation and natural selection, but an active participant in the evolutionary drama (see "4 — Epigenetics and

Developmental Dynamics: How Development Directs Evolution" in this book). In constantly responding to and transforming its environment, the organism takes part in creating the possible futures of generations to come.

The Organism is an Autonomous Coherent Whole

The concept of *coherence* has emerged within the past 20 years to describe the wholeness of the organism. The first detailed theory of coherence of the organism was presented by German-born British solid-state physicist Herbert Fröhlich (1905–1991) who argued that as organisms are made up of strongly dipolar molecules packed rather densely together (the "solid-state" cell), electric and elastic forces will constantly interact. Metabolic pumping will excite macromolecules such as proteins and nucleic acids as well as cellular membranes (which typically have an enormous electric field of some 10^7 V/m across them). These will start to vibrate and build up into collective modes — *coherent excitations* — of both phonons and photons (sound and light) that extend over macroscopic distances within the organism and perhaps also outside the organism.[39, 40] The emission of electromagnetic radiation from coherent lattice vibrations in a solid-state semiconductor has only just been experimentally demonstrated for the first time.[41] The possibility that organisms may use electromagnetic radiations to communicate between cells was already entertained by Soviet biologist Alexander Gurwitsch (1874–1954) in the 1920s.[42] This hypothesis was revived by Popp and his co-workers in the late 1970s,[26] and there is now a large and rapidly growing literature on "biophotons" believed to be emitted from a coherent photon field (or energy storage field) within the living system.[34,35]

We have indeed found that a single, one-minute exposure of synchronously developing early fruit fly embryos to white light results in the re-emission of relatively intense and prolonged flashes of light some tens of minutes and even hours after the light exposure.[43] This is reminiscent of phase-correlated collective emission or *superradiance* in physical systems, although the timescale is orders of magnitude longer. For phase correlation to build up over the entire population one must assume that each embryo has a *collective* phase of all its activities; in other words, each embryo must be considered a highly coherent domain, despite its multiplicity of activities.[44]

Actually, this is no different from the macroscopic phase correlations that are involved in the synchronous flashing of huge populations of fireflies in the wild[45] and in many physiological functions, such as precise limb coordination during locomotion[46] and coupling between heart rate and respiratory rate.[47] Under those conditions, whole limbs or entire circulatory and respiratory systems must be considered coherent domains that can maintain definite phase relationships with respect to one another.

During the same early period of development in *Drosophila*, exposure of the embryos to weak static magnetic fields also causes characteristic global transformation of the normal segmental body pattern to helical configurations in the larvae emerging 24 hours later.[48] As the energies involved are well below the thermal threshold (of a random system at thermodynamic equilibrium), we conclude that there can be no effect unless the external field is acting on a coherent morphogenetic field in which charges are moving in phase, or where magnetically sensitive liquid crystals are undergoing phase alignment globally.[49] Liquid crystals may indeed be the material basis of many, if not all, aspects of biological organization.[50]

Organisms are Polyphasic Liquid Crystals

Liquid crystals are phases of matter between the solid and the liquid states, hence the term, *mesophases*.[51] Liquid crystalline mesophases possess long-range orientational order (all the molecules pointing in the same direction), and often also varying degrees of translational order (individual molecules keeping to their positions to varying extents). In contrast to solid crystals, liquid crystals are mobile and flexible, and above all, highly responsive. They undergo rapid changes in orientation or phase transitions when exposed to electric or magnetic fields[52] or to changes in temperature, pressure, pH, hydration and concentrations of inorganic ions.[53] These properties are ideal for organisms.[54] Liquid crystals in organisms include all their major constituents: the lipids of cellular membranes, the DNA in chromosomes, all proteins (especially cytoskeletal proteins), muscle proteins, collagens and other macromolecules of connective tissues. These adopt a multiplicity of different mesophases that may be crucial for biological structure and function at all levels of organization,[50] from channelling metabolites in the cell to pattern determination and the coordinated locomotion of whole organisms.

The importance of liquid crystals for living organization was recognized by Needham among others.[32] He suggested that living systems actually *are* liquid crystals, and that many liquid crystalline mesophases may exist in the cell although they cannot then be detected. Indeed, there has been no direct evidence that extensive liquid crystalline mesophases exist in living organisms or in the cytoplasm until our recent discovery of a non-invasive optical technique.[55–57] This technique enables us to obtain high-resolution and high-contrast coloured images of live organisms based on visualizing just the kind of coherent liquid crystalline mesophases that Needham had predicted.

The technique effectively allows us to see the whole of the living organism at once from its macroscopic activities down to the phase alignment of the molecules that make up its tissues and cells. Brilliant colours are generated specific for each tissue, dependent on the molecular structure and the degree of coherent alignment of all the molecules, even as the molecules are moving about busily transforming energy. This is possible because visible light vibrates much faster than the molecules can move, so the tissues will appear indistinguishable from *static* crystals to the light passing through so long as the movements of the constituent molecules are sufficiently coherent. With this imaging technique, one can see that the organism is thick with activities at all levels, *which are coordinated in a continuum from the macroscopic to the molecular*. And that is what the coherence of the organism entails.

These images also bring out another aspect of the wholeness of the organism: all organisms, from protozoa to vertebrates without exception, are polarized along the anteroposterior axis, so that all the colours in the different tissues of the body are at a maximum when the anteroposterior axis is appropriately aligned, and they change in concert as the organism is rotated from that position. The anteroposterior axis acts as the *optical axis* for the whole organism, which behaves in effect as a single crystal. This leaves us in little doubt that the organism is a singular whole, despite the diverse multiplicity and polychromatic nature of its constituent parts.

The tissues not only maintain their crystalline order when they are actively transforming energy, but the degree of order seems to depend on energy transformation, in that the more active and energetic the organism, the more intensely colourful it is, implying that the molecular motions are all the more coherent. The coherence of the organism therefore is closely tied up with its energetic status; the coherent whole is full of energy — it is a *vibrant* coherent whole.

Quantum Coherence in Living Organisms

From the above considerations and observations, it is clear that organic wholeness is *distributed* throughout its constituent parts so that local and global, part and whole are completely indistinguishable, the organism's activities being always fully coordinated in a continuum from the molecular to the macroscopic, encompassing the most diverse collection of activities. That convinces me that there is something very special about the wholeness of organisms that is only fully captured by the concept of *quantum* coherence.[2,14,18] An intuitive appreciation of quantum coherence is to think of the "I" that each and every one of us experience of our own being. We know that our body is a multiplicity of organs and tissues composed of trillions of cells and astronomical numbers of molecules of numerous different kinds, all capable of working autonomously yet somehow cohering into the singular being of our private experience. That is just the stuff of quantum coherence. Quantum coherence does not mean that every element of the system must be doing the same thing all the time; it is more akin to a grand ballet, or better yet, a very large jazz band where everyone is doing his or her own thing while keeping perfectly in step and in tune with the whole.

A quantum coherent system maximizes both global cohesion and local freedom.[2,18] This property is technically referred to as *factorizability*, the correlations between subsystems resolving neatly into a product of self-correlations of the subsystems, so that the subsystems behave as though they are independent of one another. It enables the body to be performing all sorts of *different* but *coordinated* functions simultaneously. It also enables *instantaneous, noiseless intercommunication to take place throughout the system.*[57] As I am writing, my digestive system is working independently, my metabolism busily transforming chemical energy in all my cells, putting some away in the longer-term stores of fat and glycogen, while converting most of it into readily utilizable forms. Similarly, my muscles are keeping in tone and allowing me to work the keyboard, while, hopefully, my neurons are firing in wonderfully coherent patterns in my brain. Nevertheless, if the telephone should ring in the middle of all this, I would turn to pick it up without hesitation.

The importance of factorizability is evoked by the character Dr Strangelove memorably overplayed (and improvised) by British film actor Peter Sellers (1925–1980) in a movie of the same name — a black comedy satirizing the Cold War fears of a nuclear conflict between the Union of Soviet Socialist Republics and the United States — directed, produced and co-written by American film director and producer Stanley Kubrick (1928–1999).[58] Dr Strangelove was an unhinged ex-Nazi scientist who wanted to rule the world and a wheelchair-bound paraplegic who could not speak without raising his arm uncontrollably in a Nazi salute. That is just the symptom of the loss of factorizability, the hallmark of quantum coherence.

The coherent organism is in the ideal a quantum superposition of activities — organized according to their characteristic space-times — each itself coherent, so that it can couple coherently to the rest.[2,15,16,18] This picture is fully consistent with the earlier proposal that the organism stores energy over all space-time domains each intercommunicating (or coupled) with the rest. Quantum superposition also enables the system to maximize its potential degrees of freedom so that the single degree of freedom required for coherent action can be *instantaneously* accessed.

The Freedom of Organisms

The organism maximizes both local freedom and global intercommunication. One comes to the startling discovery that the coherent organism is in a very real sense completely and thoroughly free. Nothing is in control, and yet everything is in control. Thus, it is the failure to transcend the mechanistic framework that makes people persist in enquiring which parts are in control, or issuing instructions; or whether free will exists, and who choreographs the dance of molecules. Does "consciousness" control matter or *vice versa*? These questions are meaningless when one understands what it is to be a coherent, organic whole. An organic whole is an entangled whole, where part and whole, global and local, are so thoroughly implicated as to be indistinguishable, and each part is as much in control as it is sensitive and responsive. Choreographer and dancer are one and the same. The "self" is a domain of coherent activities, in the ideal a pure state that permeates the whole of our being with no definite localizations or boundaries, much as Bergson[3] had described.

The positing of "self" as a domain of coherent activities implies the existence of an active whole agent who is free. I must stress that freedom does *not* entail the breakdown of causality as many commentators have mistakenly supposed. On the contrary, an acausal world would be one where it is impossible to be free, as nothing would be intelligible. Nevertheless, freedom does entail a new kind of organic causality that is nonlocal, and posited with the organism itself. It is the experience of perceptual feedback consequent on one's actions that is responsible for the intuition of causality, as Freeman wrote.[59] However, it must not be supposed that the cause or consciousness is secreted from some definite location in the brain; it is distributed and delocalized throughout the entire system, brain and body included.

Freedom in the present context means being true to "self"; in other words, being coherent. A free act is a coherent act. Of course, not all acts are free, as one is seldom fully coherent. Yet the mere possibility of being unfree affirms the opposite, that freedom is real. As Bergson wrote (p.172),[3]

> "... we are free when our acts spring from our whole personality, when they express it, when they have that indefinable resemblance to it which one sometimes finds between the artist and his work."

The coherent "self" is also distributed and nonlocal — being implicated in a community of other entities with which one is entangled.[2,4] Thus, being true to self does not imply acting against others. On the contrary, sustaining others sustains the self, so being true to others is also being true to self. It is only within a mechanistic Darwinian perspective that freedom becomes perverted into acts against others.[60] The coherent "self" can also couple coherently to the environment so that one becomes as much in control of the environment as one is responsive. The organism thereby participates in creating its own possible futures as well as those of the entire community of organisms in the universe, much as Whitehead has envisaged.[4]

I venture to suggest, therefore, that *a truly free individual is a coherent being that lives life fully and spontaneously, without fragmentation or hesitation, who is at peace with herself and at ease with the universe as she participates in creating from moment to moment its possible futures.*

References and Notes

1. Freeman, W.J. *Societies of Brains. A Study in the Neuroscience of Love and Hate*. Lawrence Erlbaum Associates, Hove, 1995.

2. Ho, M.W. *The Rainbow and the Worm: The Physics of Organisms*, World Scientific, Singapore, 1993.
3. Bergson, H. *Time and Free Will: An Essay on the Immediate Data of Consciousness*, translated by F.L. Pogson. Allen & Unwin, London, 1916.
4. Whitehead, A.N. *Science and the Modern World*. Penguin Books, Harmondsworth, 1925.
5. The Theoretical Biology Club was an informal association of academics based in Cambridge University in the 1930s. Its membership was probably more extensive than I have indicated. Their project continued to some extent in a series of meetings organized by C.H. Waddington in the 1960s and 1970s. The proceedings, published under the title *Towards a Theoretical Biology* (Edinburgh University Press), were very influential among critics of mainstream neo-Darwinian theory of evolution, including myself. Four recent Waddington Memorial Conferences have been organized by Waddington's student, Brian Goodwin, and published as collected volumes: Goodwin, B.C. and Saunders, P.T., eds. *Theoretical Biology: Epigenetic and Evolutionary Order from Complex Systems*. Edinburgh University Press, Edinburgh, 1989; Stein, W. and Varela, F.J., eds. *Thinking About Biology*. Addison-Wesley, Reading, Massachusetts, (1992). These helped to keep the project of the Theoretical Biology Club alive, and I count myself among the intellectual beneficiaries.
6. Schrödinger, E. *What Is Life?* Cambridge University Press, Cambridge, 1944.
7. Ho, M.W. "Towards an Indigenous Western Science: Causality in the Universe of Coherent Space-Time Structures". In *New Metaphysical Foundations of Modern Science* (W. Harman and J. Clark, eds.), pp. 179–213, Institute of Noetic Sciences, Sausalito, 1994a.
8. Cited in Ehrenberg, W. "Maxwell's Demon". *Sci Am* 217 (1967): 103–110. (p. 103)
9. Schrödinger was criticized by both Linus Pauling and Max Perutz over his non-rigorous use of "negative entropy". The exchanges are described by Gnaiger, E. "Negative Entropy for Living Systems: Controversy Between Nobel Laureates Schrödinger, Pauling and Perutz". *Modern Trends in BioThermoKinetics* 3 (1994): 62–70.

10. Prigogine, I. *Introduction to Thermodynamics of Irreversible Processes*. John Wiley & Sons, New York, 1967.
11. Haken, H. *Synergetics: An Introduction*. Springer-Verlag, Heidelberg and New York, 1977.
12. Ho, M.W. "What Is (Schrödinger's) Negentropy?" *Modern Trends in BioThermoKinetics* 3 (1994b): 50–61.
13. Ho, M.W., ed. *Bioenergetics, S327 Living Processes: An Open University Third Level Science Course*. Open University Press, Milton Keynes, 1995a.
14. Ho, M.W. "Quantum Coherence and Conscious Experience". *Kybernetes* 26 (1997): 265–276.
15. Ho, M.W. Bioenergetics and the coherence of organisms. *Neural Network World* 5 (1995b): 733–750.
16. Ho, M.W. "Bioenergetics and Biocommunication". In *IPCAT 95 Proceedings* (R. Cuthbertson, M. Holcombe and R. Paton, eds.), pp. 251–260, World Scientific, Singapore, 1996.
17. Denbigh, K. *The Thermodynamics of the Steady State*. Methuen & Co. Ltd., London, 1951.
18. Ho, M.W. *The Rainbow and the Worm: The Physics of Organisms*, 3rd edition, World Scientific, Singapore and Imperial College Press, London, 2008.
19. Waddington, C.H. *The Strategy of the Genes*. Allen & Unwin, London, 1957.
20. I reviewed Waddington's ideas on evolutionary theory in Ho, M.W. "Evolution". In *Encyclopedia of Comparative Psychology* (G. Greenberg and M. Haraway, eds.), pp. 107–119, Garland Publishing, New York, 1998.
21. Saunders, P.T. "The Organism As a Dynamical System". In *Thinking About Biology* (W. Stein and F.J. Varela, eds.), pp. 41–63, Addison-Wesley, Reading, Massachusetts, 1992.
22. This is comprehensively described by Goodwin, B.C. "Biological Rhythms and Biocommunication". In *Biocommunication: S327 Living Processes*, pp. 183–230, Open University Press, Milton Keynes, 1995; and in accompanying video.
23. Goldberger, A.L. "Is the Normal Heartbeat Chaotic or Homeostatic"? *NIPS* 6 (1991): 87–91.

24. Kandel, E.R., Schwartz, J.H. and Jessell, T.M. *Principles of Neural Science*, 3rd edition. Elsevier, New York, 1991.
25. See Glass, L. and Mackey, M.C. *From Clocks to Chaos: The Rhythms of Life*. Princeton University Press, Princeton, New Jersey, 1988.
26. See Popp, F.A. "On the Coherence of Ultraweak Photoemission from Living Tissues". In *Disequilibrium and Self-Organization* (C.W. Kilmister, ed.), p. 207, Reidel, Dordrecht, 1986.
27. Goodwin, B.C. *How the Leopard Changed Its Spots: The Evolution of Complexity*. Weidenfeld and Nicolson, London, 1994.
28. Kacser, H. "On Parts and Wholes in Metabolism". In *The Organization of Cell Metabolism* (G.R. Welch and J.S. Clegg, eds.), pp. 327–338, Plenum Publishing Corporation, New York, 1987.
29. See Welch, G.R. and Clegg, J.S., eds. *The Organization of Cell Metabolism*. Plenum Publishing Corporation, New York, 1987.
30. Clegg, J.S. "Properties and Metabolism of the Aqueous Cytoplasm and Its Boundaries". *Am J Physiol* 246 (1984): R133–151.
31. Wheatley, D. and Clegg, J.S. "Intracellular Organization: Evolutionary Origins and Possible Consequences of Metabolic Rate Control in Vertebrates". *Am Zool* 31 (1991): 504–513.
32. Needham, J. *Order and Life*. MIT Press, Cambridge, Massachusetts, 1936.
33. Welch, G.R. and Berry, M.N. "Long-Range Energy Continua and the Coordination of Multienzyme Sequences *In Vivo*". In *Organized Multienzyme Systems* (G.R. Welch, ed.), pp. 420–448, Academic Press, New York, 1985.
34. See Popp, F.A., Li, K.H. and Gu Q. eds. *Recent Advances in Biophoton Research and its Applications*, World Scientific, Singapore, 1992.
35. Ho, M.W., Popp, F.A. and Warnke, U., eds. *Bioelectrodynamics and Biocommunications*. World Scientific, Singapore, 1994.
36. Freeman, W.J. and Barrie, J.M. "Chaotic Oscillations and the Genesis of Meaning in Cerebral Cortex". In *Temporal Coding in the Brain* (G. Bizsaki, ed.), pp. 13–37, Springer-Verlag, Berlin, 1994.
37. Ho, M.W. "Unravelling Gene Biotechnology". *Soundings* 1 (1995c): 77–98.
38. Ho, M.W. "Evolution by Process, Not by Consequence: Implications of the New Genetics on Development and Evolution". *Int J Comp Psychol* 1 (1986): 3–7.

39. Fröhlich, H. "Long Range Coherence and Energy Storage in Biological Systems". *Int J Quant Chem* 2 (1968): 641–649.
40. Fröhlich, H. "The Biological Effects of Microwaves and Related Questions". *Adv Electron El Phys* 53 (1980): 85–152.
41. Dekorsy, T., Auer, H., Waschke, C., Bakker, H.J., Roskos, H.G. and Kurz, H. "Emission of Submillimeter Electromagnetic Waves by Coherent Phonons". *Phys Rev Lett* 74 (1995): 738–741.
42. Gurwitsch, A.G. "The Mitogenic Rays". *Bot Gaz* 80 (1925): 224–226.
43. Ho, M.W., Xu, X., Ross, S. and Saunders, P.T. "Light Emission and Re-Scattering in Synchronously Developing Populations of Early Embryos: Evidence for Coherence of the Embryonic Field and Long Range Cooperativity". In *Advances in Biophotons Research* (F.A. Popp, K.H. Li and Q. Gu, eds.), pp. 287–306, World Scientific, Singapore, 1992.
44. Ho, M.W., Zhou, Y.M. and Haffegee, J. "Biological Organization, Coherence and the Morphogenetic Field". In *Physical Theory in Biology* (C. Lumsden, L.E.H. Trainor and W.A. Brandts, eds.), pp. 225–244, World Scientific, Singapore, 1998.
45. Strogatz, S.H. and Mirollo, R.E. "Collective Synchronisation in Lattices of Non-Linear Oscillators with Randomness". *J Phys A Math Gen* 21 (1988): L699–L705.
46. Kelso, J.A.S. "Behavioral and Neural Pattern Generation: The Concept of Neurobehavioral Dynamical Systems". In *Cardiorespiratory and Motor Coordination* (H.P. Koepchen and T. Huopaniemi, eds.), pp. 224–234, Springer-Verlag, Berlin, 1991; Collins, J.J. and Stewart, I.N. "Symmetry-Breaking Bifurcation: A Possible Mechanism for 2:1 Frequency-Locking in Animal Locomotion". *J Math Biol* 30 (1992): 827–838.
47. Breithaupt, H. "Biological Rhythms and Communications". In *Electromagnetic Bioinformation*, 2nd edition (F.A. Popp, U. Warnke, H.L. Konig and W. Peschka, eds.), pp. 18–41, Urban & Schwarzenberg, Berlin, 1989.
48. Ho, M.W., Stone, T.A., Jerman, I., Bolton, J., Bolton, H., Goodwin, B.C., Saunders, P.T. and Robertson, F. "Brief Exposure to Weak Static Magnetic Fields during Early Embryogenesis Cause Cuticular Pattern Abnormalities in *Drosophila* Larvae. *Phys Med Biol* 37 (1992a): 1171–1179.

49. Ho, M.W., French, A., Haffegee, J. and Saunders, P.T. "Can Weak Magnetic Fields (or Potentials) Affect Pattern Formation?" In *Bioelectrodynamics and Biocommunication* (M.W. Ho, F.A. Popp and U. Warnke, eds.), World Scientific, Singapore, 1994.
50. Ho, M.W., Haffegee, J., Newton, R.H., Ross, S., Zhou, Y.M. and Bolton, J. "Organisms As Polyphasic Liquid Crystals". *Bioelectrochem Bioenerg* 41 (1996): 81–91.
51. De Gennes, P.G. *The Physics of Liquid Crystals*. Clarendon Press, Oxford, 1974.
52. Blinov, L.M. *Electro-Optical and Magneto-Optical Principles of Liquid Crystals*. John Wiley & Sons, London, 1983.
53. Collings, P.J. *Liquid Crystals, Nature's Delicate Phase of Matter*. Princeton University Press, Princeton, 1990.
54. Gray, G. "Liquid Crystals: Molecular Self-Assembly". British Association for the Advancement of Science, Chemistry Session: Molecular Self-Assembly in Science and Life, Sept. 1, Keele, 1993.
55. Ho, M.W. and Lawrence, M. "Interference Colour Vital Imaging: A Novel Noninvasive Technique". *Microsc Anal* (1993): 26.
56. Ho, M.W. and Saunders, P.T. "Liquid Crystalline Mesophases in Living Organisms". In *Bioelectromagnetism and Biocommunication* (M.W. Ho, F.A. Popp and U. Warnke, eds.), World Scientific, Singapore, 1994.
57. Newton, R., Haffegee, J. and Ho, M.W. "Colour-Contrast in Polarized Light Microscopy of Weakly Birefringent Biological Specimens". *J Microsc* 180 (1995): 127–130.
58. "Dr Strangelove", Wikipedia, accessed 15 August 2015, https://en.wikipedia.org/wiki/Dr._Strangelove#Peter_Sellers.27s_multiple _roles.
59. Freeman, W.J. "On the Fallacy of Assigning an Origin to Consciousness". In *Machinery of the Mind: Data, Theory, and Speculations about Higher Brain Function* (E.R. John, ed.), pp. 14–26, Birkhauser, Boston, 1990.
60. Ho, M.W. "Natural Being and Coherent Society". In *Gaia in Action: Science of the Living Earth* (P. Bunyard, ed.), pp. 286–307, Floris Books, Edinburgh, 1995.

10

Quantum Coherence and Conscious Experience

Quantum coherence is the basis of living organization and can also account for key features of conscious experience — the "unity of intentionality", our inner identity of the singular "I", the simultaneous binding and segmentation of features in the perceptive act, the distributed, holographic nature of memory, and the distinctive quality of each experienced occasion.

How to Understand the Organic Whole

The 1990s was designated "the decade of the brain". Yet mid-way through the decade, British artificial intelligence and cybernetics expert Alex Andrew laments that "our understanding of brain functioning ... is still quite primitive",[1] and distinguished American neuroscientist Walter Freeman goes as far as declaring brain science "in crisis".[2] In the meantime, there is a remarkable proliferation of journals and books *about* consciousness, which brain science has so far failed to explain, at least in the opinion of those who have lost faith in the conventional mechanistic reductionist approach. One frequent suggestion is the need for quantum theory, though the theory is interpreted and used in diverse and at times conflicting ways by different authors.

I believe that the *impasse* in brain science is the same as that in all of biology: we simply do not have a conceptual framework for understanding how the organism functions as an integrated whole. Brain science has been more fortunate than many other areas in its long-established multidisciplinary approach, which is crucial for understanding the whole. In particular, the development of non-invasive, non-destructive imaging techniques has allowed access to the *living* state, which serves to constantly remind the reductionists of the ghost of the departed whole. The images obtained from ultrasensitive, and hence truly non-invasive, magnetic tomography[3] are captivating. Analysing such data presents an even greater challenge than the multichannel electroencephalography (EEG) data obtained by Freeman and his colleagues.[4] Both techniques are revealing large-scale spatiotemporal coherence of brain activities that cannot be explained by conventional mechanisms. The brain does not function as a collection of specialized brain cells, but as a coherent whole. That is surely one good reason to seek alternative perspectives that would help us understand the organic whole.

How the brain functions as a coherent whole is inseparable from how the organism functions as a coherent whole. It is the same question, stated eloquently by British biochemist/embryologist Joseph Needham (1900–1995) in *Order and Life*[5] and by Austrian Nobel laureate physicist Erwin Schrödinger (1887–1961) in *What is Life?*,[6] that has caused generations of biologists and physicists to be dissatisfied with the mechanistic approach. Inspired by a long line of distinguished dissidents, I began to work towards a theory of the organism based on empirical and theoretical findings across the disciplines, first presented in my book, *The Rainbow and the Worm: The Physics of Organisms*,[7] and extended in subsequent publications.[8–11] The theory starts from thermodynamic considerations of energy *storage* in a space-time structured system under energy flow, which, by dynamic closure, creates the conditions for *quantum coherence*. This effectively frees the organism from thermodynamic constraints so that it is poised for rapid, specific intercommunication, enabling it to function as a coherent whole. In the ideal, the organism is a quantum superposition of coherent activities, with instantaneous (nonlocal) noiseless intercommunication throughout the system.

I do not think quantum theory *per se* will lead us through the mechanistic deadlock to further understanding. Instead, we need a thoroughly *organicist* way of thinking that transcends both conventional thermodynamics and quantum theory.[7,10,11] I have focussed on quantum coherence and the attendant nonlocal intercommunication as the expression of the radical wholeness of the organism, where global and local are mutually entangled, and every part is as much in control as it is sensitive and responsive.

I shall briefly summarize the arguments for quantum coherence in the living system, then go on to explore how certain key features of conscious experience may be understood. I suggest that the wholeness of the organism *is* based on a high degree of quantum coherence. Quantum coherence underlies the "unity of intentionality" and our inner identity of the singular "I". It may account for binding and segmentation in the perceptive act, the distributed, holographic nature of memory, and the distinctive quality of each experienced occasion.

The Organism is a Quantum Coherent Whole

A Vibrant Sentient Whole

Organisms overcome the immediate constraints of thermodynamics in their capacity to *store* mobilizable energy, which circulates through a cascade of cyclic processes within the system before it is dissipated.[10,11] Dynamic closure of circulating energy gives a life cycle. Within the life cycle, coupled cyclic processes span the entire gamut of space-times from the local and fast (or slow) to the global and slow (or fast). This enables energy to be readily shared throughout the system, from local to global and vice versa, which is why we can have energy at will. But how is energy mobilization so well coordinated? That is partly a direct consequence of the energy stored, which renders the whole system *excitable*, or highly sensitive to specific weak signals. Weak signals originating anywhere within or outside the system will propagate throughout the system and become amplified, often into macroscopic action. Intercommunication can proceed very rapidly, in particular, on account of the *liquid crystalline* structure of the cells and the connective tissues in all live organisms discovered in my laboratory.[12–15]

The extracellular and intracellular matrices together constitute an excitable *continuum* for rapid intercommunication permeating the entire organism, enabling it to function as a coherent whole. The liquid crystalline continuum constitutes a "body consciousness" that precedes the nervous system in evolution; and I suggest, it still works in tandem with, and independently of, the nervous system (see next section). This body consciousness is the basis of *sentience*, the prerequisite for conscious experience that involves the participation of the intercommunicating whole of the energy storage domain. *In the limit of the coherence time and coherence volume of energy storage, intercommunication is instantaneous and nonlocal.* Because energy is stored over all modes, the organism possesses a complete range of coherence times and coherence volumes.[7]

The life cycle, with its complex of coupled cyclic processes, forms a heterogeneous, multidimensional and entangled space-time that structures experience. In the ideal, it is a quantum superposition of coherent space-time modes, constituting a pure state that maximizes both local freedom and global cohesion in accordance with the *factorizability* of the quantum coherent state, as presented by American quantum physicist Roy Glauber.[16] Quantum coherence gives rise to correlations between subsystems that resolve neatly into products of the self-correlations so that the subsystems behave as though they are independent of one another.

One can also picture the organism as a coherent quantum electrodynamical field of many modes, with an uncertainty relationship between energy and phase (of vibration), according to Italian quantum field theorist Giuliano Preparata (1942–2000):[17] $\Delta\nu\Delta\phi \geq h$, where ν is energy, ϕ the phase, and h is Planck's constant. So when phase is defined, energy is indeterminate, and vice versa; when energy is defined, phase is indeterminate.

In quantum optics and quantum electrodynamics theory, the coherent state is *asymptotically stable*,[18] meaning that the coherent state is not just stable, but it is also an attractor or ideal end-state towards which the system tends to return on being perturbed.[19] Thus, the coherent state is not an improbable state, *especially* when it is established by energy input.

There is abundant evidence of macroscopic activities with collective phases in the spectrum of biological rhythms, many of which tend towards integral phase relationships with one another.[7,20] There are also examples of collective activities that may involve phase correlations over entire populations.[21,22] As the coherence times of living processes span more than 20 orders of magnitude from 10^{-14} s for resonant energy transfer to 10^{7} s for circannual cycles, a pure coherent state for the entire system would be a many-mode quantum electrodynamical field with a collective phase potentially over *all* modes. It may be attainable only under very exceptional circumstances, as during an aesthetic or religious experience when the "pure duration" (see later) of the here and now becomes completely delocalized in the realm of no-time and no-space.[7] Nevertheless, quantum coherence can exist to different degrees or orders,[16] and I suggest that the wholeness of the organism *is* based on a high degree of quantum coherence, constituting Freeman's "unity of intentionality",[2] which is the prerequisite to conscious experience.

Quantum Coherence and Body Consciousness

From the perspective of the whole organism, the brain's primary function may be the mediation of coherent coupling of all subsystems, so the more highly differentiated or complex the system, the bigger the brain required. Substantial parts of the brain are indeed involved in integrating inputs from all over the body, and over long time scales. But not all the processing that goes on in the brain is involved in the coherent coordination of subsystems, for this coordination seems instantaneous by all accounts.

Thus, during an olfactory experience, slow oscillations in the olfactory bulb are in phase with the movement of the lungs.[4] Similarly, the coordinated movement of the four limbs in locomotion is accompanied by patterns of activity in the motor centers of the brain that are in phase with those of the limbs.[23] Those are remarkable achievements that physiologists and neuroscientists alike have taken too much for granted. The reason macroscopic organs such as the four limbs can be coordinated, is that each is individually a coherent whole, so that a definite phase relationship can be maintained among them. The hand-eye coordination required for the accomplished pianist is extremely impressive, but depends on the same inherent coherence of the subsystems that enables instantaneous intercommunication to occur. There simply isn't time enough, from one musical phrase to the next, for inputs to be sent to the brain, there to be integrated, and coordinated outputs to be sent back to the hands.

I raised the possibility that a "body consciousness" works in tandem with, but independently of, the "brain consciousness" constituting the nervous system (see above), and suggested that instantaneous coordination of body functions is mediated, not by the nervous system, but by the body consciousness inherent in the liquid crystalline continuum of the body. Working with my colleague David Knight who has studied collagenous liquid crystalline composites for many years, and following cell biologist Jim Oschman, who put energy medicine on a firm scientific footing,[24] we reviewed[25] evidence suggesting that this liquid crystalline continuum is responsible for the direct current (DC) electrodynamical field, permeating the entire body of all animals, which American orthopedic surgeon and pioneer of electro-medicine Robert Becker[26] (1923–2008) and others have detected in all organisms including humans. Becker further demonstrated that the DC field has a mode of semi-conduction much faster than nervous conduction. During a perceptive event, local changes in the DC field can be measured half a second *before* sensory signals arrive in the brain, suggesting that the activities in the brain is preconditioned by the local DC field.

Up to 70% of the proteins in the connective tissues consist of collagens that exhibit constant patterns of alignment, as characteristic of liquid crystals. Collagens have distinctive mechanical and dielectric properties that render them very sensitive to mechanical pressures, changes in pH, inorganic ions and electromagnetic fields.[25] In particular, a cylinder of bound water surrounds the triple-helical molecule, giving rise to an ordered array of bound water on the surface of the collagen network that supports rapid "jump conduction" of protons. Proteins in liquid crystals have coherent residual motions, and will readily transmit weak signals by proton conduction, or as coherent waves.[27] Thus, extremely weak electromagnetic signals or mechanical disturbances are sufficient to set off a flow of protons that will propagate throughout the body, making it ideal for intercommunication in the manner of a proton-neural network,[28] or indeed, as the acupuncture meridian system of traditional Chinese medicine as we have proposed.[25]

The liquid crystalline nature of the continuum also enables it to function as a distributed memory store. The proportion of bound versus free water on the surfaces of proteins is known to be altered by conformation changes of the proteins. Proteins undergo a hierarchy of conformational changes on a range of time scales as well as different energies. Conformers are clustered in groups that have nearly the same energy, with very low energetic barriers between them.[29] Proteins can thus be triggered to undergo global conformational changes that will, in turn, alter the structure of bound water. As the bound water forms a global network in association with the collagen, it will have a certain degree of stability, or resistance to change. In other words, it will retain tissue memory of previous experiences. The memory may consist partly of dynamic circuits, the sum total of which constitutes the DC body field. Thus, consciousness is distributed throughout the entire body, brain consciousness being embedded in body consciousness. Brain and body consciousness mutually inform and condition each other. The unity of intentionality is a complete coherence of brain and body.

Quantum Coherence and the Binding Problem

So it is that we perceive ourselves as a singular "I" intuitively, despite the extremely diverse multiplicity of tissues, cells and molecules constituting our being (c.f. Schrödinger[6]). Quantum coherence entails a plurality that is singular, a multiplicity that is a unity. The "self" is the domain of coherence,[7] a pure state or "pure duration" that permeates the whole of our being, much as French philosopher and Nobel laureate in literature Henri Bergson (1859–1941) has so vividly described in his book *Time and Free Will*.[30]

It is because we perceive ourselves as a singular whole that we perceive the real world as colour, sound, texture and smell, as a unity all at once. Sounds presented in linear sequences are recognized as speech or music, much as objects in motion are recognized as such, rather than as disconnected configurations of light and shadow. But how is this unity structured so that we can recognize whole objects *and* distinguish different objects in our perceptual field? That is the problem of binding, and reciprocally of segmentation.[31]

Detailed investigations over the past decades have revealed many brain cells that respond to isolated features such as edges or bars in the visual cortex, but no special cells have been found to respond to higher categories,[1] such as squares or cubes for example. There is simply no "grandmother cell" that integrates the separate features. So how are the separate features bound into a whole? And how is it that we can bind features correctly so that they belong to the same object in the real world? For example, how do we see correctly, a red rose in a yellow vase and not a yellow rose in a red vase? It turns out that timing is of the essence.

Freeman and his co-workers carried out simultaneous recordings with an array of 64 electrodes covering a large area of the rabbit cortex, and found oscillations that are coherent over the entire array.[2] The oscillations tend to vary continually or abruptly, but when they change, they do so in the same way over the whole area. The amplitudes will differ, but the pattern of discharges is simultaneous and uniform. Freeman concludes (p. 57):[2] "This spatial coherence indicates that the oscillation is a macroscopic property of the whole area, that all the neurons in the neuropil share it, and that the same frequency holds at each instant everywhere." Another research team recorded simultaneously from pairs of neuronal units in the cat cortex whose outputs might be subject to binding[32] in the same or different cerebral hemispheres and responded to the same or different sensory modalities. They found that throughout the wide range of situations, the characteristic feature of paired discharges that are suitable for subsequent binding is a high degree of coincidence in time. It seems that the nervous system produces "simultaneity as an aid to subsequent binding."

Neuroscientist Wolf Singer at Frankfurt Institute of Advanced Studies has also found evidence of simultaneous oscillations in separate areas of the cortex, accurately synchronized in phase as well as frequency.[33] He suggests that the oscillations are synchronized from some common source, but Freeman's group, using a large array of electrodes (see above) failed to identify any obvious source.[2] As Andrew points out,[1] the accuracy of phase agreement is far too perfect for the synchronizing to spread by normal neural transmission, and he favours some kind of optical signal transmitted by water trapped in microtubules acting as optical fibres, as proposed by American anaesthesiologist Stuart Hameroff at Arizona University and British mathematical physicist Roger Penrose at Oxford University, and others who have teamed up to study consciousness, placing great emphasis on quantum coherence in microtubules, which make up the cytoskeleton of neurons.[34,35]

If, however, the system is coherent to begin with, as I suggest, then *a genuine nonlocal simultaneity may be involved*. The present precision of recording is insufficient to distinguish between instantaneous simultaneity and propagation at the speed of light. As is well known, there is no time-like separation within the coherence volume, and no space-like separation within the coherence time, so apparent "communication" is instantaneous, and synchrony can be established with no actual delay. This simultaneity may be mediated and gated by the DC body field mentioned above. That can easily be tested by repeating the measurements carried out by Becker.[26]

British cyberneticist Peter Marcer has proposed a "quantum holographic" model of consciousness in which perception involves the conversion of an interference pattern (between a coherent wave-field generated by the perceiver and the wave-field reflected off the perceived) to an object image that is coincident with the object itself.[36,37] This is accomplished by a process known as *phase conjugation*, whereby the wave reflected from the object is returned (by the perceiver) along its *original* path to form an image where the object is situated. The perceiving being is *into* the act of perceiving, as Freeman observes.[2]

If Marcer is right, endogenously generated coherent waves or activities can function as precise gating, on the basis of phase information, to bind and segment features as appropriate. Furthermore, in the act of perceiving, the organism also perceives itself situated in the environment, through active phase conjugation. As American psychologist James Jerome Gibson (1904–1979) points out in his book *The Ecological Approach to Visual Perception*,[38] perception and proprioception are one and the same. Within the perceptive realm of the organism, there will always be an image of the self as the focus of "prehensive unification" (in the words of English mathematician philosopher Alfred North Whitehead (1861–1947)),[39] to which all features in the environment are related. Marcer's quantum holographic model of self-consciousness[34,35] would involve an image of the self that is coincident with the organism itself, so "self" and "other" are simultaneously defined. What is the source of the coherent wave-field generated by the perceiver? Could it be the body field itself? Or the body field as modulated by the nervous system? Again, this could be subject to empirical investigation.

One thing seems clear. Quantum coherent systems can bind and segment simultaneously and nonlocally by virtue of their factorizability (see above), which is how living processes are organized. Circulation, metabolism, muscular and nervous activities all go on simultaneously and independently, yet nevertheless cohering into a whole. A multitude of bound and segmented simultaneities are created in the act of experiencing, which define the *here* and *now*. These simultaneities are nonlocal and heterogeneous. They contain further simultaneities within and become entangled as they cascade through a quasi-continuum of space-times. The here and now is, therefore, not a flat instantaneity, nor "a travelling razor blade dividing past from future (c.f. Gibson)".[38] Instead, it is the grain of experiencing — a labyrinth of commuting and non-commuting simultaneities within simultaneities out of which hesitations we weave our futures.

Coherent Information Storage and Quality of Perception

The conscious being initiates and gates experience and determines the content of the experience, so it is that two people can experience the same music simultaneously, one with the highest rapture, and the other, the utmost indifference. According to Gibson,[38] objects in the environment provide "affordances", which are selected by the subject in the act of perceiving. The information goes into "resonant circuits in the brain" from which "effectivities" flow, ultimately as "object-oriented actions" complementary to the affordances. Thus, the quality of each perception is coloured by all that has gone before, as Freeman states.[2] The brain does more than coordinate subsystems of the body — it forms images (or, at any rate, takes part in forming images), *and* stores them for future reference.

The stored information, or memory system, is generally found to be distributed over the entire brain, perhaps in the form of "reverberations" or circuits that "mediate" responses to stimuli and initiate actions. Thus, in contrast to the rapidity with which simultaneity can be established in different parts of the brain, half a second is required for the subject's brain to become "aware" (as evidenced by its electrical activities) that something has happened, although the subject automatically back-dates it to make up for the delay (see earlier). Freeman's view is that the delay is the time needed for (p. 83)[2] "propagation of a global state transition through a forebrain to update the state of the intentional structure by learning". In other words, that is the time taken to reorganize the whole system.

There does appear to be a circulating activity in a network consisting of different brain structures, and transmitted between various regions in a highly organized fashion. These circulating activities, modified by sensory inputs, are thought to be responsible for "short-term" memory, which becomes long-term memory by causing structural chemical changes.[2] However, it would be a mistake to suppose that memory is thereby "fixed" once and for all. Molecules in the brain, as in all of the rest of the body, are subject to metabolic turnover. So, it is more realistic to suppose that so-called long-term memory is subject to the same dynamic modification and reconstitution as short-term memory, and that short-term and long term are simply the ends of a continuum that extends from the most microscopic "here and now" to the individual's entire life-span and beyond. It is this dynamic information store, distributed over a whole gamut of timescales that underlies the distinctive quality of each experience, for the experiencing being is constantly being renewed and updated.

Thus, Freeman and his co-workers found that rabbits trained to distinguish odours have patterns of brain activities for each odour that are never twice the same in any one session for any animal.[2] And each animal has its own repertoire of patterns that evolve in successive trials. Far from being disconsolate, Freeman gained new insights from the experiments into the unity of intentionality in that every perception is influenced by all that has gone before. Constant stimulus-response relationships are not mediated by correspondingly constant cause-and-effect associations of brain activities. In contrast to the microscopic patterns carried by a few sensory neurons that differ consistently with each smell, the macroscopic spatial patterns in the olfactory bulb are distributed over the entire bulb for every odour, and (p. 59)[2] "did not relate to the stimulus directly but instead to the meaning of the stimulus". So when reward was switched between two odours, the patterns of activities changed for both odours, as also did the control patterns without odour in background air. The patterns changed whenever a new odour was added to the repertoire. There is no mosaic of compartments in the olfactory memory in the bulb. It is a seamless information store.

Furthermore, all the evidence points to a dynamic maintenance and recreation of memory over all timescales.[40] There is a transfer of information to ever longer and longer timescales exactly in the way that energy gets transferred in cascades of processes of increasingly larger space-times.[7] In the transfer of memory, different memories also become entangled in the reconstitution of the whole, thus continually redefining a unique here and now. One never ceases to write and overwrite one's biography — it is a tissue of reconstructions. There is no sharp distinction between the here and now and what has gone before. "Past" simultaneities overarch the "now" and extend beyond while further simultaneities are seeded within the "now".

Strong evidence that memory storage is delocalized at least over the whole brain is the finding that it is able to survive large brain lesions. This has already led a number of people to suggest that memory storage is holographic in the same way that perception is holographic, so that the whole can be reconstructed from even a small part, albeit with less detail. As neurobiologist Philip Landfield at University of California Irvine points out, holography enables complex information to be retrieved simply by generating a regular wave without any informational content.[41] Of course, the same regular or coherent wave is instrumental in creating and coding the complex information in the first place. Likely candidates for coherent reference waves are considered to include alpha waves and waves generated by the hippocampus. Landfield has proposed a model in which memory is encoded by coherent waves from the hippocampus interacting with sensory inputs and undergoing a phase change. These modulated "object" waves are then recombined with the reference waves to form an interference pattern in the pyramidal cells of the hippocampus, from which a "reconstructed wavefront" is projected to other parts of the brain to generate the circulating patterns of activity that constitute "short-term" memory. This short-term memory is thought to be consolidated during sleep, whereas the alpha rhythms occurring during states of relaxation are believed to play a special role in memory retrieval.

Holographic memory storage is orders of magnitude more efficient than any model that makes use of "representations" because, as Marcer points out, holographic memory employs actual physical simulations of processes and does not require lengthy sequences of *arbitrary* coding and decoding of isolated bits.[34,35] He suggests that the brain stores experienced holographic spatiotemporal patterns and compares stored with new patterns directly, recognition and learning being reinforced in "adaptive resonance", thus also making for much faster processing. As mentioned before, the liquid crystalline continuum supporting the body field may also take part in memory storage, although this possibility has never been seriously considered. Hungarian philosopher and systems theorist Ervin Laszlo goes even further to suggest that much of memory may be stored in an ambient, collective holographic memory field delocalized from the individual; and that memories are only accessed by the brain from the ambient field.[42]

Quantum Coherence and the Macroscopic Wave Function of the Conscious Being

If quantum coherence is characteristic of the organism as a conscious being as I have argued here, then the conscious being will possess something like a macroscopic wave function. This wave function is ever evolving, entangling its environment, transforming and creating itself anew.[7] I agree with the ontological interpretation of quantum theory[43] by British quantum physicists David Bohm (1917–1992) and Basil Hiley at Birbeck College, London, to the extent that there is *no* collapse of the wave function. The Bohm and Hiley wave function, with quantum potential playing the role of *active information* to guide the trajectories of particles, simply changes after interaction to become a new one. The possibility remains that there is no resolution of the wave functions of the quantum objects after interacting. So one may remain entangled and indeed, delocalized over past experiences (i.e., in Laszlo's ambient field.[42] Some interactions may have timescales that are extremely long, so that the wave function of interacting parties may take a correspondingly long time to become resolved, and large-scale nonlocal connectivity may be maintained.

What would our wave function look like? Perhaps it is an intricate supramolecular orbital of multidimensional standing waves of complex quantum amplitudes. It would be rather like a beautiful, exotic flower, flickering in and out of many dimensions simultaneously. That would constitute our quantum holographic self, created from the entanglements of past experiences, the memory of all we have suffered and celebrated, the totality of our anxieties and fears, our hopes and dreams.

References and Notes

1. Andrew, A.M. "The Decade of the Brain: Some Comments". *Kybernetes* 24 (1995): 54–57.
2. Freeman, W.J. *Societies of Brains: A Study in the Neuroscience of Love and Hate*. Lawrence Erlbaum Associates, Hove, 1995.
3. Iaonnides, A. A. "Estimates of Brain Activity Using Magnetic Field Tomography and Large Scale Communication within the Brain". In *Bioelectrodynamics and Biocommunication* (M.W. Ho, F.A. Popp and U. Warnke, eds.), pp. 319–354, World Scientific, Singapore, 1994.
4. See Freeman, W.J. and Barrie, J.M. "Chaotic Oscillations and the Genesis of Meaning in Cerebral Cortex". In *Temporal Coding in the Brain* (G. Bizsaki, ed.), pp. 13–38, Springer-Verlag, Berlin, 1994.
5. Needham, J. *Order and Life*. MIT Press, Cambridge, Massachusetts, 1936.
6. Schrödinger, E. *What Is Life?* Cambridge University Press, Cambridge, 1944.
7. Ho, M.W. *The Rainbow and the Worm: The Physics of Organisms*. World Scientific, Singapore, 1993.
8. See Ho, M.W. "What Is (Schrödinger's) Negentropy?" *Modern Trends in BioThermoKinetics* 3 (1994): 50–61.
9. Ho, M.W. "Bioenergetics and the Coherence of Organisms". *Neural Network World* 5 (1995): 733–750.

10. Ho, M.W. "The Biology of Free Will". *J Conscious Stud* 3 (1996a): 231–244.
11. Ho, M.W. "Bioenergetics and Biocommunication". In *IPCAT 95 Proceedings* (R. Cuthbertson, M. Holcombe and R. Paton, eds.), pp. 251–260, World Scientific, Singapore, 1996b.
12. Ho, M.W and Lawrence, M. "Interference Colour Vital Imaging: A Novel Noninvasive Technique". *Microsc Anal* (1993): 26.
13. Ho, M.W. and Saunders, P.T. "Liquid Crystalline Mesophases in Living Organisms". In *Bioelectromagnetism and Biocommunication* (M.W. Ho, F.A. Popp, and U. Warnke, eds.), pp. 213–228, World Scientific, Singapore, 1994.
14. Newton, R., Haffegee, J. and Ho, M.W. "Colour-Contrast in Polarized Light Microscopy of Weakly Birefringent Biological Specimens". *J. Microsc* 180 (1995): 127–130.
15. Ho, M.W., Haffegee, J., Newton, R.H., Ross, S., Zhou, Y.M. and Bolton, J. "Organisms As Polyphasic Liquid Crystals". *Bioelectrochem Bioenerg* 41 (1996): 81–91.
16. Glauber, R.J. "Coherence and Quantum Detection". In *Quantum Optics* (R.J. Glauber, ed.), pp. 15–60, Academic Press, London, 1969.
17. Preparata, G. "What is Quantum Physics? Back to the QFT of Planck, Einstein and Nernst. Lecture given at IX Winter School on Hadron Physics, Folgaria, Italy, 1994.
18. Goldin, E. *Waves and Photons: An Introduction to Quantum Optics*. John Wiley & Sons, New York, 1982.
19. See Ho, M.W. "Towards a Theory of the Organism". *Integr Physiol Behav Sci* 32 (1997): 343–363.
20. Breithaupt, H. "Biological Rhythms and Communications". In *Electromagnetic Bioinformation,* 2nd edition (F.A. Popp, U. Warnke, H.L. Konig and W. Peschka, eds.), pp. 18–41, Urban & Schwarzenberg, Berlin, 1989.
21. Strogatz, S.H. and Mirollo, R.E. "Collective Synchronisation in Lattices of Non-Linear Oscillators with Randomness". *J Phys A Math Gen* 21 (1988): L699–L705.
22. Ho, M.W., Xu, X., Ross, S. and Saunders, P.T. "Light Emission and Re-Scattering in Synchronously Developing Populations of Early

Embryos: Evidence for Coherence of the Embryonic Field and Long Range Cooperativity". In *Advances in Biophotons Research* (F.A. Popp, K.H. Li and Q. Gu, eds.), pp. 287–306, World Scientific, Singapore, 1992.
23. See Kelso, J.A.S. "Behavioral and Neural Pattern Generation: The Concept of Neurobehavioral Dynamical Systems". In *Cardiorespiratory and Motor Coordination* (H.P. Koepchen and T. Huopaniemi, eds.), pp. 224–234, Springer Verlag, Berlin, 1991.
24. Oschman, J. *Energy Medicine: The Scientific Basis.* Churchchill Livingstone, *New York*, 2000.
25. Ho, M.W. and Knight, D. "Liquid Crystalline Meridians". *Am J Chin Med* 26 (1998): 251–263.
26. Becker, R.O. *Cross Currents: The Promise of Electromedicine, the Perils of Electropollution.* Jeremy P. Tarcher, Inc., Los Angeles, 1990.
27. Mikhailov, A.S. and Ertl, G. "Nonequilibrium Structures in Condensed Systems". *Science* 272 (1996): 1596–1597.
28. Welch, G.R. and Berry, M.N. "Long-Range Energy Continua and the Coordiantion of Multienzyme Sequences *In Vivo*". In *Organized Multienzyme Systems* (G.R. Welch, ed.), Academic Press, New York, 1985.
29. See Welch, G.R., ed. *The Fluctuating Enzyme.* John Wiley & Sons, New York, 1986.
30. Bergson, H., translated by Pogson, F.L. *Time and Free Will: An Essay on the Immediate Data of Consciousness.* George Allen & Unwin, Ltd., New York, 1916.
31. Hardcastle, V.G. "Psychology's 'Binding Problem' and Possible Neurobiological Solutions". *J Conscious Stud* 1 (1994): 66–90.
32. Gray, C., Konig, P., Engel, A.K. and Singer, W. "Oscillatory Responses in Cat Visual Cortex Exhibit Inter-Columnar Synchronization Which Reflects Global Stimulus Properties". *Nature* 33 (1989): 334–337.
33. Singer, W. "Self-Organization of Cognitive Structures". In *The Principles of Design and Operation of the Brain* (J. Eccles and O. Creutzfeld, eds.), pp. 119–135, Springer, Berlin, 1990.

34. Hameroff, S. and Penrose, R. "Orchestrated Reduction of Quantum Coherence in Brain Microtubules: A Model of Consciousness". *Neural Network World* 5 (1995): 793–812.
35. Jibu, M., Hagan, S., Hameroff, S.R., Pribram, K.H. and Yasue, K. "Quantum Optical Coherence in Cytoskeletal Microtubules: Implications for Brain Function". *Biosystems* 32 (1994): 95–209.
36. Marcer, P.J. "Designing New Intelligent Machines: The Huygens' Machine". *CC-AI Journal* 9 1992: 373–394.
37. Marcer, P.J. "The Need to Define Consciousness: A Quantum Mechanical Model". Symposium (P.J. Marcer and A.M. Fedorec, eds.), pp. 23–15, University of Greenwich, Greenwich, 1995.
38. Gibson, J.J. *The Ecological Approach to Visual Perception*, MIT Press, Massachusetts, 1966.
39. Whitehead, A.N. *Science and the Modern World*. Penguin Books, Harmondsworth, 1925.
40. Verzeano, M. "The Activity of Neuronal Network in Memory Consolidation". In *Neurobiology of Sleep and Memory* (R.R. Drucker and J.L. McGaugh, eds.,), pp. 75–97, Academic Press, New York, 1977.
41. Landfield, P.W. "Synchronous EEG Rhythms: Their Nature and Their Possible Functions in Memory, Information Transmission and Behaviour". In *Molecular and Functional Neurobiology* (W.H. Gispen, ed.), pp. 390–424, Elsevier, Amsterdam, 1976.
42. Laszlo, E. *The Interconnected Universe*. World Scientific, Singapore, 1995.
43. Bohm, D. and Hiley, B.J. *The Undivided Universe*. Routledge, London, (1993).

11

Why Say "Consciousness"?

Why say "consciousness"?
In this ever-present knowing-
experiencing-growing of being alive
One cannot grab hold of this "consciousness" thing
This thing you reified
Wrenched from the tangled whole
with roots, tendrils, raw and torn
Sadly dangling, weeping sap till life is drained
Then wonder: does it exist, does one exist
Does the world exist?
Or ask of this withered thing,
"Can it be caused or causal?"
"Can it control or be controlled?"

This is where madness begins
where Western civilization sins.

Where is "consciousness" bred?
In the heart, or in the head?
In the neurons of the brain
where homunculus divides pleasure from pain
Or in some universal cosmic ether
Unsullied by gross bodily matter
So say the lotus-eaters, the body-snatchers
Both renouncing the real entanglement
Of experiencing-feeling-loving
For the thin veneer of "ep-ist-em-ology"
Or is it "phen-om-en-ology"?
To ask of this withered thing,

"Can it be causal or caused?"
"Is it controller or controlled?"

This is where madness lies
where Western civilization dies.

This "consciousness" stance
The composure of those who will not dance
When all nature dances
Who will not engage
in knowing-experiencing-transforming
Instead, you hold steadfast yourself,
too steadfast an "observer", jealously guard
your "subjectivity"
against the onslaught
of the "objective" mean
You "introspect" so circumspect-ly
disturbing not even a fly
While worlds swirl stormily by

And from your stately throne
in suspended animation,
you bid nature, "hold still!"
Still you try
to stem the tide
To make of nature the "observed".
And ask why
"Why are we here?"
What is the purpose of it all?
And where,
where is the unmoved mover?
The unknown knower
The unthought thought that is
"Consciousness"?

The dance of life revolves
evolves from galaxies to the tiniest atom all
Pirouetting, pulsating

as we turn and leap, each
to her own exquisite music
embodying, enacting the whole
Weaving the web of eternity
To no cosmic choreographer
nor grand design
Yet each infinitesimal gesture
within gesture cascading
freely dropping,
Drops like a miraculous prophesy
Of ever more exotic figures to be

22 November, 1994

12

Art and Science

12

Significant Form in Science and Art

The creation of significant form is the basis of knowledge, possibly for all living species; it holds the key to aesthetic experience in science and art, and depends on the inextricable entanglement of all beings in nature.

Feeling for the "Sublime"

Many years ago, I attended for the first time a performance of *The Magic Flute*, the last opera written by German composer Wolfgang Amadeus Mozart (1756–1791), a genius and child prodigy of the classical era. The electrifying moment came when the Queen of the Night launched into her aria.[1] I found myself bolt upright on the edge of the seat, and must have held my breath for the entire duration. My heart ached and tears welled up in my eyes. Her voice rang through me everywhere as though I had dematerialized into an exquisitely sensitive ethereal being that filled the auditorium. There was intense excitement, but also something supremely joyful and serene. No words can capture that charged moment but that I was in the presence of the "sublime".

I have experienced similar moments on very different occasions: in the theatre or cinema, while listening to music, reading, and once, during a lecture on mathematics that I barely understood at the time. These moments are never *passively* aroused, but involve an intensely *active* engagement; so they can also happen when I am performing a scientific experiment, writing, painting, sculpting, or simply thinking and dreaming while awake or asleep. Always, there is something familiar, recognizably the same, even though the onrush of feelings and imageries that fill the moment to overflowing never fails to colour each occasion uniquely.

One of my first encounters with the sublime was also perhaps the most significant, as it shaped the rest of my life. It happened when, as a young undergraduate, I came across a remark attributed to the Hungarian-born American biochemist Albert Szent-Györgyi (1893–1986) that life is "interposed between two energy levels of an electron."[2] I was so smitten with the poetry in the idea that I spent the next 30 years searching for it, wandering in and out of diverse fields; though it was the same poetry in an ever-changing guise that led me on.

The experience of the sublime is not exclusively provoked by "works of art" in the conventional sense, but also by "works of science". Volumes have already been written on aesthetics, and I am inclined to agree with what Derek Jarman put in the mouth of Ludwig Wittgenstein in his film about the Austrian philosopher: that there can be nothing better said on aesthetics as saying nothing.[3] So I am saying nothing on aesthetics. Instead, I want to explore that kernel of the sublime that resides in all those special occasions.

"Significant Form"

Clive Bell (1881–1964) was an English art critic, and one of the Bloomsbury literati around the English writer Virginia Woolf (1882–1941). Bell attempted to revive and revitalize what he saw as the dwindling creative spirit in Western art increasingly preoccupied with illusionism and the mechanical representation of natural forms. To that end, he stressed the universal, timeless aspects of art:

> "What quality is shared by all objects that provoke our aesthetic emotions? ... What quality is common to Sta Sophia and the windows at Chartres, Mexican sculpture, a Persian bowl, Chinese carpets, Giotto's frescoes at Padua and the masterpieces of

Poussin, Piero della Francesca, and Cézanne? Only one answer seems possible — significant form."[4]

In other words, all works of art produced anywhere at any time whatsoever, are capable of arousing our aesthetic emotion because they possess *significant form*.

But what *is* significant form? Bell referred to it as a "moving combination of lines and colours", a quality distinct from the surface appearance of the artwork itself. Hence, significant form is not the same as the beautiful form, say, of a man or woman, a flower or a butterfly. It is supposed to be a pure, abstract quality. The emotion it arouses is not ordinary emotion but aesthetic rapture, given only to a chosen few. The effete elitism implied in those remarks probably aroused more hostility against the idea than anything else. Although what Bell said contains a germ of truth, I feel he has quite misrepresented the case.

To me, the aesthetic experience is intuitive and primitive, and hence universal to all human beings. I would guess that animals, too, have aesthetic experiences. For many of us bird-song and whale-song do touch the sublime, and so why not for individuals of their own species or genera? The neo-Darwinists' tedious obsession with "explaining" everything — including why birds and whales sing — in terms of its "selective advantage" in the struggle for survival simply misses the point, and has produced nothing convincing (see "4 — Epigenetic and Developmental Dynamics: How Development Directs Evolution" in this book). More importantly, the significant form in aesthetic experience is by no means abstract. On the contrary, the more significant the form, the more *concrete* it has to be, as I shall show later. Aesthetic emotion can be developed to great depths, but it can also be suppressed and obliterated, particularly in the fragmented, formless society we now inhabit. Bell's invectives were directed, after all, against the philistines on the one hand and the academicians on the other, both equally lacking in artistic sensibility, but nevertheless dominating the art world then, as they do today.

I discovered Bell's idea just as I was becoming quite convinced through my own activities and experiences of the symmetry between science and art as ways of getting to know nature intimately. Science, like art, *creates* the significant form that gives rise to all aesthetic experiences.

A scientific theory is above all, a form or a pattern that connects seemingly unrelated or disparate phenomena, and therein lies its ability to arouse aesthetic feelings. It is surely the stuff of poetry that an apple falling to the earth in our garden should have reference to the motion of planets and stars. Equally so, is the realization that all living things, from the tiniest microbes to human beings and whales, are animated by the same infinitesimal quanta of sunlight captured one at a time to raise electrons from the ground state to the excited state; and that within the duration of the electron falling back to the ground state, the whole of biological creation is poised. This is in essence the process of photosynthesis in green plants which capture carbon and energy for practically all living things on earth (see "15 — Water the Means, Medium and Message of Life" in this book).

The significance of an authentic scientific theory thus depends on its richness of content and also in its "ring of truth": it is what *we feel to be consonant with our most intimate experience of nature.* Can we say the same about artistic form? Can we judge the significance of artistic form according to its richness of content and its consonance with our most intimate experience of reality? I suggest we can. I make no claims to scholarship, or to being anything like a *connoisseur*. Instead, I am literally an *amateur* who loves both science and art, and practises both to some extent. Inevitably, I shall be drawing mainly on my own experiences, and no one should take what I say to be a pronouncement on which particular works are significant or on how science and art ought to be experienced by everyone.

Form and Wholeness

What is form? "Form" is a web of interrelationships making a *whole*, more importantly, *apprehended* as a whole. A pure form is nothing if not concentrated interrelationship. The intuition of form is the prerequisite to knowledge; hence it is common to all ways of knowing, in science as in art. For a form to be significant requires something in addition. A significant form is never just the superficial form of any object or work of art as such, nor is it merely a certain abstract formal combination of lines and colours. It is a form that *signifies* some deep interrelationships in nature, *to which the apprehending being is connected*. Without this connection, there can be no significance in the content, and hence, no significant form. The significant form is a *conduit* to the nexus of interrelationships beneath the surface appearance of things. One is suddenly drawn into a catenated flux of associations, propagating and circulating endlessly in a subterranean sea of meaning. For a fleeting yet eternal moment, we lock onto the pulse of some timeless universal being.

"Form" is the irreducible coherence of part and whole. A random collection of bricks is construed as a work of art precisely because in its very formlessness, it challenges each of us, the "spectator", to participate and create a form, if not a *significant* form. We cannot help but see faces and castles in clouds, monsters in inkblots and exotic shapes in random dots. Form is so central to human perception that I am told it is extremely difficult to prove something random or formless.

The intuition of form and wholeness is the basis of perception, and all the more so, of artistic perception. It is by no means restricted to visual art. Mozart is reputed to have "seen" a symphony all at once, before he wrote it out in its entirety without a single mistake, much to the chagrin of the Italian court composer Antonio Salieri (1750–1825). Salieri's scourge in life was that he clearly recognized in Mozart a genius towering above his own.[5] Music can never be seen literally. Just as our eyes seek out and create spatial patterns, our ears assemble and weave parallel strings of sounds into temporal patterns that make a symphony. A significant temporal form emerges from an intimate communion between the artist and nature.

Music conjures a wealth of rhythmic patterns that are the fabric of natural processes. Day alternates with night as the moon waxes and wanes and the seasons follow one after the other in regular succession. The heavens are thick with the tangled paths of stars and galaxies encircling one another in an intricate cosmic dance. Our bodies, similarly, are replete with rhythms, from the infinitesimal vibrations of molecules to the thump-thumping of the beating heart, all pulsating in complex harmonies to the music of the earth and heavenly spheres.[6]

Rhythmic elements are very prominent in the "decorative art" of all indigenous cultures, which include textiles, rugs, pottery, basket-work, body-painting and motifs applied to habitations, clothing, shields, weapons, utensils, jewellery and practically every article of use. An embroidery from the Koryaks of Eastern Siberia reproduced by German-American anthrologist Franz Boas (1858–1942) looks at first glance, remarkably like a musical score for an ensemble of ten or more (Fig. 12.1).[7] None of the repeated patterns in "decorative art" realistically represent anything in nature, however. The artists have created a pattern, a significant form, capturing what they intuitively feel and perceive, and communicable through a certain indefinable, universal syntax of form.

Is that not so for science and scientist?

Mephistopheles in the epic poem *Faust* by the great German polymath Johann Wolfgang von Goethe (1749–1832) scoffs at the

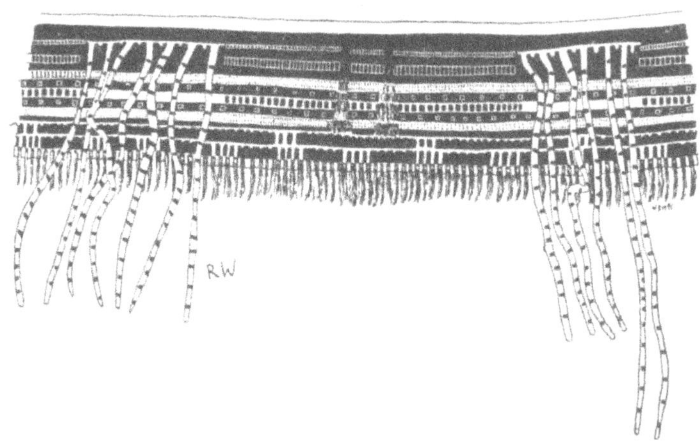

Figure 12.1 Embroidery from the Koryaks of Eastern Siberia.

scholars who try to understand a living organism by the detailed description of its parts:

"Dann hat er die Theile in seiner Hand
Fehlt leider! nur das gestige Band."[8]
(Then he has all the parts within his hand
Except sadly, the living bond.)

Goethe might well have been describing the project carried out to its logical conclusion by molecular biologists of the present day. Goethe himself, both poet and scientist, knew more than anyone else that the artist who recognizes wholes makes a better scientist than the analyst who breaks up the whole into parts:

> "In all ages even among scientific men, there can be discerned the urge to apprehend the living form as such, to grasp the connections of their external visible parts; to take them as intimations of inner activity, and so to master, to some degree, the whole in an intuition."[9]

Universal Wholeness and the Apprehension of Significant Form

The intuition of form, or wholeness, is the prerequisite to knowing, in science as in art. The British mathematician/philosopher Alfred North Whitehead (1861–1947) referred to the primitive act of an organism "getting hold" of the world "prehensive unification".[10] One must realize that Whitehead's organism refers to any and every entity in nature, from an elementary particle to what we would call an organism, and all the way up to much larger things such as Earth itself, or a galaxy. The word "prehensive" is used deliberately, to include the non-cognitive perception of organisms other than human beings, or perhaps also the intuitive perception of human beings.

Is there evidence that organisms other than human beings perceive patterns and forms? There certainly is. Bees appear to have the ability to create a map of their surroundings *and* to communicate it as such. Austrian ethologist Karl von Frisch (1886–1982) had discovered in the late 1940s that when a bee scout finds a source of food, she returns to the hive and performs an elaborate dance to tell her hive-mates where the food is located.[11] Actually, a song generated by the beating of her wings accompanies the dance, and both song and dance are necessary to convey the precise information. This is an astonishing feat; not only does the bee have a map of its world as a whole; it can transcribe this map into the significant form of a dance and communicate that to its hive-mates.

Mathematician Barbara Shipman at the University of Texas Arlington in the United States was introduced to the bees' dance as a child, and while doing her PhD in the 1990s, discovered that the bees' dance could be modelled as transformations in a six-dimensional space called a "flag manifold" that quantum physicists use to describe the behaviour of quarks.[12] The bees seem to have discovered flag algebra long before humans did; or else, as suggested by Shipman, they are directly sensitive to the quantum field of quarks. Bees and clever human beings, it seems, have a common grammar of form and transformation, for whatever underlying reason. Could it be that this grammar of form and transformation is universal?

Ants live in huge colonies and super-colonies with cities, gardens and many technologies. Biologists have found that ants, too, can tell each other complicated stories with something pretty close to a language.[13] An ant scout was made to go through a maze to find food. On returning to the nest, it huddled together for a while with a group of workers, touching antennae and mouthparts. The scout was then removed from the nest. The workers subsequently found their way to the food through the maze in a much shorter time than when the scout was removed before it had the chance to confer with them. So ants also have the ability to perceive patterns, make maps of their environment, *and* communicate them in some form of language. These amazing capabilities depend, as Whitehead rightly said, on perceiving the whole in the most primitive act of "prehensive unification".[10] Whitehead's organic vision is one of universal wholeness, in which everything is entangled with everything else through individual acts of prehensive unification (see *The Rainbow and the Worm: The Physics of Organisms*[14]).

The paradoxical conclusion to centuries of mechanical reductionism in Western science is to rediscover the organic reality that we are all entangled with and within nature, from the infinitesimal quantum of light to stars and galaxies. *This natural entangled state is the only possible ground for the creation and apprehension of significant form and hence of authentic knowledge.*

Significant Form is Deep and Dynamic

Significant form is deep and dynamic. It is not to be found in the surface appearance of things, but in their reference to realms of reality not immediately before us. A beautiful woman is not a significant form as such, but becomes so in the immortal lines of English poet Lord Byron (1788–1824):

She walks in beauty like the night
Of cloudless climes and starry skies;
And all that's best of dark and bright
Meet in her aspect and her eyes;[15]

The significance lies neither in the form of the woman nor in the night, but in the dynamic transference of sympathetic resonance between the two: the clear starlit night and her dark, shimmering, mysterious beauty, each reflecting and heightening the qualities of the other in our mind, intensifying their simultaneous presence.

In just the same way, a significant form in science is as deep as it is dynamic. The search for natural order in 17th century Europe[16] was nothing if not a quest for a deep, significant form. In biology, this began as the idea of "the unity of type" among organisms superficially different, yet sharing a common structure or "body plan". Not only is the organism perceived as a whole, a form in itself, but as a community of forms united by dynamic transformation.

The science of biological form, or morphology, is not about the study of *Gestalt*, or *fixed* form. A Gestalt is but an instantaneous snapshot of the organic *process* of transformation and development. Form, to Goethe, is "the intimation of inner process",[9] that displays itself fully only in the transformations of becoming. In a community of organisms, this dynamic form captures the convergence of resonance, affinities and sympathies, *and at the same time*, the creative divergence of individualities, multiplicities and diversities, rather like endless variations on a theme. In Goethe's view, living things in their totality strive to manifest an idea. They are nature's works of art, and so incidentally, they require an artist to understand and a poet to interpret them.

The Significant form and Poetic Imagination

The significance of a form lies in its ability always to transport us away from the here-and-now in a wide sweep of the imagination that returns only to be led away again and again. The moment expands and grows with each cycle around the ever-widening circuit of signification, and so one seems to dwell in the moment forever. *It is for this reason that significant forms are often figurative or non-representational.* A "realistic" work can inhibit those flights of the imagination by focussing attention ever back onto itself until one is overwhelmed with a sense of oppression.

Stage sets and productions in the theatre are most suggestive when they are spare and simple. One particularly memorable example is a 1990 production of *Travels with My Aunt* by British actor/director Giles Havergal, an adaptation of a novel with the same title by English novelist Graham Greene (1904–1991) in which three men, dressed in identical brown suits, take turns playing aunt and nephew as well as all the other characters, including a dog.[17] This so effectively underlines the irony and pathos in the humour that one begins to thoroughly identify with the everyman bank clerk, who, in the drab-brown dullness of his uneventful humdrum existence, nevertheless harbours a romantic fantasy of bohemian life in the retold (perhaps imagined) adventures of his anarchic, eccentric aunt. In a more "realistic" production, one's imagination cannot *participate* to the same degree, and hence partake of the significance of the occasion.

A significant form always invites participation, as it used to be in the days of William Shakespeare (1564–1616), English poet, playwright and actor. Theatre was not a spectacle — the stage sets being always minimal. Members of the audience were not spectators, but active participants in a timeless drama of the imagination. As Shakespeare wrote in the prologue to Henry V:

Think, when we talk of horse,
that you see them
Printing their proud hoofs
i' the receiving earth.
For 'tis your thoughts that
now must deck our kings,
Carry them here and there,
jumping o'er times;
Turning the accomplishments
of many years
into an hour glass.[18]

One can see the parallels in the development of the science of biological form. For Goethe, the unity of the biological world is a manifestation of some deep natural order. The attraction of a *seemingly* abstract, transcendental primeval form or archetype can be understood in the same way, for it invites our imagination to actively participate. I stress "seemingly" because I shall presently show why this position is in reality, the most *concrete*.

Form and Transformation

Darwin and practically all post-Darwinian systematists (those who study classification of living things) regard the unity of type as implying nothing else than the community of descent. The significant form loses significance as its content collapses into one dimension. It comes to signify only one thing: heredity, connection through the bloodline. There is no deep, transformational order encompassing the biological world, in such a way that a multitude of apparently disparate forms can be made simultaneously present to our mind. That is why I often experience an unbearable sense of oppression whenever I come up against a neo-Darwinist who sees biological forms as nothing more than an imperfect record of evolutionary history. Indeed, most neo-Darwinists interested in evolutionary history have given up studying biological forms to concentrate on comparing the DNA of different species instead.

As I have demonstrated in a series of papers,[19–23] form and transformation are independent of heredity. The dynamics generating form in development naturally give rise to a transformation set depicting how different forms are related by transformation, and therefore, also providing a *natural system* for classifying the forms. Swedish botanist Carl Linnaeus (1707–1778), the father of taxonomy, was searching for such a natural system. Linnaeus realized that the system he invented — which biologists have inherited to the present day — was a stopgap, and not the natural system intended.[22]

A range of body patterns in the first instar fruit fly larva was obtained by exposing the embryos briefly at different times during early development to an atmosphere saturated with ether.[22] From this, a tree of transformation could be produced (see "4 — Epigenetics and Developmental Dynamics: How Development Directs Evolution" in this book, for the transformation trees of fruit fly larva body pattern and for leaf arrangement around the stem in plants mentioned below.) The main sequence, going up the trunk of the tree, is the normal transformational pathway, which progressively divides up the body into domains, ending up with 16 body segments of the normal larva. This transformational tree reveals how different forms are related to one another. It gives the logic, or grammar of transformation, showing how superficially similar forms are quite far apart on the tree, while forms that look most different are neighbours. Fig. 12.2 depicts the model of successive bifurcation and the embryos arrested at different stages in the main sequence.

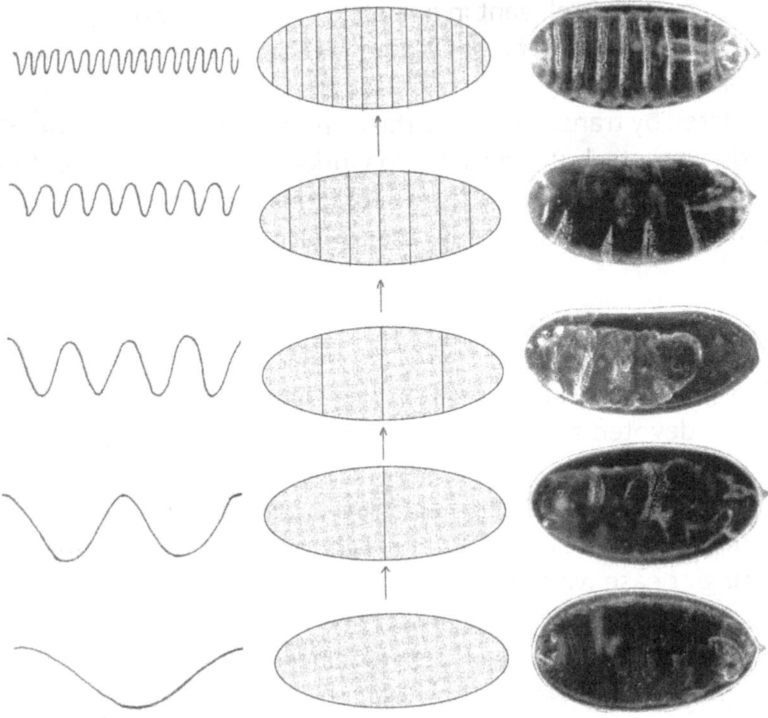

Figure 12.2 Model of successive bifurcation and actual embryos arrested in the main sequence.

There are 676 possible forms, according to the rule or grammar of transformation that gives rise to the 16 segments. If all the body segments were free to vary independently, the number of possible forms would have been 2^{16}, or more than 60 000.

A transformation tree for all possible ways leaves are arranged around the stem in plants is also produced[23] based on the dynamics that generate the patterns discovered by French mathematical physicists Stephane Douady and Yves Couder at Laboratoire de Physique Statistics in Paris.[24] Leave arrangement, *phyllotaxis,* has been a long-standing problem in biology. For decades, numerous neo-Darwinian "just-so stories" have been invented to account for different leaf arrangements in terms of "selective advantage", and all were proven irrelevant in one stroke. The power of dynamics — the syntax of form — is that it predicts the set of possible transformations, *excluding all others*. It also tells us how the possible forms are related by transformation. The transformational trees are scientific documents, but they are also works of art giving access to the natural process — the *tao* in ancient Chinese philosophy or natural grammar — that connects apparently disparate forms.

Participation in Significance

In the indigenous Taoist tradition[25] of Chinese art and poetry, great effort is devoted to *cultivating* spontaneity.[26] Spontaneity has the quality of free flow, of being both innocent, the Chinese for which is *tiān zhēn*, heaven-true, and natural, the Chinese for which are both *tiān rán*, heaven-being, and *zi rán*, self-being, with the connotation of being at ease with heaven and with oneself. It is, of course, also a state of freedom: *zi yóu*, self-sourced.

It is significant that "self", *zi*, in Chinese does not mean the isolated individual; rather, it has the sense of a being located by its specific relationships to the cosmos. The self, as it were, is held and supported by a myriad of specific entanglements. Thus, whereas the predominant trend in Western culture has been to sever the connections between self and nature, and to fragment the self into a pure ("objective") intellect divorced from all bodily feelings, indigenous Chinese culture, as indeed, indigenous cultures all over the world, simply regard the unity of nature and the integrity of self as a matter of immediate experience that needs no special pleading.[27] Consequently, any person, or "self", is empowered to participate in nature's process.

Furthermore, acting spontaneously and freely is also acting in accordance with the cosmos. This may be compared to the quantum coherent state that maximizes both local freedom and global cohesion.[14] In order to attain true spontaneity, it is necessary to cultivate a heightened awareness of one's entanglement with the whole.

Traditional Chinese artists spend a long time meditating and attending to the object, which may be a landscape, or flowers, or some other living beings (it is also highly significant that there is no category of "still life" or "*nature morte*" in Chinese painting, for everything is alive), and will pick up the brush only when the moment is ripe: when the will of all nature, centred at that moment on the artist, becomes concentrated in one unbroken gesture. The work of art is a unity, formed "in one breath" in a single duration enfolding a multitude of durations, when artist and nature are mutually coherent and transparent.

The same sense of participation in the significance of the occasion is responsible for the extraordinary power of so-called primitive art to move us. In Palaeolithic cave paintings and petroglyphs (Fig. 12.3), neolithic Chinese jade sculptures (Fig. 12.4) and African masks (Fig. 12.5), we perceive the archetype of a multitude of significant forms and transformations rooted in the cultural histories of peoples living fully *within* nature. Their works of art are hymns to the creativity of nature herself.

Participating in the significance of the occasion is a *concrete* act, both for the artist and the amateur. The most significant form is

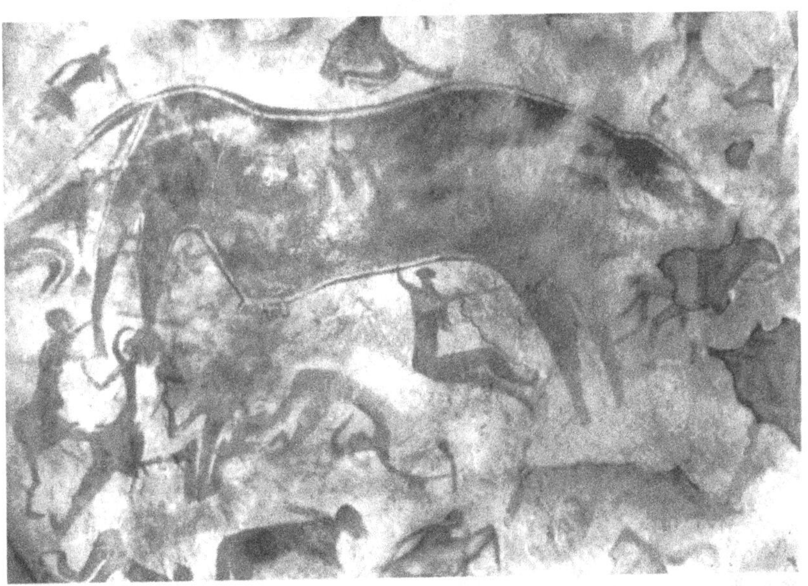

Figure 12.3 Palaeolithic rock painting courtesy of Prague Museum.

Figure 12.4 Neolithic Chinese jade phoenix.

Figure 12.5 Mask, Ivory Coast.

hence also the most concrete because it signifies ultimately all of nature, all of reality by the dynamic transference of signification. This recalls what British philosopher and poet Owen Barfield (1898–1997) said of language.[28] In the beginning, the meanings of words were concrete, because they were the sign to things and the invisible, inextricable links between them. Later on, meaning became *abstract* and subject to definitions, denuded of all associations and feelings. So language suffers a loss of significance. Words become mere conventional symbols, *representing* things and ideas we no longer *feel.*

I touched the sublime the first time I heard French mathematician René Thom's lecture on catastrophe theory and morphogenesis (form generation) more than 30 years ago.[29] Here was a theory that concretely signified to me all forms in nature, offering a tantalizing glimpse of *the* universal generative grammar of form, the *tao* of nature beyond the archetype whereby the multiplicity of things may converge and diverge, transmute and commute in weird and wonderful ways. Mathematics can indeed be a deep and significant form encapsulating the dynamic transference between forms. It is by no means abstract or Platonic, but can be the most concrete and complete apprehension of nature's unity.

References and Notes

1. The Queen of the Night aria, "*Der Hölle Rache Koch in menem Herzen*" ("Hell's vengeance boils in my heart") is actually the second aria sung by the Queen of the Night. See "*Der Hölle Rache Koch in menem Herzen*", Wikipedia, accessed 23 July 2015, https://en.wikipedia.org/wiki/Der_H%C3%B6lle_Rache_kocht_in_meinem_Herzen.
2. Although I am sure that Szent-Györgyi actually made the remark, I have been unable to track it down. It was reported by Alan Marshall in one of his lectures that I attended at Hong Kong University.
3. This was said in the film *Wittgenstein* made in 1993 by English director Derek Jarman (1942–1994), https://www.youtube.com/watch?v=6WzqyO-wIMI&list=PLFCA7FB1ED3C1F5C0.

4. Bell, C. *Art*, p. 3. Frederick A. Stokes Company, New York, 1914, http://rci.rutgers.edu/~tripmcc/phil/poa/bell-art.pdf.
5. This was admirably portrayed in the play *Amadeus* by English playwright Peter Shaffer, first performed in 1979, and turned into a film in 1984. See "Amadeus", Wikipedia, accessed 24 September 2015, https://en.wikipedia.org/wiki/Amadeus.
6. See Ho, M.W. "Quantum Jazz: The Meaning of Life, the Universe and Everything". *Sci Soc* 32 (2006): 11–14.
7. Reproduced from Boas, F. *Primitive Art*, p. 43. Dover Publications, Toronto, 1927, 1955.
8. Goethe, J.W., 1807. Cited and translated by E.S. Russell. In *Form and Function*, p. 50, John Murray, London, 1916; retranslated by the author.
9. Goethe, J.W., 1807. Cited and translated by R.H. Brady, "Form and Cause in Goethe's Morphology", p. 273. In *Goethe and the Sciences: A Re-Appraisal* (F. Amrine, F.T. Zucker and H. Wheeler, eds.), pp. 257–300, Boston Studies in the Philosophy of Science, Volume 97, D. Reidel Publishing Company, Dordrecht, 1987.
10. Whitehead, A.N. *Science and the Modern World*. Penguin Books, Harmondsworth, 1925.
11. "Bee Learning and Communication", Wikipedia, accessed 22 February 2008, http://en.wikipedia.org/wiki/Bee_learning_and_communication.
12. Frank, A. (1997). "Quantum Honey Bees". *Discover*, pp. 80–87, November 1997.
13. Michie, D. "Look Who's Talking". *The Independent on Sunday* magazine, pp. 7–10, 15 November, 1998.
14. Ho, M.W. *The Rainbow and the Worm: The Physics of Organisms*, 3rd edition. World Scientific, Singapore and London, 2008.
15. Lord Byron, 1814. "She Walks in Beauty", http://www.poetryfoundation.org/poem/173100.
16. See Russell, E.S. *Form and Function*. John Murray, London, 1916.
17. There have been several different productions of the play in which different numbers of actors were used, from two to four. See "Travels with My Aunt", Wikipedia, accessed 24 May 2015, https://en.wikipedia.org/wiki/Travels_with_My_Aunt_(play).

18. The complete text of the prologue is an apology for the "unworthy scaffold" serving as the stage for the great drama. See Shakespeare-online here: http://www.shakespeare-online.com/plays/henryv_1_1.html.
19. Ho, M.W. "Where Does Biological Form Come From"? *Revista di Biologia* 77 (1984): 147–179.
20. Ho, M.W. "How Rational Can Rational Morphology Be? A Post-Darwinian Rational Taxonomy Based on a Structuralism of Process". *Rivista di Biologia*, 81 (1988): 11–56.
21. Ho, M.W. "An Exercise in Rational Taxonomy". *J Theor Biol* 147 (1990): 43–57.
22. Ho, M.W. "Development, Rational Taxonomy and Systematics". *Rivista di Biologia* 85 (1992): 293–311.
23. Ho, M.W. and Saunders, P.T. "Rational Taxonomy and the Natural System: Segmentation and Phyllotaxis". In *Models in Phylogeny Reconstruction* (R.W. Scotland, D.J. Siebert and D.M. Williams, eds.), pp. 113–124, The Systematics Association Special Volume No. 52, Oxford Science, Oxford, 1994.
24. Douady, S. and Couder, Y. "Phyllotaxis As a Physical Self-Organized Growth Process". *Phys Rev Lett* 68 (1992): 2098–2101.
25. See Addiss, S. transator. *Tao Te Ching*. Hackett Publishing Company, Indianapolis/Cambridge, 1993.
26. Chang, C.Y. *Creativity and Taoism: A Study of Chinese Philosophy, Art and Poetry*. Singing Dragon, London, 2011.
27. See Ho, M.W. "Toward an Indigenous Western Science: Causality in a Universe of Coherent Space-Time Structures". In *New Metaphysical Foundations of Modern Science* (W. Harman and J. Clark, eds.), pp. 179–213, Institute of Noetic Sciences, Sausalito, 1994.
28. Barfield, O. *Poetic Diction*. Faber and Faber, London, 1951.
29. See Thom, R. *Structural Stability and Morphogenesis* (D.H. Fowler, trans.), W.A. Benjamin, Inc., Reading, Massachusetts, 1976.

13

Why Beauty is Truth and Truth Beauty

Recovering beauty and truth from corporate obfuscation is the most urgent task in our bid to save people and planet.

Is Beauty Truth in Science?

Scientists, especially the greatest scientists, are motivated by the beauty of the natural order of things. So intensely felt is the love for the beauty of a scientific theory that some scientists are unconcerned as to whether the theory happens to be true. Fortunately, really beautiful theories tend to be true, in the sense that their predictions can be tested and confirmed empirically. That's what Indian-born American astrophysicist Subrahmanyan Chandrasekhar (1910–1995), recipient of the 1983 Nobel Prize for his work on the evolution of stars, argued in his book *Truth and Beauty: Aesthetics and Motivation in Science*[1] published in 1987.

It is important to distinguish between mathematics and science here, even though they are closely linked. Mathematics is *always* true in the sense that the deductions follow from the axioms, and the theorems have been proven (although my mathematician husband Peter Saunders tells me that some proofs are controversial and not universally accepted as such). The big difference is that unlike science, it need not apply to nature. Indeed, it could be misapplied, resulting in scientific theories full of *ad hoc* elements that are neither true nor beautiful. A case in point is Big Bang cosmology based on the general relativity theory of Albert Einstein (1879–1955), which has come to depend on hypothetical entities such as black holes, dark matter and dark energy largely because it has failed to explain the deluge of empirical astronomical observations from powerful telescopes and dedicated space missions.[2] So although Einstein's theory of general relativity is beautiful, the Big Bang theory is not.

In his book, *Why Beauty Is Truth: A History of Symmetry*,[3] published in 2007, mathematician Ian Stewart at Warwick University extols the beauty of mathematical symmetry. But does it give us beautiful science?[4] Mathematical symmetry has spawned a series of string theories and superstring theories, 10^{500} to be precise, to explain ultimate physical reality.[5,6]

Many scientists are unimpressed, some declaring string theory outright ugly[7] because it is strongly reminiscent of the epicycles that described the paths of the planets around the earth in the earth-centred theory of the solar system before it was replaced by the much simpler, more elegant, and hence genuinely beautiful, heliocentric theory.

Beauty is pivotal to the scientist's judgment as to whether a theory is true. More than that, the quest for beauty is central to the life of the scientist. Henri Poincare (1854–1912), French mathematician-physicist described as the Last Universalist excelling in all fields of mathematics in his time, once wrote that the scientist does not study nature because it is useful to do so, but because he takes pleasure in it; and he takes pleasure in it because it is beautiful. "If nature were not beautiful, it would not be worth knowing and *life would not be worth living* [emphasis added]."[8]

I add emphasis to "life would not be worth living" because that's exactly how I feel. The greatest gift a human being can have — and everyone has it — is the capacity to be inspired by beauty; it is the fount, if not the *raison d'etre* of all creation. To be inspiring, one must have the capacity to be inspired.

Poincare went on to refer to ultimate beauty as[8] "the harmonious order of its parts [to the whole]; and the scientist delights in both the "vastness" and the "prodigious smallness" of things — domains that transcend everyday experience — thereby inviting us to reach for the deeper mysteries of nature.

Transcendence of the mundane is a hallmark of beauty, as it is of science and art. And I hasten to say, the beauty I am talking about has nothing to do with the superficial appearance of things. That's why one should be wary of educationalists who insist on reducing science and mathematics to prosaic everyday experience. For the same reason, one should hold onto that dumb inspiration, the sublime beauty that leaves one lost for words. You will find yourself returning to it again and again, until perhaps a great scientific theory or work of art drops like a ripe fruit from the tree of creativity that grows out of the fertile ground of the imagination.

I wholeheartedly agree with JWN Sullivan (1886–1937), London-born journalist and biographer of Newton and Beethoven among others, who regarded that the measure of success of a scientific theory is a measure of its aesthetic value, and[9] "The measure in which science falls short of art is the measure in which it is incomplete as science." Implicit in what Sullivan said is that art is a measure of beauty.

In response to Sullivan, artist and art critic Roger Fry (1866–1934) observed that there is no reason why a beautiful scientific theory has to "agree with facts".[10] Like Poincare, who posited "pleasure" in the beauty of nature as the motivation for science, Fry laid great store by "emotional pleasure" in the pursuit of art. But is it the same "pleasure" in both cases? Fry pointed out that in art, the emotional pleasure comes from "the recognition of relations" that is "immediate and sensational", and "curiously akin to those cases of mathematical geniuses who have immediate intuition of relations which it is beyond their powers to prove". In that sense, art is more akin to mathematics than science.

A famous case was the Indian mathematical genius Srinivasa Ramanujan (1887–1920), who left a large number of notebooks that recorded several hundred formulae and identities; many were proven decades after his death by methods that Ramanujan could not have known.

British mathematician G.N. Watson (1886–1965) spent several years proving Ramanujan's identities, and vividly described how coming across those identities gave him "a thrill" indistinguishable from that which he felt on seeing Michelangelo's sculptures "Day", "Night", "Evening" and "Dawn" over the Medici tombs in Florence, Italy.[11]

Ramanujan inspired the recent multi-award winning play *A Disappearing Number*[12] by Théâtre de Complicité and British playwright Simon McBurney; and it *is* the most beautiful play I have seen in ten years.

But it is all the more thrilling, I think, when nature appears to conform to an elegant mathematical theory. German theoretical physicist Werner Heisenberg recalled his extreme elation on discovering the laws of quantum mechanics:[13] "I had the feeling that, through the surface of atomic phenomena, I was looking at a strangely beautiful interior, and felt almost giddy at the thought that I now had to probe this wealth of mathematical structure nature had so generously spread out before me."

Is Beauty and Truth Relevant to Art?

Heisenberg was anticipated by English romantic poet John Keats (1795–1821) in the enigmatic last lines of his poem "Ode on a Grecian Urn", which have been debated by generations of poets and critics since:[14]

Beauty is truth,
truth beauty — that is all
Ye know on earth,
and all ye need to know.

A commentator tells us that[7] the American poet T.S. Eliot said the lines were "meaningless" and "a serious blemish on a beautiful poem". And famously aggressive American critic John Simon opened a movie review with "one of the greatest problems of art — perhaps the greatest — is that truth is not beauty, beauty not truth. Nor is it all we need to know."

Are artists motivated by the quest for beauty and truth? What would "truth" mean in art?

Or is the quest for beauty and truth in both science and art no longer relevant in the present day, having been overtaken by the profit imperative? Arthur Danto, Emeritus Professor of Philosophy at Columbia University and art critic remarks:[15] "A century ago, beauty was almost unanimously considered the supreme purpose of art and even synonymous with artistic excellence. Yet today beauty has come to be viewed as an aesthetic crime. Artists are now chastised by critics if their works seem to aim at beauty."

But the pendulum is swinging back. Since the early 1990s there has been a rising chorus to bring beauty back to art,[16–18] if not to science. Danto, for example, said that the modernists were right to exclude beauty from art, but also that beauty is essential to human life, and need not always be excluded from art.[15]

Somewhat surprisingly, Danto never defined beauty in his book, but conflates aesthetic beauty, say, with the superficial appearance of a woman "beautified" with cosmetics. Again, I have to object that the aesthetic beauty of science and art has nothing to do with the surface appearance of things. No one, I think, would fail to understand what I meant when I said *A Disappearing Number* is a beautiful play, but there was nothing overtly, or cosmetically beautiful about the play. The beauty I have in mind is a transcendent quality more akin to the sublime (see later).

If beauty (and truth) is essential for human life, then beauty and truth are central to art and science, and recovering them is the most urgent task facing humanity as corporate manipulation of truth and beauty threatens the survival of people and planet. That is the project we have taken on at the Institute of Science in Society.

Recovering Beauty in its Organic Form

What does it mean to recover beauty and truth in science and art? Is there an agreed, universal concept of beauty? I cannot think of a universal *concept* of beauty, although I believe the *sense* of beauty is universal.

Of course, there is no end of apologists who tell us that the concept of beauty changes through the ages, and what was thought ugly when first perceived, such as Andy Warhol's *Brillo Box*, or Marcel Duchamp's *Fountain*, a urinal, is now revered as high art. And there are those who claim that beauty is culture-bound, and what is beautiful in one culture is abomination in another.

Yet, beneath it all, we know that beauty is timeless and universal. Witness that collective sharp intake of breath when successive prehistoric cave paintings were unveiled to the world, the most recent spectacle being in the Chauvet cave of Ardeche, France (Fig. 13.1) dating from 30 000 BC.[19,20] Or marvel at the astonishing creative transgressions of cultural/ethnic/genre boundaries in contemporary music and art, albeit mostly away from the mainstream.[21]

The year 2011 happened to be the 12th anniversary of the Institute of Science in Society, co-founded by me and Peter Saunders, with a mission to reclaim science for the public good through providing accessible and critical scientific information to the public, insisting

Figure 13.1 A painting of horses from the Chauvet cave of Ardeche.

on sustainability and accountability in science, and especially, promoting holistic organic science in place of the mechanistic. And we celebrated it in style with an art/science event.[22,23] Holistic science is ultimately a way of knowing and understanding that engages *all* our faculties, both rational and aesthetic, and hence recognizes no separation between science and art. So we are really recovering the values of the pre-modernist romantic era, with one important difference. We need to transcend traditional concepts of harmony and elegance that are static and mechanical, hence quite inadequate to encompass the protean, shimmering splendor of nature's organic beauty.

In science, the static harmony of the "golden mean" has blossomed into the mathematics of chaos and strange attractors, and fractal geometry, much closer to how nature expresses herself, though not exactly; and that is important in real science, as opposed to mathematics.

Mathematics and science have caught up to the capricious volatility of natural processes and the endless diversity of natural forms. Fractal mathematics reveals that the variability of the healthy heartbeat is a sign that the heart is in coherent intercommunication with the rest of the body, and hence reflects the entire range of the body's biological rhythms[24,25] with a characteristic self-similar mathematical structure over a range of time scales. The dynamic structure is absent from the unhealthy and superficially much more regular heartbeat.

Even more fundamentally, fractal processes and structures are ubiquitous in organisms, and this accounts for the allometric scaling of physiological processes with body size that has long puzzled biologists (Fig. 13.2). This new discovery deserves to be called perhaps "biology's theory of everything".[26]

Nevertheless, nature does not conform to ideal mathematics, which succeeds only in approximating it.[27] The contingent, unpredictable, unrepeatable freedom of the moment is what makes life and art. In my essay "The Biology of Free Will"[28] I show how, in liberating itself from mechanical determinism and mechanistic control, the organism becomes a sentient, coherent being that is free,

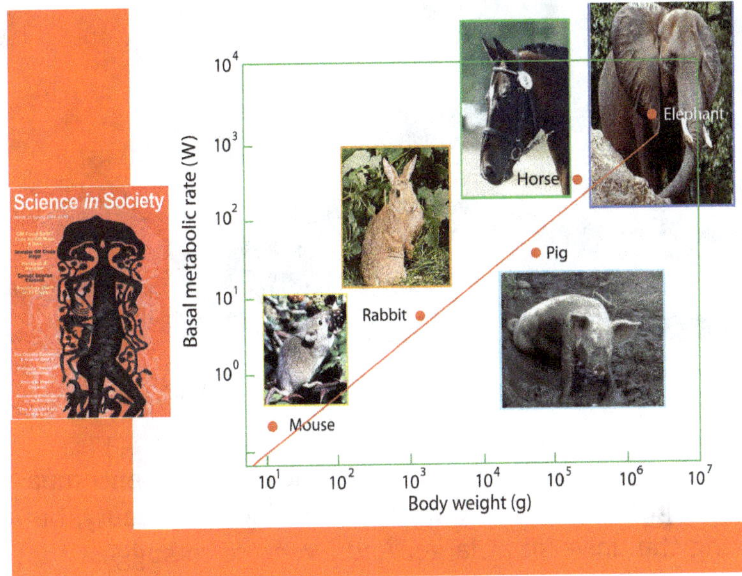

Figure 13.2 The mouse to elephant allometric scaling of metabolic weight and body weight.

from moment to moment, to explore and create its possible futures. A work of art is unique for the occasion, as are all organic forms and organisms, as opposed to mechanically manufactured objects. A later essay ("18 — Space-Time Is Fractal and Quantum Coherent in the Golden Mean" in this book) shows how and why "golden mean" fractals are the fabric of organic space-time and consciousness.

So perhaps Andy Warhol was saying something critical and profound with his piles of identical Brillo boxes and Campbell tinned soups.

Mechanical uniformity has given way to dynamic coherence.

Beauty is Truth in Art

Beauty in science and art comes in endlessly diverse organic forms, surprising and unpredictable, and always *sublime* in arousing that

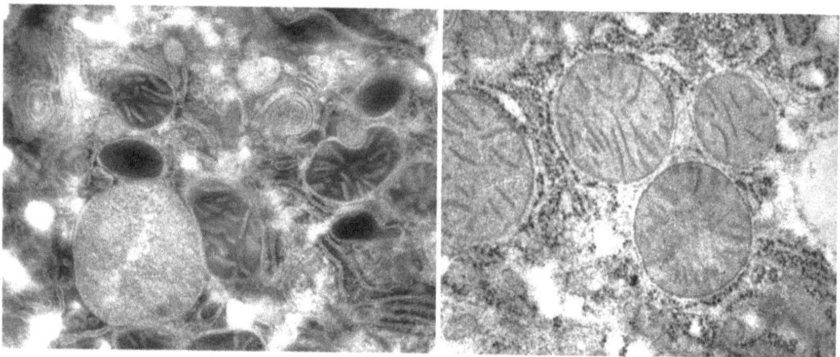

Figure 13.3 Which is the more beautiful?

rapturous "thrill" that I can only describe as an all-encompassing *love*, not just directed at the particular object of beauty, but permeating the universe at large. And I venture to suggest that the same "truth" or authenticity underlies beauty in art as in science, in that it resonates with some universal, timeless aspects of nature to which we are connected, and in which we are utterly immersed.[28]

Let's do a little experiment to illustrate what I am saying about beauty and authenticity. Fig. 13.3 shows two electron micrographs of the same kind of cells prepared using two different procedures. Which is the more beautiful, the one on the left, or the one on the right?

I have carried this out on at least two different small audiences, one of ~50, the other 70; on both occasions *all* chose the one on the left.

Now, which is the more true to life, the one on the left, or the one on the right?

The one on the right is the state-of-the-art conventional electron microscopy, involving fixation and dehydration before staining. The one on the left, however, was done without dehydration. Instead, it was freeze-dried, then sectioned, and the water sublimated away through a very long, laborious process, which the German scientist Ludwig Edelman believes, is crucial in capturing what the cell is really like in life, when it is hydrated with 70–80% by weight of water.[29,31] There are more structures preserved in the non-dehydrated cell processed by Edelman, so it is reasonable to conclude that it represents more closely what the cell is like in life. Beauty and truth coincide in the electron micrograph, which is as much a work of art as a work of science.

From other research including my own that I shall presently describe, we now know that the water in the cell and extracellular matrix is in a very special, liquid crystalline state that forms dynamically coherent units with the macromolecules embedded within it, and without which, the enzyme proteins, the DNA and RNA would not work at all.

Water has very special properties, even outside the body. It has fascinated and intrigued generations of scientists, including me. For the past five years, more and more scientists are saying that the properties of water cannot be explained except in terms of quantum theory. There is even evidence that water may be quantum coherent at ambient conditions. That is the subject of another essay (see "15 — Water the Means, Medium, and Message of Life" in this book).

Quantum Jazz Biology and Art

It was 1992 when I peered down the microscope and saw the little fruit fly larva hatch from its egg, dancing in all the colours of the rainbow, *which no one else had seen before*. Those are real interference colours that some artists would die for.

Later, the true significance of this vision dawned on me: it would not have been possible unless *all* the molecules in the cells and tissues within the body of the little larva are aligned as liquid crystals *and* moving coherently together, all including the water molecules that form dynamically coherent units with the macromolecules embedded in it, enabling them to function as quantum molecular machines with efficiencies close to 100%. That water I call "liquid crystalline water", without which the molecular machines cannot work and life would be impossible.

The fruit fly larva is not unique; all living cells and organisms display themselves like that under the polarized light microscope geologists use to look at rock crystals (Fig. 13.4).

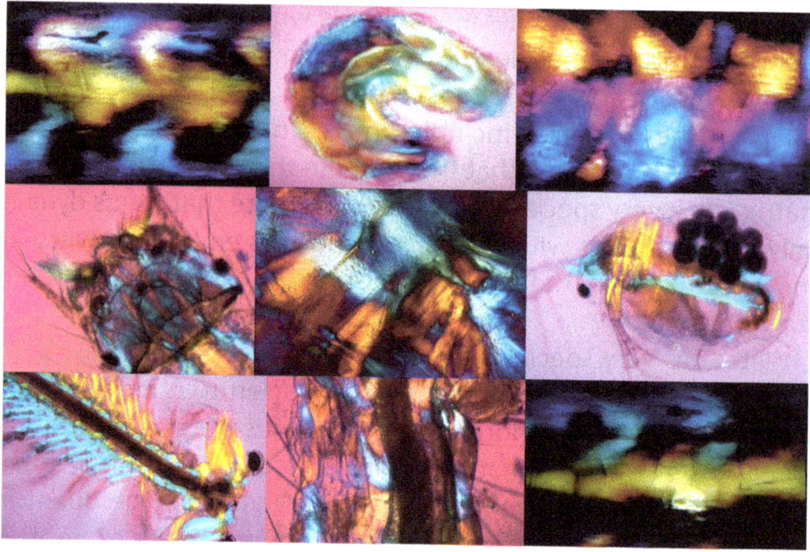

Figure 13.4 All organisms are liquid crystalline.

These stunning images of live organisms are among the key evidence that convinced me organisms are coherent to a high degree, from the macroscopic level down to the molecular and below.

The other important reason we can see such images at all is that we have used a non-destructive technique to allow the organisms to show us what they are really like, and we are richly rewarded for this. Again, these are beautiful and authentic works of science *and* art.

In the book named after the fruit fly larva,[31] I presented theoretical arguments and empirical evidence to support the idea that the organism is quantum coherent, and depends on individual molecular energy machines embedded in the "liquid crystalline water" matrix working seamlessly together to transform material and energy without loss.

An organism going round its business of living is doing quantum jazz inside: an incredible dynamic light-and-sound dance show involving astronomical numbers of the most diverse players, coordinated from sub-atomic dimensions to the macroscopic performing over a frequency range of 70 octaves, each improvising spontaneously and freely yet keeping perfectly in step and in tune with the whole.

When I first proposed that the organism is quantum coherent in the early 1990s, only a handful of exceptional scientists thought quantum theory had anything to do with biology at all. The situation has greatly changed since then. Google ran a workshop on quantum biology in October 2010[32] where various scientists spoke about quantum coherence at the micron scale for photosynthesis, and about the "collapse" of the quantum wave function as the basis of consciousness and a kind of quantum computing in our brain. Obviously, they have yet to catch up with the evidence suggesting (to me at least) that the organism is completely quantum coherent, and breathtakingly beautiful to behold.

References and Notes

1. Chandrasekhar, S. *Truth and Beauty: Aesthetics and Motivation in Science*. University of Chicago Press, Chicago and London, 1987.
2. Ho, M.W. "Electric Plasma Universe Arrives". *Sci Soc* 68 (2015): 8–11, 23.
3. Stewart, I. *Why Beauty Is Truth: The History of Symmetry*. Basic Books, New York, 2007.
4. Levy, D. "String Theorist Explores Dark Energy and Our Unique 'Pocket' of the Universe". Stanford news service, 15 February 2005, http://news.stanford.edu/pr/2005/pr-aaas_susskind-021605.html.
5. Chu-Caroll, M. The mapping of the E8 Lie Group (minor update). Good Math, Bad Math, 19 March 2007, http://scienceblogs.com/goodmath/2007/03/19/the-mapping-of-the-e8-lie-grou/.
6. Stewart, I. "Symmetry, String Theory and an Equation for Everything (Plus — Could We Be Living in an Asymmetrical Universe?)". Warwick University podcast, 2007, http://www2.warwick.ac.uk/newsandevents/podcasts/media/more/symmetry/.
7. Gardner, M. "Is Beauty Truth and Truth Beauty?" *Scientific American*, 18 March 2007, http://www.scientificamerican.com/article.cfm?id=is-beauty-truth-and-truth&page=2.

8. Poincare, H. *Science and Method* (Francis Maitland, trans. 1914), p. 19, Cosmo Classics, New York, 1908.
9. Sullivan, J.W.N. *Athenaeum*, 1919. Cited in Chandrasekhar, S. *Truth and Beauty: Aesthetics and Motivation in Science*, p. 60, University of Chicago Press, Chicago and London, 1987.
10. Fry, R. "Science and Art". Cited in Chandrasekhar, S. *Truth and Beauty: Aesthetics and Motivation in Science*, p. 60, University of Chicago Press, Chicago and London, 1987.
11. Watson, G.N. Cited in Chandrasekhar, S. *Truth and Beauty: Aesthetics and Motivation in Science*, p. 61, University of Chicago Press, Chicago and London, 1987.
12. "A Disappearing Number", Wikipedia, accessed 6 November 2010, http://en.wikipedia.org/wiki/A_Disappearing_Number.
13. Heisenberg, W. *Physics and Beyond*, p. 61. Allen & Unwin, London, 1971.
14. "Ode on a Grecian Urn" and "Beauty and Truth Debate", Wikipedia, accessed 22 February 2011, http://en.wikipedia.org/wiki/Ode_on_a_Grecian_Urn#Beauty_is_truth_debate.
15. Danto, A.C. *The Abuse of Beauty: Aesthetics and the Concept of Art*, back cover. The Paul Carus Lectures 21, Open Court Publishing, Chicago, 2003.
16. Scruton, R. "What Has Art Got to Do with Beauty?" *The Sunday Times*, 16 March 2009, http://www.timesonline.co.uk/tol/comment/columnists/guest_contributors/article5913530.ece.
17. Gablik, S. "The Nature of Beauty in Contemporary Art". *New Renaissance magazine* Volume 8, no. 1, 1998, http://www.ru.org/81gablik.html.
18. Strickland, C. "Does Beauty Still Belong in Art?" *The Christian Science Monitor*, 20 December 2007, http://www.csmonitor.com/2007/1220/p09s03-coop.html.
19. Le web pedagogique, accessed 12 January 2011, http://lewebpedagogique.com/notreprof/.
20. *Quantum Jazz Art*, Virtual Art Exhibition DVD, ISIS, London, 2011, http://www.i-sis.org.uk/onlinestore/av.php#315.
21. Ho, M.W. (ed). *Celebrating ISIS, Quantum Jazz Biology*Medicine* Art*, ISIS, London, 2011, http://www.i-sis.org.uk/onlinestore/books.php#312

22. *Celebrating ISIS Event*, DVD of Art Exhibitions, Lectures, Musical Performances, Interviews, and other activities at the weekend event 26–27 March 2011, http://www.i-sis.org.uk/onlinestore/av.php#315.
23. Ho, M.W. "The Heartbeat of Health". *Sci Soc* 35 (2007a): 10–13.
24. Ho, M.W. "Happiness Is a Heartbeat Away". *Sci Soc* 35 (2007b): 14–19.
25. Ho, M.W. "Biology's Theory of Everything". *Sci Soc* 21 (2004a): 46–47.
26. "Fractal". Wikipedia, accessed 4 March 2011, http://en.wikipedia.org/wiki/Fractal.
27. Ho, M.W. "The Biology of Free Will". *J Conscious Stud* 3 (1996): 231–244.
28. Ho, M.W. "In Search of the Sublime: Significant Form in Science and Art". *Sci Soc* 39 (2008a): 4–14.
29. Edelman, L. "Freeze-Dried and Resin-Embedded Biological Material Is Well-Suited for Ultrastructure Research". *J Microsc* 207 (2002): 5–26.
30. Ho, M.W. "What's the Cell Really Like?" *Sci Soc* 24 (2004b): 46–47.
31. Ho, M.W. *The Rainbow and the Worm: The Physics of Organisms*, 3rd edition. World Scientific, Singapore, 2008b.
32. Google workshop on quantum biology, 22 October 2010, http://sitescontent.google.com/google-workshop-on-quantum-biology/.

14

Thermodynamics of Sustainability

14

Sustainable Cities as Organisms: A Circular Thermodynamics Perspective

Urbanization is among the most pressing problem facing the world, but it is possible to make the most of the process by looking at sustainable cities as organisms in a circular thermodynamics perspective.

Urbanization Challenges and Opportunities

Urbanization has been expanding rapidly within the past 200 years, emerging as the most pressing global problem of this century. Since the year 2006, those of us who live in cities have exceeded 50% of the world population. Today, more than 80% of people in developed countries such as the United States are urbanized, and by 2050, more than 75% of the world population will live in cities.

Geoffrey West at Santa Fe Institute in the United States summed up the challenges of urbanization:[1] "Cities are the origins of global warming, impact on the environment, health, pollution, disease, finance, economies, energy — they're all problems that are confronted by having cities." On the other hand, he also noted that cities have proven to be "engines of creativity",[2] wealth creation and economic growth.

He called for a "Grand Unified Theory of Sustainability."[1,3]

Sustainable Cities and Fractal Cities as Organisms

It has been long recognized that cities can save energy on mass and food transport, and facilitate the provision of essential services for health, education, sanitation and energy, as well as building communities.[4] The idea of sustainable cities goes back at least to the 1980s. For example, Richard Register coined the term "ecocity" in his book *Ecocity Berkeley: Building Cities for a Healthy Future*.[5] Architect Paul F. Downton founded the company Ecopolis in 1990, specializing in ecological architecture and bio-urban design[6] and published a book on the subject[7] in 2009.

There is no agreed definition for a sustainable city. Nevertheless, the following would probably be generally acceptable: a sustainable city should be able to feed and power itself using land and energy efficiently with the least environmental pollution and impact on climate change, minimizing waste, and recycling and reconverting wastes into resources. A long list of sustainable cities or ecocities is found in countries across the world,[4] with indicators for sustainability typically in three areas — economy, environment and social — the details depending on local conditions.[8] There is strong emphasis on green spaces, renewal energies/energy efficiencies, reduced carbon emissions, transport, water availability and quality, air quality, and minimizing and recycling waste for the environment; complete neighbourhood, equity and affordable housing as well as education, sanitation and health for the social; and employment rates especially in green jobs and economic growth for the economy. Surprisingly there is nothing on food and its availability, considering that 842 million people worldwide do not have enough to eat.[9] I shall show why these indicators make sense, especially in the context of the circular thermodynamics of organisms.

Since the 1990s, stimulated by the science and mathematics of complex systems, there has been substantial effort dedicated to understanding cities in terms of fractal geometry and dynamics.[10] Fractals are geometrical or mathematical objects with fractional dimensions instead of the usual integers 1, 2, 3 or 4. They also have

Figure 14.1 Fractal city of Siena.

self-similar properties, i.e., they appear similar on different scales. Siena in Italy is a wonderful example of a fractal city (Fig. 14.1).

Fractals are also the key to living organization and characteristic of living organisms (as will be made clear later) and this led to the transition in thinking of cities as machines to cities as organisms.[11] While living cities have fractal properties that extend downwards to human scales, these are lacking or destroyed in modernist cities built to accommodate the automobile and population growth. Fractal properties are being recreated and recovered[12] in pedestrian walkways and precincts such as those along and across the River Thames in London.[13] A related move is "smart growth"[14] in urban planning to avoid urban sprawl. It advocates compact, transit-oriented walkable and bicycle-friendly land use, with neighbourhood schools, complete streets that include safe access for bicycles, pedestrians as well as for delivering goods, and mixed-use development that blends a combination of residential, commercial, cultural, institutional or industrial uses with a range of housing choices, all aimed at restoring connectivity and structures on human scales. Much attention is obviously being paid to the social aspects of sustainability.

Despite these promising developments, cities are still poorly understood. There is as yet no real science of cities,[10] let alone a grand unified theory of sustainability.[1,3] Nevertheless there are hints that a theory of sustainable cities may exist, and fractals very much part of it.

Universal Scaling of Organisms and Cities Based on Thermodynamics

One of the most intriguing and best known physiological relationships is allometric scaling, especially of resting metabolic rate with size,

$$I = I_0 M^{\beta} \tag{1}$$

where I is the resting metabolic rate, M the body mass, and both I_0 and β are constants. Plotting log I against log M gives a straight line with slope β and intercept log I_0. The value of β is ¾, and it applies universally across the living world, from giant redwood trees and whales all the way down to cells, mitochondria and the respiratory molecular complex, spanning 27 orders of magnitude.[15] This sublinear scaling ($\beta < 1$) means there is economy of scale; larger organisms have proportionately lower resting metabolic rates, using energy more efficiently than smaller organisms. West and colleagues produced the first overarching theory[16–18] on how this scaling depends on fractal networks optimized for transport, such as the circulatory system in mammals, trachea in insects and xylem and phloem in plants. These fractal structures maximize the area across which they can take up and release resources and minimize the energy required to deliver those resources throughout the organism. They are space-filling in three dimensions, but not quite three-dimensional, and hence have a fractional dimension between 2 and 3. They are similar over many scales, as characteristic of fractals, but unlike mathematical fractals that are self-similar over infinite scales, real live fractals terminate in finite units, as for example, actual metabolic units of the respiratory complexes.

The metabolic theory applies not only to the physiology of organisms, but also to ecology. As American ecologist James Brown wrote in his MacArthur Award Lecture (p. 1771):[19] "Metabolic theory predicts how metabolic rate, by setting the rates of resource uptake from the environment and resource allocation to survival, growth and reproduction, controls ecological processes at all levels of organization from individuals to the biosphere."

It is well known that metabolic rate increases exponentially with temperature, as described by the Boltzmann factor, $e^{-E/kT}$, where E is the activation energy, k is Boltzmann's constant and T is the absolute temperature. This relationship holds over the range of normal temperatures between 0°C and 40°C. Combining the effects of body size and temperature gives[19]

$$I = i_0 M^{3/4} e^{-E/kT} \tag{2}$$

where i_0 is a normalization constant independent of body size and temperature. Equation 2 predicts that the natural logarithm of the mass-corrected whole organism metabolic rate should be a linear function of the inverse absolute temperature ($1/kT$), the slope of which yields the activation energy of metabolism. The prediction proved to be true for birds and mammals, fish, amphibians, reptiles, invertebrates, unicellular organisms and plants. The slope (activation energy) was the same over all the groups at 0.69 eV. This is within the range of activation energy (0.60–0.70 eV) reported for aerobic respiration.

Most intriguingly, West and others also observed allometric scaling in cities, raising the possibility that cities too, may be considered as organisms.[20] For example, cities seem to scale sublinearly with size as measured by population for infrastructures such as number of petrol stations, length of roads, electric lines, etc., with economy of scale about 15% (scaling constant 0.85). A similar sublinear scaling has been claimed for carbon footprints, suggesting that larger cities are greener than small ones. More surprisingly, cities also scale superlinearly by approximately the same amount ($\beta \sim 1.15$) in gross domestic product, wages, income, creativity (number of patents), etc. However, a new analysis using state-of-the-art clustering technique for identifying cities and the most detailed available data shows that large cities are much less green than small ones. Carbon emissions scale superlinearly, with $\beta = 1.46 \pm 0.02$. Income level, on the other hand, scales superlinearly only for those with per capita incomes >$37 235, whereas for those with incomes lower than that, the scaling is sublinear; in other words, the low-paid get less as the size of the city increases.[21,22]

West and Brown are right in linking the impressive universality of allometric scaling to thermodynamics,[15] which in general terms is the transformation of energy and materials in metabolism that enables organisms to develop and grow (accumulate biomass). But without a clear thermodynamic theory, they appealed erroneously to Darwinian natural selection for optimization.

My own conjecture is that this can be explained in terms of a circular thermodynamics of organisms and sustainable systems that results in dissipationless (zero entropy) energy transfer and transformation in the ideal; and that is also implied in the quantum coherence of organisms. The detailed empirical evidence and theoretical arguments are presented in my book *The Rainbow and the Worm: The Physics of Organisms*,[23] and elsewhere.[24–26] In *Living Rainbow* H_2O[24] I elaborate on the universal metabolism in living organisms based on splitting and recombining water in photosynthesis and respiration that underlies the thermodynamics of universal allometric scaling.

To think of cities as organisms,[1–3] we must concentrate on the thermodynamics — the transformation of energy and resources — which is nothing but the economy of cities; and it could be a good guide to making cities (and economies) sustainable.

Circular Thermodynamics of Sustainable Cities

The most important thing to note is that organisms are *not* heat engines and do not make their living by heat transfer. Instead they are isothermal systems maintained far from thermodynamic equilibrium and depend on the direct transfer of molecular energy by proteins and other macromolecules acting as "quantum molecular energy machines". For isothermal processes the change in Gibbs free energy ΔG (thermodynamic potential for doing work at constant temperature and pressure) is given as

$$\Delta G = \Delta H - T\Delta S \tag{3}$$

where ΔH is the change in enthalpy (heat content), T is temperature in deg K, and ΔS is the change in entropy, the measure of dissipation or loss of energy.

In a thermodynamically efficient process, ΔS approaches 0 (least dissipation) and $\Delta H = 0$; or else $\Delta G = 0$ via entropy-enthalpy compensation, i.e., entropy and enthalpy changes cancelling each other out. We shall see how the organism manages to do that.

Importance of Energy Capture and Storage

For a system to keep far away from thermodynamic equilibrium — death by another name — it must capture energy and materials from the environment to develop, grow and recreate itself from moment to moment during its lifetime, and also to reproduce and provide for future generations, all part and parcel of sustainability. The key to understanding the thermodynamics of the living system is not so much energy flow as energy capture and storage under energy flow to create a reproducing life cycle (see "9 — The Biology of Free Will", Fig. 9.1, in this book).

The analogy with a city is clear: the city too is far from thermodynamic equilibrium; it captures and stores resources to build physical structures that keep resources circulating within and the city alive. Figure 14.2 is a remarkable fresco of the idealized 14th century autonomous Siena city-state by Lorenzetti, representing the peace and prosperity of a well-governed city.[27] The thin city wall is emblematic of the fragile physical closure against invaders. Physical closure is perhaps much less important for the survival and autonomy of the city than dynamic closure due to the *characteristic spacetimes of processes* that create the fractal dynamics of a living city.

Figure 14.2 Lorenzetti's idealized Siena city-state representing peace and prosperity.

Characteristic Space-Times and Fractal Dynamics

Characteristic space-time is a very important concept all too often ignored. All real processes and objects have characteristic space-times. In the organism, the heart (10^{-1} m) beats in a second, nerve cells (10^{-4} m) fire in a tenth of a second or faster, and protons (10^{-15} m) and electrons (10^{-17} m) move in 10^{-12} to 10^{-15} s. Cells divide in minutes, and physiological processes have longer cycles of hours, a day, a month or a year.

Living activities are organized by their characteristic space-times in a coherent fractal hierarchy. Processes with matching space-times interact most strongly with one another through resonance, but also link up to the entire hierarchy.

Similarly, cities have fractal hierarchies of activities in buying and selling, manufacturing and scavenging, imports and exports, borrowing and lending, construction and deconstruction, etc., *which must also match and link up.*

Cycles are the Key to Fractal Space-Times

Cycles are the key to the fractal space-time organization of living systems. The way to match space-times and link up the entire hierarchy is through cycles. Practically all living activities come in cycles, i.e., biological rhythms. Cycles are also ubiquitous in the physical universe, and some cosmologists believe that even the universe itself is cyclic.[28] The possibility for cycles in the living world coupling and linking up to those in the physical universe is surely why life exists, and indeed some would argue, suggests that the entire universe may be alive.

The coupled cycles form a nested fractal self-similar structure: the life cycle is made up of small cycles; each small cycle has similar smaller cycles within, spanning characteristic space-times from subnanometre to metres and from 10–15 seconds to hours and years. Cycles enable the activities to be coupled together, so that energy-yielding processes can transfer energy directly to those requiring energy. *And the direction can be reversed when necessary.* Cooperativity and reciprocity are the hallmark of a sustainable system. This is an extended Onsager reciprocity relationship in thermodynamics that strictly applies only to the steady state.[23] What it means in practice is that energy can be concentrated to any local point where it is needed, and conversely spread globally from any local point. In that way, the fractal hierarchy of coupled cycles maximizes both local autonomy and global cohesion (see "9 — The Biology of Free Will", Fig. 9.2, in this book).

(To get an idea of such coupled cycles, one needs to look no further than charts of biochemical metabolic pathways;[29] most if not all of the reactions are reversible, depending on the local concentrations of reactants and products. Biochemical recycling is ubiquitous; there are numerous scavenging or salvaging pathways for the recovery of building blocks of proteins, nucleic acids, glycolipids and even entire proteins.)

The fractal hierarchy of cycles confers dynamic stability as well as autonomy to the system on every scale. Thermodynamically, no net entropy is produced in the case of perfect cycles; hence the system can maintain its organization.

The fractal structure effectively partitions the organism into a hierarchy of systems within systems *defined by the extent of equilibration of thermal (dissipated) energies*. Thus, energies thermalized or equilibrated within a smaller space-time will still be out of equilibrium in the larger system encompassing the first, and hence capable of doing work.

Consequently, there are two ways to mobilize energy efficiently with entropy change approaching zero: very slowly with respect to the characteristic time so it is reversible at every point, or very rapidly with respect to the characteristic time, so that in both cases the energy remains stored (in a coherent non-degraded form) as it is mobilized. It is that which enables the organism to *simultaneously* achieve the most efficient equilibrium *and* far-from-equilibrium energy transfer.

This nested dynamical structure also optimizes the kinetics of energy mobilization. For example, biochemical reactions depend strictly on local concentrations of reactants, which are very high, as their extent of equilibration is typically at nanometre dimensions (in a nanospace).

Space-Time Differentiation Directly Proportional to Energy Storage and Resource Residence Time

The degree of space-time differentiation in the living system is directly proportional to its energy storage capacity or energy residence time, which is also proportional to the material residence time; in ecosystem/sustainable system terms, these translate to biodiversity (species richness), and biomass or productivity. There is now abundant evidence that biodiversity and productivity go together in agricultural systems.[30] Biodiverse agricultural systems based on the circular economy of nature are many times more productive than industrial monocultures in which all fractal space-time differentiation has been destroyed. Monocultures need chemical fertilizers on account of barren soils that cannot retain or recycle nutrients, and pesticides for want of pollinators and other beneficial animals and plants that could control pests.

For a city, the fractal structure maximizes the retention of resources (and wealth) within the city, where they circulate in a hierarchical way, benefiting local economies, encouraging diversity and diversification, which in turn feeds back to benefit local economies.

Re-use and recycling make perfect sense in prolonging the retention of useful resources in the system. So do cooperation and reciprocity in the coupled cycles of giving and taking among local businesses and people in general, in place of the dominant misguided culture of relentless competition.

Fractals Mathematically Isomorphic to Quantum Coherence

The coherent fractal structure maximizes both global connectivity and local autonomy, the hallmark of quantum coherence.[23] Coincidentally, there is now a formal connection established between fractals and quantum coherence. Italian quantum field theorist Giuseppe Vitiello at Salerna University has proven mathematically that fractals are isomorphic to quantum coherence.[31] This is very significant and will be further discussed in "18 — Space-Time is Fractal and Quantum Coherent in the Golden Mean" in this book.

The Zero-Entropy Ideal

In the ideal — approached most closely by the healthy mature organism and the healthy mature ecosystem — an overall internal conservation of energy and compensation of entropy ($\Sigma\Delta S = 0$) is achieved. In this state of balance, the system organization is maintained and dissipation minimized; i.e., the entropy exported to the environment also approaches zero, $\Sigma\Delta S \geq 0$ (Fig. 14.3).

Internal entropy compensation (and energy conservation) implies that there is free variation in microscopic states within the macroscopic system; i.e., *the internal microscopic detailed balance at every point of classical steady-state theory is violated.*

For an organism, it means that detailed energy balance is neither necessary nor possible at every point within the system. Most often, parts of it are in deficit, and severely so when one needs to run from a tiger, with the knowledge that the energy can be replenished after a successful escape. The same applies to ecosystems: all species are in a sense storing energy and resources (nutrients) for

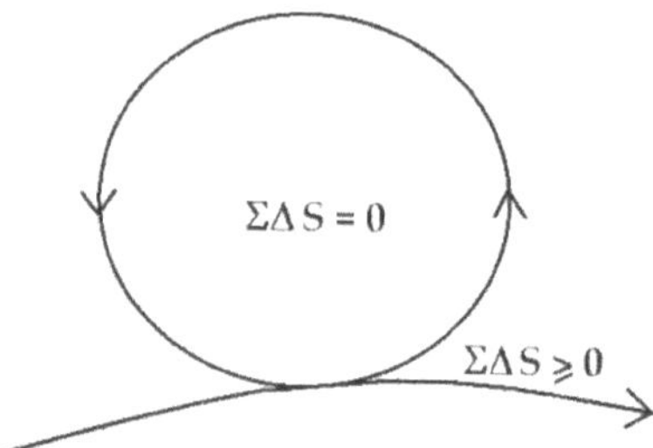

Figure 14.3 The zero-entropy ideal of circular thermodynamics.

every other species via ecomplex food webs and other symbiotic relationships. That is why biodiverse ecosystems and agricultural systems are more resistant and resilient to stress and environmental extremes.

For a sustainable city, all wealth and other resources are shared, either directly or indirectly. A proper city bank will see to it that excess wealth generated somewhere in the city can be directed in timely fashion to help those in need, or to finance key innovations and diversifications.

The above considerations give rise to the prediction that a sustainable system maximizes cyclic, non-dissipative flows while minimizing dissipative flows, i.e., it tends towards *minimum entropy production* as conjectured by Ilya Prigogine.[32] (This also accords with the asymptotic stability of the quantum coherent state[23] in which entropy is effectively zero.)

In other words, minimum entropy production requires abandoning the principle of microscopic detailed balance of the classical steady state, which applies at near to equilibrium condition. This is most crucial for sustainability.

Sustainable Systems Need to Sustain the Ecosystem in which it is Embedded

Minimum entropy production means that little or no entropy is exported to the environment. As the system depends on environmental input, entropy and wastes exported to the environment will

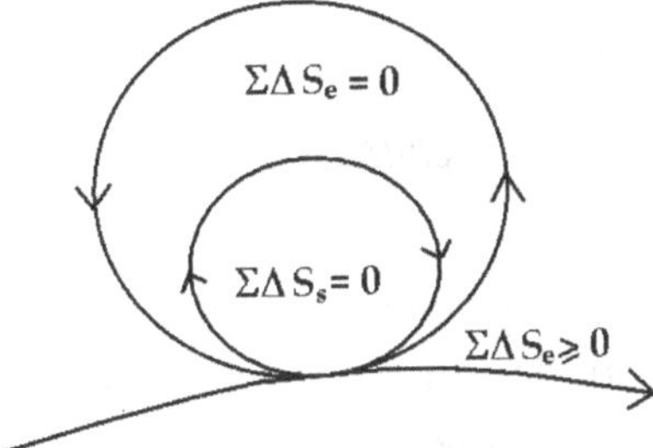

Figure 14.4 The coupled flows of system and ecological cycles in a sustainable system.

simply mean diminished environmental input. Yet, generations of economists persist in speaking of economic growth as though the environment does not exist. Figure 14.4 makes it explicit that an organism or a sustainable system is coupled to its ecological environment, hence the importance of minimum entropy export to the environment. In practice, this means the organism or sustainable system must not overexploit the ecological environment. Again this diagram is a fractal, so that smaller ecosystems are embedded and coupled to larger and larger ones ending with the entire Earth ecosystem.

For a city with ready access to air travel in a globalized world, the implications are profound. Not only does its immediate ecological environment need to be conserved, but a sustainable city must take care not to overexploit ecosystems in other countries especially in the developing world, because ultimately, the entire global fractal ecosystem is one. This also shows why exploitation and unequal exchanges in general are bad for sustainability. Overexploitation of people leads to overexploitation of the environment both because it drives the exploited to faster deplete and degrade Earth's resources, and/or creates false "wealth" for the exploiter to overconsume. Equity and self-sufficiency make perfect sense in this context. Cities can and should produce as much of their own food as possible in urban and peri-urban green spaces and revive its local markets, which are great for rebuilding and revitalizing local communities.

Some Applications

Circular Economy in Business

Circular economy, if not circular thermodynamics, has been enthusiastically championed by Dame Ellen MacArthur since 2010.[33] The Ellen MacArthur Foundation has produced reports with analyses from McKinsey highlighting an annual trillion-dollar opportunity globally in net material cost savings for companies making the transition to circular economy. And big companies and governments all over Europe have indeed signed up to the Circular Economy 100 that the Foundation has initiated.

The Dyke-Pond System and Dream Farm 2

I started thinking about cycles and circular thermodynamics in the first edition of *Rainbow Worm*[23] published in 1993, and used the term "circular economy" for the dyke-pond system of the Pearl River Delta in China[34] which I visited in 2006, and discovered that "circular economy" is already in mainstream and official discourse in China, albeit also restricted to the manufacturing and service industries sector.[35]

Leaving out agriculture — the primary production of the economy — does not make sense, especially for the "green economy" on everyone's lips, and certainly runs counter to circular thermodynamics. Industrial monoculture is completely non-circular; it is also decidedly not green as it pollutes the environment, requires massive inputs in energy and is responsible for huge amounts of carbon emissions.[30] It is now generally recognized that a paradigm change from industrial to agro-ecological farming is urgently needed to feed the world that at the same time can best mitigate and adapt to climate change.[36]

The dyke-pond system evolved over 2 000 years, perfected by generations of Chinese farmers into a paradigmatic circular economy of intensive agriculture. It depends on maximizing internal inputs between land and water, optimizing the efficient use of

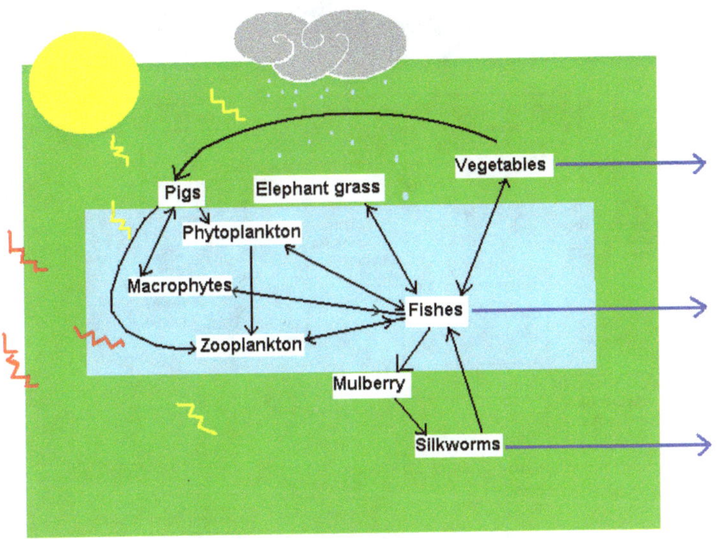

Figure 14.5 The circular economy of a dyke-pond system in China's Pearl River Delta.

resources while minimizing wastes, and transforming wastes into resources. One version is shown in Fig. 14.5.

Numerous cycles (I counted 15 in all) are obvious involved in keeping resources within the system to minimize dissipation and preventing pollution of the general environment. Dyke-pond systems are so productive that they supported an average of 17 persons per hectare compared with the "carrying capacity" of about two persons per hectare stipulated by the United Nations Food and Agriculture Organization.[37]

I extended the dyke-pond system by incorporating renewable energies including biogas from anaerobic digestion of wastes into a Dream Farm 2, which could form part of peri-urban agriculture (Fig. 14.6).[30,37,38]

The diagram is colour-coded. Pink is for energy, green for agricultural produce, blue is for water conservation and flood control, black is waste in the ordinary sense of the word, which soon gets converted into food and energy resources. Purple is for education and research into new science and technologies. It is estimated to

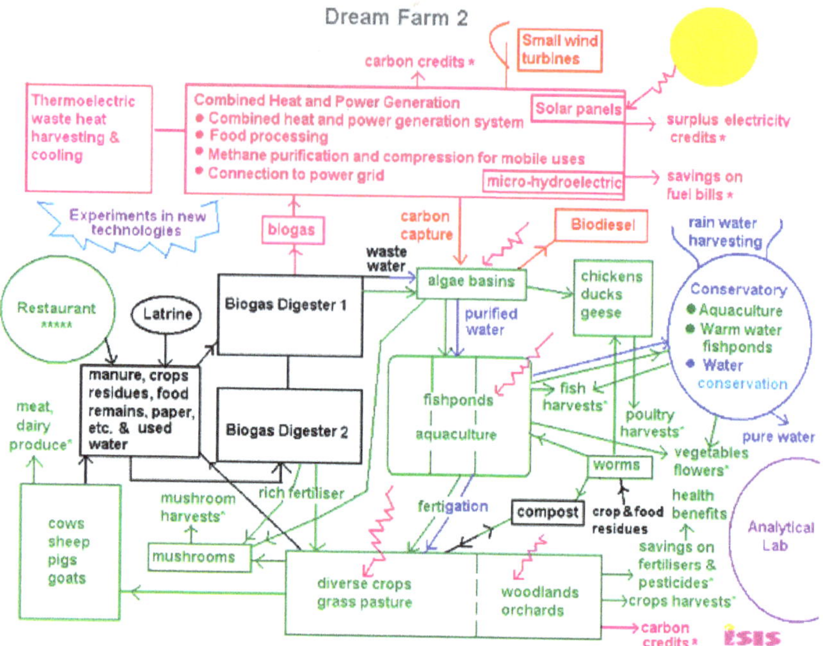

Figure 14.6 A food and energy self-sufficient farm run on the circular economy of nature.

save more than 40% of energy consumption, only counting anaerobic digestion for a country like China.[38] With the addition of solar, wind or micro-hydroelectric technologies as appropriate, such farms could provide more than enough energy for the entire farm (while also compensating substantially if not completely for carbon emissions). As a peri-urban farm, it not only helps provide food security, but also healthy nutrition for the inhabitants (another indicator of sustainability). In addition, it frees up land in rural areas to support wild life to protect and increase natural biodiversity.

Dream Farm 2 is an excellent project for a university or a research institute because engineers, architects, scientists, artists, medical doctors, sociologists, economists and business can all work together across the disciplines to realize the closed loop model in design, energy flow, architecture, marketing, etc., while providing huge opportunities for education, research and innovation.

To Conclude

The principles of circular thermodynamics can help cities become as sustainable as organisms. A sustainable city has a dynamic fractal hierarchy of activities that bridges all space-times. The coherent fractal hierarchy maximizes cyclic activities that enable the most efficient transformation and exchanges of resources, goods and services over all space-times on a symbiotic and reciprocal basis. It maximizes both global cohesion and local autonomy, such that energy and resources could be readily shared from local to global and vice versa. It minimizes wastes and recycles wastes into resources, thereby achieving the state of minimum dissipation that protects its own organization as well as safeguarding the ecological environment on which it depends for input. These circular thermodynamic principles are not difficult to implement, and are already widely implemented in sustainable cities all over the world. Now is the time to move ahead to meet both the challenges and opportunities in making cities sustainable as urbanization is one of the most pressing problems of our time.

Acknowledgment

I am very grateful to Carlos Brebbia and the Committee of the Wessex Institute for awarding me the 2014 Prigogine Medal, and to the University of Siena for hosting the conference on Sustainable Cities, which provided the major impetus for my venturing into this topic. Many thanks also to my son Adrian Ho and my husband Peter Saunders for their help in preparing this report. Robert Ulanowicz has provided many astute comments and suggestions for improving the essay.

References and Notes

1. West, G. (2011). "The Surprising Math of Cities and Corporations". TEDGlobal 2011, July 2011, https://www.ted.com/talks/geoffrey_west_the_surprising_math_of_cities_and_corporations/transcript.
2. Bettencourt, L. and West, G. A Unified Theory of Urban Living. *Nature* 447 (2010): 912–913.

3. West, G. (2013). "Life from Cells to Cities: Are They Sustainable?" Boston University lecture, 8 May 2013, https://www.youtube.com/watch?v=etfRE5-YlXs.
4. "Sustainable City", Wikipedia, accessed 4 July 2014, http://en.wikipedia.org/wiki/Sustainable_city.
5. Register, R. *Ecocity Berkeley: Building Cities for a Healthy Future*. Atlantic Books, North Berkeley, 1987.
6. "Paul F. Downton", Wikipedia, accessed 31 May 2014, http://en.wikipedia.org/wiki/Paul_F_Downton.
7. Downton, P.F. *Ecopolis: Architecture and Cities for a Changing Climate*. Springer, Berlin and CSIRO, Australia, 2009.
8. *Indicators for Sustainability: How Cities Are Monitoring and Evaluating Their Success*. Sustainable Cities International, Vancouver, November 2012.
9. "10 Hunger Facts for 2014". World Food Programme, 30 December 2013, https://www.wfp.org/stories/10-hunger-facts-2014.
10. Batty, M. and Longley, P. *Fractal Cities: A Geometry of Form and Function*. Academic Press, London, 1994.
11. Batty, M. "Building a Science of Cities". 30 October 2011, http://www.complexcity.info/.
12. Salingaros, N.A. "Connecting the Fractal City". Keynote speech, 5th Biennial of Towns and Town Planners in Europe, Barcelona, April 2003. Published in *Planum the European Journal of Planning Online*, March 2004. Reprinted as Chapter 6 of *Principles of Urban Structure*, Techne Press, Amsterdam, 2005.
13. "Thames Path Walk". Walk London, accessed 29 August 2014, http://www.walklondon.org.uk/route.asp?R=6.
14. "Smart Growth", Wikipedia, accessed 10 July 2014, http://en.wikipedia.org/wiki/Smart_growth.
15. West, G.B. and Brown, J.H. "The Origin of Allometric Scaling Laws in Biology from Genomes to Ecosystems: Towards a Quantitative Unifying Theory of Biological Structure and Organization". *J Exp Biol* 208 (2005): 1575–1592.
16. West, G.B., Brown, J.H. and Enquist, B.J. "A General Model for the Origin of Allometric Scaling Laws in Biology". *Science* 276 (1997): 122–126.

17. Enquist, B.J., West, B.G., Charnov, E.L. and Brown, J.H. "Allometric Scaling of Production and Life-History Variation in Vascular Plants". *Nature* 401 (1999): 907–911.
18. Gilloly J.F., Brown J.H., West G.B., Savage V.M. and Charnov E.L. "Effects of Size and Temperature on Metabolic Rate". *Science* 293 (2001): 2248–2251.
19. Brown, J.H. Toward a Metabolic Theory of Ecology". *Ecology* 85 (2004): 1771–1789.
20. Ho, M.W. "'Grand Unified Theory of Sustainability' For Cities?" *Sci Soc* 64 (2014a): 32–36.
21. Oliveira, E.A., Andrade, J.S., Jr and Makse, H.A. "Large Cities Are Less Green", 2014, http://arxiv.org/abs/1401.7720.
22. Ho, M.W. "Larger Cities in USA Less Green Than Small Ones". *Sci Soc* 64 (2014): 32–36.
23. Ho, M.W. *The Rainbow and the Worm: The Physics of Organisms*, 3rd edition. World Scientific, Singapore, 2008.
24. Ho, M.W. *Living Rainbow* H_2O. World Scientific, Singapore, 2012.
25. Ho, M.W. and Ulanowicz, R.E. "Sustainable Systems As Organisms"? *BioSystems* 82 (2005): 39–51.
26. Ho, M.W. "Circular Thermodynamics of Organisms and Sustainable Systems". *Systems* 1 (2013a): 30–49.
27. Harskamp, J. and Dijstelberge, P. "Piazza del Campo (Siena) or Rather: Not the Piazza Del Campo", 17 November 2013, http://abeautifulbook.wordpress.com/2013/11/17/piazza-del-campo-siena-or-rather-not-the-piazza-del-campo/.
28. Ho, M.W. "Golden Cycles of Organic Spacetimes". *Sci Soc* 62 (2014b): 32–35.
29. "Metabolic Pathways". Sigma Aldrich, Sigmaaldrich.com/content/dam/sigma-aldrich/docs/Sigma/General_Information/metabolic_pathways_poster.pdf, accessed 27 October 2014.
30. Ho, M.W., Burcher, S., Lim, L.C. *et al.* "Food Futures Now". ISIS/TWN report, 2008, http://www.i-sis.org.uk/foodFutures.php.
31. Vitiello, G. "On the Isomorphism Between Dissipative Systems, Fractal Self-Similarity and Electrodynamics: Toward an Integrated Vision of Nature". *Systems* 2 (2014): 203–216.

32. Prigogine, I. "Time, Structure and Fluctuations". Nobel Prize lecture, 8 December 1977, http://www.nobelprize.org/nobel_prizes/chemistry/laureates/1977/prigogine-lecture.pdf.
33. The Circular Economy 100, Ellen MacArthur Foundation, accessed 27 August 2014, http://www.ellenmacarthurfoundation.org/business/ce100.
34. Ho, M.W. "Circular Economy of the Dyke-Pond System". *Sci Soc* 32 (2006): 38–41.
35. Ho, M.W. "Sustainable Agriculture, Green Energies and the Circular Economy". *Sci Soc* 46 (2010): 8–13.
36. Ho, M.W. "Paradigm Shift Urgently Needed in Agriculture: UN Agencies Call for an End to Industrial Agriculture and Food System". *Sci Soc* 60 (2013b): 4–10.
37. Ho, M.W. Chapter 1 Commentary XIII: "Sustainable Agriculture and Off-Grid Renewable Energy". In UNCTAD, *Wake Up Before It Is Too Late: Make Agriculture Truly Sustainable Now for Food Security in a Changing Climate*, Trade and Environment Review 2013, pp. 19–21, UNCTAD, Geneva, 2013.
38. Ho, M.W. "Sustainable Agriculture Essential for Green Circular Economy". ISIS lecture in Ten+One Conference, Bradford University, Bradford, 29 November–1 December 2010, http://www.i-sis.org.uk/sustainableAgricultureEssentialGreenCircularEconomy.php.

15

Quantum Electrodynamics of Water and Life

15

Water the Means, Medium and Message of Life

Water is essential for life. Yet this simple, ubiquitous chemical compound has remained completely mysterious for centuries until quite recently. New evidence indicates that liquid water may be quantum coherent even at ordinary temperatures and pressure. It associates with macromolecules and membranes in a liquid crystalline configuration that enables enzymes and nucleic acids to function as quantum molecular machines that transform and transfer energy at close to 100% efficiency. Liquid crystalline water at interfaces also provides the excitation energy that enables it to split into hydrogen and oxygen in photosynthesis, simultaneously generating electricity for intercommunication and for the redox chemistry that ultimately powers the entire biosphere. Water is the means, medium and message of life — "the rainbow within that mirrors the one in the sky".

More than 20 years ago, I peered down a polarizing light microscope and saw a rainbow dancing inside a worm (see "13 — Why Beauty Is Truth and Truth Beauty", Fig. 13.4, in this book). It took a while to decipher the meaning of that vision.

Typically, colours are generated by birefringent crystals such as quartz with ordered arrangement of atoms, or liquid crystals with ordered alignment of electrically polarized molecules. Birefringent crystals split plane-polarized white light — containing all the frequencies in the visible spectrum from red to violet — into two orthogonally oriented rays, one propagating more slowly than the other. When the two rays are recombined with a second polarizer (analyser), the two rays interfere, and that is how colours are generated. To amplify the effect, a full wave plate — the wavelength of green light — is added, hence the characteristic pink background.

We had stumbled upon a new setting for the wave plate that is especially good for biological liquid crystals with low birefringence. Instead of positioning the vibrating directions of the wave plate at the usual 45° to the polarizers, we placed it at a small angle of 7.5°, which greatly improved the contrast.[1–3]

The rainbow in the worm means that the organism is liquid crystalline *and* coherent to a high degree, even quantum coherent, as consistent with other evidence (see below). It means that the entire organism is electrically polarized from head to tail, like a single uniaxial crystal. Not only are the macromolecules in all the tissues and cells perfectly aligned, but also the 80% by weight of water in the organism.[4] It is actually the water that makes the entire organism liquid crystalline because this water is liquid crystalline, as others also discovered later. To see the rainbow colours in the living organism, the liquid crystalline molecules not only have to be fully aligned, but also moving coherently, macromolecules *and* water molecules together. (This has been demonstrated recently by nuclear magnetic resonance (NMR) measurements for a population of water molecules associated with a single molecule of ubiquitin protein trapped inside a water droplet in oil, which moves in concert with different parts of the protein at rates varying by more than ten orders of magnitude from one region to another.[5]) As coherent molecular motions are much slower than visible light vibrations, the ordered alignment of molecules will still be registered by the light transmitted. The most active parts are always the brightest, indicating that the molecular motions are the most coherent. The effect is similar to that obtained using a very sensitive camera with ultra-short exposure time to capture sharp images of a fast-moving object. In other words, the colour images of live organisms are direct evidence of coherent molecular motions in the organism that depend absolutely on the liquid crystalline water.

Water the Medium of Life

The liquid crystalline living water enables macromolecules to function as quantum molecular machines that transfer and transform energy at close to 100% efficiency. If not for that, life would be impossible. Water is the supreme medium of life. The archetypal quantum molecular energy machine is the enzyme. Enzymes speed up chemical reactions in organisms[6] by a factor of 10^{10} to 10^{23} and they cannot do that without water.[7] However, the role of water is still hardly recognized in the conventional biochemical community. Water gives flexibility to proteins, reduces the energy barrier between reactants and products, and increases the probability of quantum tunnelling by a transient compression of the energy barrier.

The rainbow worm was the immediate inspiration for *The Rainbow and the Worm: The Physics of Organisms*,[8] first published in 1993 and now in its third edition. It presents empirical evidence and theoretical arguments in support of the idea that the organism is quantum coherent, and that liquid crystalline water plays the lead in creating and maintaining the coherence of organisms. In it, I define quantum coherence after American quantum physicist and Nobel Laureate Roy Glauber[9] in terms of *factorizability*, which is later expressed as follows:[10] a system is quantum coherent if its parts are so perfectly correlated that their cross-correlations factorize exactly as the product of the individual self-correlations, so that each appears paradoxically as though totally uncorrelated with the rest. It is a state of maximum local freedom *and* global cohesion; something that is impossible in a classical mechanical system.

Quantum coherence is a sublime state of wholeness; a superposition of coherent activities over all space-times, a pure (ideal) dynamic state towards which the system tends to return asymptotically (approach it closely). I call it "quantum jazz" to highlight the immense diversity and multiplicity of supramolecular, molecular and submolecular players, the complexity and the coherence of the performance, and above all, the freedom and spontaneity, with each and every player improvising from moment to moment yet keeping in step and in tune with the whole.

Quantum coherent energy storage and mobilization over all space-times dovetails with a circular thermodynamics of organisms (presented in the same book) that approaches the zero-entropy ideal simultaneously under both equilibrium and far-from equilibrium conditions[11] and is a good model also of sustainable ecological/economic systems (see "14 — Sustainable Cities As Organisms: A Circular Thermodynamics Perspective" in this book).

But the full extent to which life, the universe, and everything depends on water is still unfolding. Astronomers now think water is actually the most abundant substance in the universe, and was present at the birth of the universe.[12] Perhaps the universe too, is powered by water electricity, just as the organism is powered by water electricity, as I shall presently show.

I wrote *Living Rainbow* H_2O[10] focussing on water in living organisms as a sequel to the *Rainbow Worm*.[8] Water scarcity is now threatening the survival of people and planet, especially in times of climate change.[13] Yet, most people still don't know why it is so essential for life, and hence how precious it really is, and why the quality of water is so important. This new book is a synthesis of the recent findings in the quantum physics and chemistry of water that have unlocked some of the mysteries of why it is so fit for life. It is "the means, medium and message of life, the beautiful rainbow within that mirrors the one in the sky".

Water is Weird and Wonderfully Fit for Life

The water molecule is a dipole with separated positive and negative charges associated with the oxygen and the two hydrogen atoms respectively, so it can engage in dipole interactions with other molecules of water or other dipoles. However, it prefers to hydrogen-bond whenever possible, where the hydrogen atom of one molecule is shared between two oxygen atoms in neighbouring molecules. The favoured configuration is a tetrahedron in which a molecule accepts two hydrogen atoms and donates two hydrogen atoms to neighbouring molecules. It is estimated that at ordinary temperatures and pressures, up to 90% of the water molecules are hydrogen-bonded,[14] although the hydrogen bonds flicker on and off randomly in a matter of pico (10^{-12}) seconds.

Water is notorious for a host of anomalous properties (some listed in Box 15.1) attributed to its propensity to form hydrogen bonds. The same anomalies are widely regarded as precisely the qualities that make water fit and essential for life.

Box 15.1

Major Anomalies of Water[15,16]

- Neighbours of oxygen form gases with hydrogen at ordinary temperatures and pressures, but water boils at 100°C and only freezes at 0°C under standard atmospheric pressure, which means organisms are composed of and bathed in liquid water on earth.
- Water has a high heat capacity and high thermal conductivity, thereby preventing temperature fluctuations, enabling organisms to control their body temperature; in large bodies of water such as oceans and seas, the high heat capacity enables them to serve as heat reservoirs, thereby moderating our climate.
- Other liquids increase in density on becoming solid, but ice is lighter than water and floats on it, most fortunately for fish and other aquatic inhabitants.
- Liquid water can be supercooled below 0°C without freezing, but on heating, the supercooled liquid does not expand like other liquids; instead it contracts to a maximum density at about 4°C; this is very important for the hydrological cycle[17] as it plays a key role in rain water percolating underground to refill the aquifers.
- Water's compressibility atypically decreases with increasing temperature reaching a minimum at about 46.5 °C.
- At ordinary temperatures below 35°C, increasing pressure results in decreased viscosity, which facilitates flow, again at odds with other liquids.

Quantum Delocalization of the Hydrogen Bond

The key to water's remarkable properties is the hydrogen bond interconnecting water molecules. It is usually regarded as classical and electrostatic, but many observations are inconsistent with that picture.

Nobel laureate American chemist Linus Pauling (1901–1994) first suggested in 1935 that the hydrogen bond and covalent bond in ice may switch places between two neighbours on either side in view of residual entropy (randomness) existing even at very low temperatures[18] and thus, the hydrogen bond must be at least partly covalent.

In 1999, the hydrogen bond in ordinary ice was probed with inelastic X-ray scattering at the European Synchrotron Radiation Facility in Grenoble, France.[19] Beams of X-rays are bounced off electrons so both the energy of the electron and the X-ray are changed. The intensity of scattering was measured as a function of energy or momentum at different orientations of a carefully prepared slab of ice and the anisotropy plotted. The results were in good agreement with the predictions based on a fully quantum mechanical model, while predictions based on the classical electrostatic model did not agree with the data at all (Fig. 15.1).

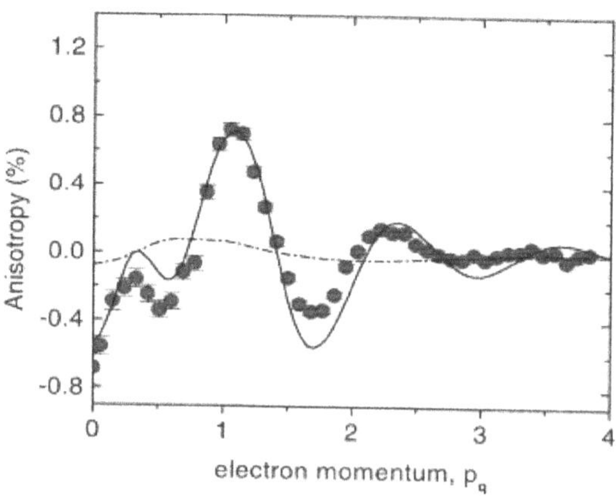

Figure 15.1 Anisotropy versus electron momentum; solid line quantum model, dotted line classical model, filled circles data.

In the same year, time-resolved pump-probe laser spectroscopy on *liquid* water at room temperature and pressure revealed resonant intermolecular transfer of OH-stretch excitations mediated by dipole-dipole interactions that was substantially faster than the classical Förster mechanism would predict,[20] which again suggesting a quantum mechanism. Further experiments a few years later showed that not only did the electron of the hydrogen bond fail to conform to the classical electrostatic model but that the proton too was quantum mechanical. Using ultrafast femto-second (10^{-15} s) pulses of infrared light to excite and probe the O-H covalent bond vibration in liquid water,[21] researchers found that only quantum mechanical calculation of the vibrational wave functions could reproduce the experimental absorption spectrum. The excited proton was simultaneously (delocalized) at the O-H bond distance from either of two neighbouring oxygen atoms (belonging to two different water molecules) (Fig. 15.2).

The excited v = 2 state for proton delocalization is 6 500 cm^{-1} (0.82 eV), less than 20% the O-H bond energy of 38 750 cm^{-1} (4.8 eV). The delocalization of protons in water increases the probability of proton transfer, which may play a crucial role in intercommunication (see later).

Quantum Coherent Water Makes Life on Earth

Standard quantum theory does not predict quantum coherence for liquid water, largely because it ignores both quantum fluctuations and the interaction between matter and electromagnetic field; these are only taken into account in quantum electrodynamics field theory. But conventional quantum electrodynamics field theory applies only to gases.

Italian quantum field theorists Giuliano Preparata (1942–2000) and Emilio Del Giudice (1940–2014) and their colleagues extended conventional quantum electrodynamics theory to the condensed phase of liquids. They showed that interaction between the vacuum (or ambient) electromagnetic field and liquid water induces the formation of large, stable coherent domains (CDs) of about 100 nm

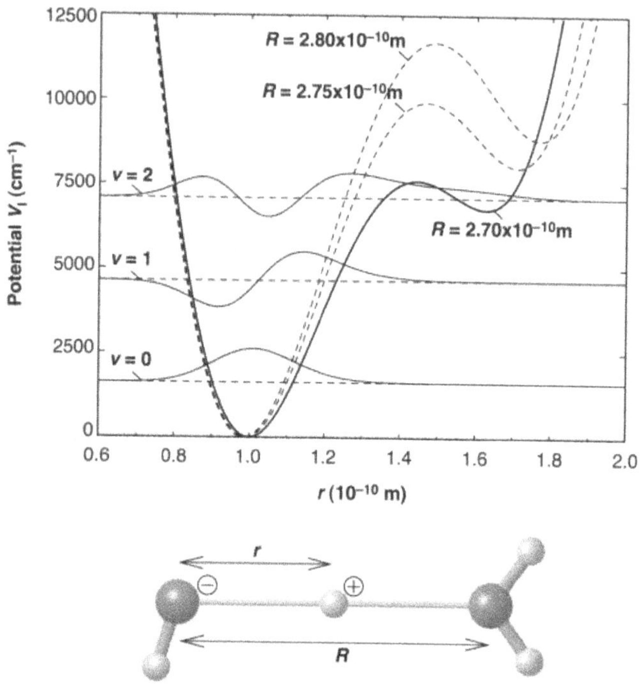

Figure 15.2 Excitation of the O-H bond and delocalization of the proton. The three lowest energy (v = 0, 1, 2) vibrational wave functions in the potential for R (O-O distance) = 2.7 Å.

in diameter at ordinary temperature and pressure, and these CDs may be responsible for all the special properties of water including life itself.[22–26] In particular, the propensity of water molecules to form tetrahedral directed hydrogen bonds is a *consequence* of the excited state of water in the coherent domains that would not happen otherwise.

The CD is a resonating cavity produced by the electromagnetic field that ends up trapping the field because the photon acquires an imaginary mass in water, so the frequency of the CD electromagnetic field becomes much smaller than the frequency of the free field with the same wavelength. In other words, as water is a much denser medium than the vacuum or air, electromagnetic fields of the same wavelength will vibrate at a much lower frequency.

Under ambient conditions, water is an approximately equal mixture of coherent domains surrounded by incoherent regions. This picture, according to Del Giudice and colleagues, is reflected in the many observations supporting a two-state model of liquid water.[15,27,28] (It is perhaps more accurate to say that the water molecules are dancing between the tetrahedral configuration corresponding to the CDs and non-tetrahedral configuration, so both the CD and non-CD molecules are interchangeable.)

What is special about water is that the coherent oscillation occurs between the ground state and an excited state at 12.06 eV, just below the ionizing threshold of water at 12.60 eV. In liquid water, the CD of about 100 nm in diameter contains millions of water molecules, and hence close to a million almost free electrons (the fraction of molecules in the excited state is estimated to be 0.13,[25] which can be donated readily to electron acceptors). It is this property that crucially makes water the means of life and the message of life.

Water the Means of Life

Water is the basis of the energy metabolism that powers all living processes; it is the chemistry, the means of life. Water is also the electricity of life for intercommunication; it is both the message and messenger (see next section).

The abundant life on earth, including the human species, depends on photosynthesis in green plants, algae and cyanobacteria that traps the energy of sunlight by means of chlorophyll to split water into hydrogen, electrons and oxygen (Eq. 1), giving life access to an enormous energy source, and more importantly, liberating oxygen for the evolution of teaming millions of air-breathing species that fill the earth. (If water is the most abundant substance in the universe, could there be abundant extraterrestrial beings not too different from earthlings in the rest of the universe?)

$$H_2O \rightarrow 2\,H^+ + 2\,e^- + O \qquad (1)$$

The hydrogen ion (protons) and electrons go to reduce (fix) carbon dioxide into carbohydrates and the biomass of photosynthetic organisms, which serve as food for herbivores, and down the food web to include the vast majority of air-breathers. During respiration, carbohydrates are broken down by oxidation (with oxygen) in mitochondria to release energy for growth and reproduction, regenerating carbon dioxide and water.[10] This completes the living dynamo of photosynthesis and respiration, the magic roundabout that turns inanimate substances into living organisms (Fig. 15.3).

However, it takes 12.6 eV to split water, an energetic photon in the soft X-ray region that would destroy life, and is not what green plants and cyanobacteria use. They use mainly red and to some extent, blue light in the visible spectrum.

More than 50 years ago, Nobel Laureate Hungarian-born American biochemist Albert Szent-Györgyi (1893–1986) had already suggested[29] that water at interfaces was the key to life. He proposed that water next to membranes is in the excited state, which requires considerably less energy to split than water in the ground state. Most, if not all, water in living organisms is interfacial water, as it is almost never further away from surfaces such as membranes

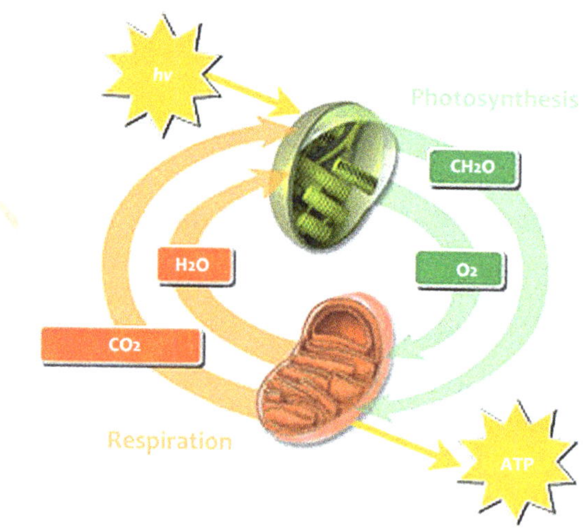

Figure 15.3 The dynamo of life running on water.

or macromolecules than a fraction of a micron. This is the "liquid crystalline water" mentioned at the beginning of this essay, which is present in all organisms, animals and plants. And chloroplasts where water is split are particularly rich in membrane surfaces, as are mitochondria, where water is regenerated.

A sign that interfacial water is excited is a voltage difference at the boundary between interfacial water and bulk water, as predicted by Szent-Györgyi[29] and observed not long afterwards, as Del Giudice pointed out.[23]

A vivid demonstration of interfacial water hundreds of microns thick next to the surface of a hydrophilic gel was presented by Gerald Pollack and his team at University of Washington Seattle,[30–33] which excludes microspheres as well as other solutes such as proteins and dyes, and hence referred to as an "exclusion zone" (EZ).

Del Giudice and colleagues[24] suggested that EZ water is in fact a giant coherent domain stabilized on the surface of the hydrophilic gel. As coherent water is excited water with a large collective of almost free electrons, it can easily transfer electrons to molecules on its surface. The boundary between fully coherent interfacial water and normal bulk water becomes a "redox pile". In line with this proposal, EZ water does have a potential difference of –100 mV to –200 mV with reference to the bulk water, and can act as a battery.[33,34] Pollack's explanation for the formation of EZ is based purely on classical physics, and although it converges in some respects with Del Giudice's quantum field theoretical explanation, there are key differences (for details see "16 — Illuminating Water and Life: Essay in Honour of Emilio Del Giudice" in this book).

Water the Message of Life: Superconducting Proton Currents

The core chemistry of life is reduction-oxidation or redox reactions that transfer electrons between chemical species. The movement of electrons is nothing if not an electric current. However, water electricity is special in that it also involves the movement of positive charges, i.e., protons.[35,36] Water conducts protons by "jumping"

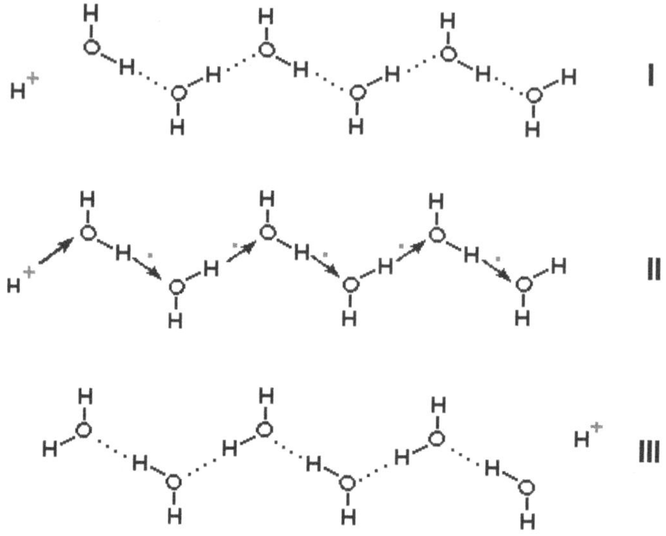

Figure 15.4 Proton jump conduction down a string of hydrogen-bonded water molecules.[8]

down a chain of water molecules connected by hydrogen bonds (Fig. 15.4). A proton leaps on one end of the chain, and a second leaps off at the other end, while electrons are displaced in the other direction. This is much faster than an ordinary electric current involving the flow of electrons, and the total current is twice that due to electrons alone.

Structured water confined in carbon nanotubes less than 5 nm in diameter was first demonstrated by researchers at Drexel University Philadelphia in the United States using high resolution transmission microscopy accompanied by parallel modelling with a Hyperchem software package. The water appeared completely different from that confined in larger nanotubes[37,38] (Fig. 15.5). I suggested that water confined in the small diameter nanotube, being more ordered, could be superconducting because proton jump-conduction could occur simultaneously down multiple chains of hydrogen-bonded water molecules.[39]

Later, American biochemist Gary Fullerton (now at University of Colorado Denver) and colleagues offered a convincing model

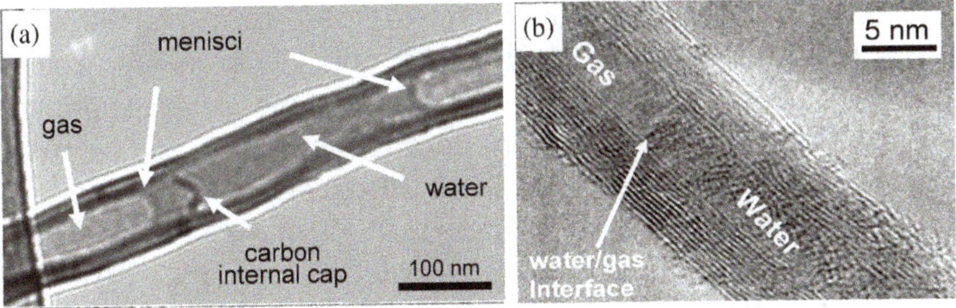

Figure 15.5 Structured water confined in narrow nanotube (right) compared with ordinary water in wide nanotube (left).

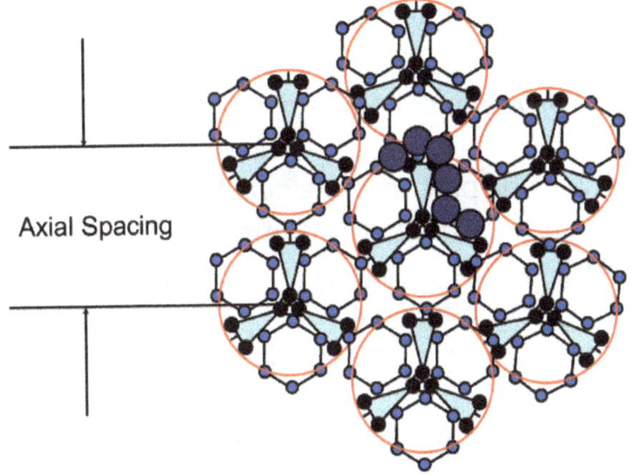

Figure 15.6 Collagen water structure revealed: cross-section of a fibril consisting of seven triple helical collagen molecules with six-member water nanotubes fitting within the grooves of the triple helices.

of liquid crystalline six-member diameter nanotubes of water interwoven with the triple-helix molecules of collagen molecules in the collagen fibres[40] (Fig. 15.6) reminiscent of those identified in the small-diameter carbon nanotubes, suggesting that these water structures in the extracellular matrix could also be superconducting.[41] That is just the beginning of a story yet to be told by anyone else.

It has been known since the 1970s that collagen supports jump-conduction of protons, and proton conductivity goes up exponentially with water content. Dutch biophysicist G.H. Bardelmeyer[42] found that electrical conductivity in bovine Achilles tendon is fully determined by hydration, and the current is primarily carried by protons at water contents up to 45%, and by small ions at water contents beyond 65%. Between water contents of 8.5% and 126%, conductivity went up by eight orders of magnitude. He estimated that pure water's dissociation constant is 10^{-5} that of absorbed water; i.e., adsorbed water is more likely to let go of protons. More than ten years later, Japanese biophysicist Naoki Sasaki at Hokkaido University found that the conductivity of collagen increased markedly with water absorbed — at an exponent of 5.1–5.4 — between a water content of 0.1 and 0.3 g/g,[43] suggesting that continuous chains (of five or more ordered water molecules) adsorbed in collagen enable proton jump-conduction to take place.

As seen earlier, the proton is in a delocalized quantum state even in bulk water under ambient conditions. This delocalization has been confirmed recently for water confined in a range of hydrophilic gels, as reviewed elsewhere.[44] Delocalized protons mean jump-conduction can be *very* fast indeed. We shall see exactly how fast it could be.

Nafion is a synthetic polymer used as a proton exchange membrane. Its chemical structure is a hydrophobic backbone with hydrophilic side chains. When hydrated, it forms channels of inverse micelles with hydrophilic groups facing the cavity and hydrophobic groups facing out. This structure is a good model of the living cell. The interstices between fibres of the cytoskeleton and cytoplasmic membranes form inverse micelle nanospaces and channels that drastically alter enzyme/substrate relationships, often greatly enhancing enzyme activity compared to bulk phase thermodynamic models that still dominate conventional cell biology (see Chapter 18 of *Living Rainbow* H_2O[10]).

When Nafion was drawn out into fibres by electrospinning, the proton conductivity of fibres with diameters >2 μm was similar to the bulk Nafion film (~0.1 S/cm). However, when the fibre diameter was <1 μm, proton conductivity rose sharply with decreasing fibre diameter and reached 1.5 S/cm for the 400-nm diameter fibre, an order of magnitude higher than the bulk Nafion film, or silicon, a semiconductor. Conductivity of the fibre also increased a hundred-fold as relative humidity rose from 50% to 90%; in comparison, conductivity of the bulk film increased only ten-fold.[45]

The inverse micelle model may be even more relevant to the extracellular *milieu* of multicellular animals, which is traversed by collagen fibres consisting of fibrils interwoven with nanotubes of water (Fig. 15.6). These water channels aligned with collagen fibres are most likely the anatomical correlates of the acupuncture meridians of traditional Chinese medicine, as I and David Knight first suggested in 1998;[46] the hypothesis is still much alive and untested.[47]

What one needs to imagine are proton and electron currents flowing inside the cells and over extracellular distances, delivering physical and chemical messages concerning the redox status, setting in motion the requisite core chemical reactions that restore local and global energy balance (and also the peripheral chemistry that forms the basis of the highly nuanced passions and feelings that make life so exciting for organisms).

I hinted that protons could move even faster than simple jump conduction along the aligned water molecules. There is new evidence for that.

Protons in Quantum Jazz Concert

Multiple protons in water clusters have been caught quantum-tunnelling in concert, resulting in superfast and accurately directed current flow, leading us to ask if this could be happening in our body.[48]

Chinese researchers Wang En-Ge and Jiang Ying at Peking University in Beijing and their colleagues designed experiments to catch four protons switching partners *simultaneously* in a cyclic water tetramer under a cryogenic scanning tunnelling microscope (STM). The tetramers were carefully constructed so they are *chiral*, i.e., with their hydrogen bonds all pointing in one direction, either clockwise, or anticlockwise.[49] Each proton in this four-membered ring is covalently bonded to an oxygen on the left (or right), and hydrogen-bonded to the oxygen on another water molecules on the right (or left) for the clockwise (anticlockwise) state. To convert between the clockwise and anticlockwise states the protons essentially change partners, from a hydrogen bond to a covalent bond and from acovalent bond to a hydrogen bond, and they all have to do it *at the same time.*

Ab initio (starting from first principles) quantum mechanical density function theory calculations (for the electronic structure of the many body systems) show that the energy barrier to switching is lowest when the four protons hop in concert from the hydrogen-bond donor to the acceptor molecule. Any conceivable sequential or stepwise rearrangement of the protons would result in significantly higher energy barriers.

The mechanism proposed is similar to that suggested earlier for the concerted tunnelling of six protons within a chiral hexamer in ordinary ice at 50 K carried out by Christof Dreschsel-Grau and Dominik Marx at Rhur-University Bochum in Germany[50] in a similar quantum mechanical simulation. They found unexpectedly, that only a moderate contraction of the oxygen skeleton of the hexamer is sufficient to enable quasiparticle-like concerted tunnelling of all six protons. Commenting on the new results from Beijing, Dreschsel-Grau and Marx speculate that the same can occur also in larger water clusters, or maybe chirality could be transferred between individual clusters, and that it might also occur at higher temperatures.[51] Taking this speculation a bit further, I suggested that concerted quantum tunnelling of many protons may indeed happen in our body, through the nanotubes of liquid crystalline water molecules associated with collagen (see Fig. 15.6).[48] If so, proton conduction through liquid crystalline water may be infinitely faster (in a time period during which a single proton can tunnel), more substantial, and precisely directed than previously thought.

Collagen, composed of three polypeptide chains wound together in a right-handed triple-helix, is intrinsically chiral. Chiral changes are conventionally studied by optical circular dichroism (CD) spectroscopy, due to the unequal absorption of right and left plane-polarized light. However, the contrast in CD is typically less than 1%. Second harmonic generation (SHG) is a nonlinear optical phenomenon in which a birefringent crystalline medium combines two photons to generate a new one with double the frequency (and energy). Collagen exhibits exceptionally strong SHG, as discovered more than 30 years ago, and has been used in imaging ever since. It has also been shown that chirality can give rise to different efficiency of SHG for left circularly polarized and right circularly polarized light, resulting in SHG-CD responses. These considerations prompted researchers at National Taiwan University in Taiwan and Tampere University of Technology in Finland to provide definitive evidence that the chirality of collagen can indeed give rise to strong SHG-CD responses, resulting in 100% contrast with submicron resolution of individual collagen fibres in a laser scanning microscope.[52]

The imaging was done on sliced ligament of a freshly slaughtered young pig ~10 μm thick sealed on the microscope slide with "abundant water". The reason is simple. There is already evidence that the SHG signal depends almost entirely on the liquid crystalline water associated with the collagen fibres.[47] It is therefore highly likely that the CD signal also comes from the liquid crystalline water. This liquid crystalline water is most likely chiral based on its probable structure in the form of a nanotube six water molecules in diameter nested in each groove of the triple helix (Fig. 15.6).

Quantum tunnelling proton currents within the body would indeed enable "every single molecule to intercommunicate [instantaneously] with every other", as they are all performing the most exquisite quantum jazz[8,10] within the liquid crystalline medium.

Electromagnetic Language of Cells and Organisms

Finally, how do cells and molecules actually find one another? Conventional wisdom says this is through hormones and receptors, cell-cell recognition molecules, and a lock-and-key principle for molecules that somehow bump into each other at random (see current authoritative textbook *Molecular Biology of the Cell*.[53]

Actually, there is substantial evidence that molecules find each other by electromagnetic fields, by resonating to common frequencies.[8,10] Molecules that react together were found to share a common frequency; which is thought to be how they can attract each other.[54] This makes even more sense in the context of quantum coherent water.

Del Giudice and colleagues[22–25] argue that water CDs can be easily excited, and are able to capture surrounding electromagnetic fields to produce coherent excitation in the frequencies of the external fields. This in turn enables selective coherent energy transfer to take place. All molecules have their own spectrum of vibrational frequencies. If the molecule's spectrum contains a frequency matching that of the water CD, it would get attracted to the CD, become a guest participant in the CD's coherent oscillation, and settle on the CD's surface. Furthermore, the CD's excitation energy would become available to the guest molecules as activation energy for the chemical reactions.

Is it possible that cells or organisms as a whole also intercommunicate by means of electromagnetic and electric signals? This is completely uncharted territory as far as conventional cell biology is concerned. It is the water in us that gives us life; and makes us sensitive to electromagnetic fields; there is a distinct possibility that we are sensitive to the fields of other organisms, as we are sensitive to fields of the sun and the earth, and possibly also from distant stars; all without our conscious knowledge.[55]

To Conclude

Life appears quantum electrodynamical through and through, and water is at the heart of it all. A whole new vista has opened up, thanks to all the dedicated water scientists who have contributed to this vision and whose work I have cited, and especially the pioneers Albert Szent-Györgyi[29] and Chinese-born American physiologist Gilbert Ling,[56] both of whom recognized the central role of water in producing and animating life more than half a century ago. Our adventures have only just begun.

References and Notes

1. Ho, M.W. and Lawrence, M. Interference Colour Vital Imaging: A Novel Noninvasive Technology. *Microscopy Analy* (September 1993): p. 26.
2. Ho, M.W. and Saunders, P.T. "Liquid Crystalline Mesophases in Living Organisms". In *Biodynamics and Biocommunication* (M.W. Ho, F.A. Popp and U. Warnke, eds.), pp. 213–228, World Scientific, Singapore, 1994.
3. Ross, S., Newton, R.H., Zhou, Y.M., Haffegee, J., Ho, M.W., Bolton, J. and Knight, D. "Quantitative Image Analysis of Birefringent Biological Materials". *J Microscopy* 187 (1997): 62–67.
4. Ho, M.W., Zhou, Y.M., Haffegee, J., Watton, A., Musumeci, F., Privitera, G., Scordino, A. and Triglia, A. "The Liquid Crystalline Organism and Biological Water". In *Water in Cell Biology* (G. Pollack, I.L. Cameron and D.N. Wheatley, eds.), pp. 219–251, Springer, Berlin, 2006.
5. Nucci, N.V., Pometun, M.S. and Wand, A.H. "Site-Resolved Measurement of Water-Protein Interactionb Solution NMR". *Nat Struct Mol Biol* 18 (2010): 245–250.
6. Kraut, D.A., Carrol, K.S. and Herschlag. D. "Challenges in Enzyme Mechanism and Energetics". *Ann Rev Biochem* 72 (2003): 517–571.
7. Ho, M.W. "Dancing with Macromolecules". *Sci Soc* 48 (2010): 11–15.
8. Ho, M.W. *The Rainbow and the Worm: The Physics of Organisms*, 3rd edition. World Scientific, Singapore, 2008.
9. Glauber, R.H. "The Quantum Theory of Optical Coherence". *Phys Rev* 130 (1963): 2529–2539.
10. Ho, M.W. *Living Rainbow H2O*. World Scientific, Singapore and Imperial College Press, London, 2012a.
11. Ho, M.W. "Circular Thermodynamics of Organisms and Sustainable Systems". *Systems* 1 (2013a): 30–49.
12. "Astronomers Find Largest, Most Distant Reservoir of Water". *Mission News*, NASA, 22 July 2011, http://www.nasa.gov/topics/universe/features/universe20110722.html.
13. Ho, M.W. "World Water Supply in Jeopardy". *Sci Soc* 56 (2012b): 38–43.

14. Bakker, H.J. and Skinner, J.L. "Vibrational Spectroscopy As a Probe of Structure and Dynamics of Water". *Chem Rev* 110 (2010): 1498–1517.
15. Chaplin, M. "What Is Liquid Water?" *Sci Soc* 58 (2013): 41–45.
16. Vedamuthu. M., Singh, S. and Robinson, C.W. "Properties of Liquid Water: Origin of the Density Anomalies". *J Phy Chem* 98 (1994): 2222–2230.
17. Coat, C. *Living Energies*, Gateway Books, Bath, 1996.
18. Pauling, L. "The Structure and Entropy of Ice and Other Crystals with Some Randomness of Atomic Arrangement". *J Am Chem Soc* 57 (1935): 2680–2684.
19. Isaacs, E.D., Shukla, A., Platzman, P.M., Hamann, D.R., Barbiellin, B. and Tulk, C.A. "Covalency of the Hydrogen Bond in Ice: A Direct X-Ray Measurement". *Phys Rev Lett* 82 (1999): 600–603.
20. Woutersen, S. and Bakker, H.J. "Resonant Intermolecular Transfer of Vibrational Energy in Liquid Water". *Nature* 402 (1999): 507–509.
21. Bakker, J.H. and Nienhuys, H.K. "Delocalization of Protons in Liquid Water". *Science* 297 (2002): 587–590.
22. Arani, R., Bono, I., Del Giudice, E. and Preparata, G. "QED Coherence and the Thermodynamics of Water". *Int J Mod Phys B* 9 (1995): 1813–1841.
23. Del Giudice, E. Old and New Views on the Structure of Mattera the Special Case of Living Matter". *J Phys Conf Ser* 67 (2007): 012006.
24. Del Giudice, E., Spinetti, P.R. and Tedeschi, A. "Water Dynamics at the Root of Metamorphosis in Living Organisms". *Water* 2 (2010): 566–586.
25. Del Giudice, E. and Pulselli, R.M. "Formation of Dissipative Structure in Liquid Water". *Int J Des Nat Ecodyn* 5 (2010): 21–26.
26. Ho, M.W. "Quantum Coherent Water and Life". *Sci Soc* 51 (2011): 26–28.
27. Ho, M.W. "Two-States Water Explains All?" *Sci Soc* 32 (2006a): 17–18.
28. Huang, C., Wikfeldt, K.T., Tokushima, T. and Nilsson, A. "The Inhomogeneous Structure of Water at Ambient Conditions". *Proc Natl Acad Sci* 106 (2009): 15214–15218.

29. Szent-Györgyi, A. *Introduction to a Supramolecular Biology*. Academic Press, London, 1960.
30. Zheng, J.M. and Pollack, G.H. "Long-Range Forces Extending from the Polymer Gel Surfaces". *Phys Rev E* 68 (2003): 314–318.
31. Ho, M.W. "Water Forms Massive Exclusion Zones". *Sci Soc* 23 (2004): 50–51.
32. Pollack, G.H. "Water, Eenergy and Life: Fresh Views from the Water's Edge". *Int J Des Nat Ecodyn* 5 (2010): 27–29.
33. Pollack, G.H. *The Fourth Phase of Water*. Ebner & Sons, Seattle, 2013.
34. Ho, M.W. "Liquid Crystalline Water at the Interface". *Sci Soc* 38 (2008): 37–39.
35. Riistama, S., Hummer, G., Puustinen, A., Dyer, R.B., Woodruff, W.H. and Sikatrom, M. "Bound Water in the Proton Translocation Mechanism of the Haem-Copper Oxidation". *FEBS Lett* 414 (1997): 275–289.
36. Ho, M.W. "Positive Electricity Zaps Through Water Chains". *Sci Soc* 28 (2005a): 49–50.
37. Ye, H., Naguib, N. and Gogotsi, Y. "TEM Study of Water in Carbon Nanotubes". *JEOL News* 39 (2004): 2–7.
38. Naguib, N., Ye, H., Gogotsi, Y., Yazicioglu, A.G., Megaridis, C.M. and Yoshimura, M. "Observation of Water Confined in Nanometer Channels of Closed Carbon Nanotubes". *Nano Lett* 4 (2004): 2237–2243.
39. Ho, M.W. "First Sighting of Structured Water". *Sci Soc* 28 (2005b): 47–48.
40. Fullerton, G.D. and Amurao, M.R. "Evidence That Collagen and Tendon Have Monolayer Water Coverage in the Native State". *Int J Cell Biol* 30 (2006): 56–65.
41. Ho, M.W. "Collagen Water Structure Revealed". *Sci Soc* 32 (2006b): 15–16.
42. Bardelmeyer, G.H. "Electrical Conduction in Hydrated Collagen: I. Conductivity Mechanisms". *Biopolymers* 12 (1973): 2289–2302.
43. Sasaki, N. "Dielectric Properties of Slightly Hydrated Collagen: Time-Water Content Superposition Analysis". *Biopolymers* 23 (1984): 1725–1734.

44. Ho, M.W. "Superconducting Quantum Coherent Water in Nano-space Confirmed". *Sci Soc* 55 (2012c): 48–51.
45. Dong, B., Gwee, L., Salas-de-la-Cruz, D., Winey, K.I. and Elabd, Y.A. "Super-Proton Conductive High-Purity Nafion Nanofibers". *Nano Lett* 10 (2010): 3785–3790.
46. Ho, M.W. and Knight, D.P. "The Acupuncture System and the Liquid Crystalline Collagen Fibers of the Connective Tissues". *Am J Chin Med* 26 (1998): 251–263.
47. Ho, M.W. "Super-Conducting Liquid Crystalline Water Aligned with Collagen Fibres in the Fascia As Acupuncture Meridians of Traditional Chinese Medicine". *For Immunopathol Dis Therap* 3 . (2012d): 221–236.
48. Ho, M.W. "Protons in Quantum Jazz Concert". *Sci Soc* 66 (2015): 46–47.
49. Meng, X., Guo, J., Peng, J., Wang, Z., Shi, J.R., Li, X.Z., Wang, E.G and Jiang, Y. "Direct Visualization of Concerted Proton Tunneling in a Water Nanocluster". *Nat Phys*, published online 16 February 2015, DOI: 10.1038/NPHYS3225.
50. Dreschsel-Grau, C. and Marx, D. "Quantum Simulation of Collective Proton Tunneling in Hexagonal Ice Crystals". *Phys Rev Lett* 112 (2014): 148302.
51. Dreschsel-Grau, C. and Marx, D. "Protons in Concert: News and Views". *Nat Phys*, 16 February 2015, advance online publication, www.nature.com/naturephysics.
52. Lee, H., Huttunen, M., Hsu, K.J., Partanen, M., Zhuo, G.Y., Kauranen, M. and Chu, S.W. "Chiral Imaging of Collagen by Second-Harmonic Generation Circular Dichroism". *Biomed Opt Express* 4 (2013): 909–916.
53. Lewis, J. and Raff, M. *Molecular Biology of the Cell*. Garland Science, New York, 2008.
54. Ho, M.W. "The Real Bioinformatics Revolution". *Sci Soc* 33 (2007): 42–45.
55. Ho, M.W. "Life is Water Electric". *Sci Soc* 57 (2013b): 43–47.
56. Ling, G. *Life at the Cell and below Cell Level*. Pacific Press, Nampa, 2001.

16

Illuminating Water and Life: Essay in Honour of Emilio Del Giudice

The quantum electrodynamics theory of water put forward by Emilio Del Giudice and his colleagues provides a useful foundation of a new science of water for life.

Emilio Del Giudice

Emilio Del Giudice (1940–2014) was a prolific and original scientist whose intellectual legacy needs to be fully assessed in decades to come, being so far ahead of his time. He is best known for pioneering the quantum field theory of condensed soft matter, especially water.[1] I confess I only began to fully appreciate his work after becoming convinced from examining the more conventional literature that water is quantum coherent even under ambient conditions.[2] Quantum coherent water underlies the quantum coherence of organisms;[3] it is the medium of life in that it enables quantum macromolecular machines (proteins and nucleic acids) to work at close to 100% efficiency and with precise coordination down to molecular and sub-molecular levels, which I call "quantum jazz". Inspired by Del Giudice, I began to see clearly how quantum coherent water also fuels the biosphere, and provides the electricity that animates life and makes quantum jazz possible. That led to my recent book *Living Rainbow* H_2O,[4] which presents quantum coherent water as "the means, medium and message of life."[5] Del Giudice has given us a unifying basis for a new science of water for life. In this essay, I shall give an overview of how I see the subject in the light of his contribution; and more importantly, how it could be taken forward.

Illuminating Water Creates Coherence

Del Giudice has literally brought light into the quantum physics of water, thereby also illuminating life. Standard quantum theory does not predict quantum coherence for liquid water, because it ignores both quantum fluctuations and the interaction between matter and light, the electromagnetic vacuum field. These are only taken into account in quantum electrodynamics field theory. But conventional quantum electrodynamics field theory applies only to gases.

Working with his close associate Giuliano Preparata (1942–2000), Del Giudice and other colleagues, especially Giuseppe Vitiello, Antonella De Ninno and Alberto Tedeschi, extended conventional quantum electrodynamics theory to the condensed phase of liquids. They showed that interaction between the vacuum electromagnetic field and water induces coherent excitations that lead to the formation of large, stable coherent domains (CDs) about 100 nm in diameter, resulting in the condensation to the liquid phase at some critical density. These CDs are present in liquid water at ordinary temperature and pressure, and may be responsible for all the special properties of water including life itself.[6–10]

Each CD of water is effectively a resonating cavity produced by the electromagnetic field that ends up trapping the field because the photon acquires an imaginary mass, so the frequency of the CD electromagnetic field becomes much smaller than the frequency of the free field with the same wavelength.

Under ambient conditions, water is an approximately equal mixture of CDs surrounded by incoherent regions. (More accurately, the water molecules are dancing between the CD and non-CD configurations, and both the CD and non-CD molecules are interchangeable.) This picture is reflected in the many observations supporting a two-state model of liquid water, a dense state and less dense state coexisting simultaneously. The less dense state with tetrahedral-directed fully hydrogen-bonded water molecules is the excited state and hence corresponds to water in the CD, while the non-CD water molecules represent the dense state in which the hydrogen bonding is not so regular.

Coherent Water at Interfaces Makes Life Possible

The water molecules in the CD are oscillating between the ground state and an excited state of 12.06 eV, just below the first ionization potential of 12.56 eV, and therefore contain close to a million almost-free electrons (the proportion of excited-state molecules within the CD is estimated to be 0.13). (In a 2012 publication[11] these values are re-estimated from empirical data to be 12.07 eV and 12.62 eV respectively, but the main argument remains unchanged.) That means CDs are mostly likely negatively charged at the periphery close to or at the surface of the sphere. (At the same time, positively charged protons are probably extruded just outside the coherent domain.) The surface of the CD is "a redox pile" where the almost-free electrons can be readily donated to electron acceptors. Excited coherent water is the basis of the oxidation-reduction energy metabolism that powers all living processes; it is both the chemistry and the electricity of life, and as we shall see, it orchestrates all the necessary chemical reactions.[4]

The abundant life on earth, including the human species, depends ultimately on photosynthesis by green plants, algae and cyanobacteria. In the process, sunlight is trapped by chlorophyll (the green pigment in chloroplasts) to split water into hydrogen, electrons and oxygen (Eq. 1), giving life access to an abundant energy source, and more importantly, liberating oxygen for the evolution of air-breathing organisms that filled Earth with the teaming millions of species.

$$H_2O \rightarrow 2\,H^+ + 2\,e^- + O \tag{1}$$

The hydrogen ion (protons) and electrons go to reduce carbon dioxide into carbohydrates and biomass of photosynthetic organisms, which feed herbivores, and down the food web, the vast majority of animal species. The air-breathers break down carbohydrates by oxidizing them with oxygen in the mitochondria of cells to obtain energy for growth and reproduction, regenerating carbon dioxide and water. This completes the living dynamo of photosynthesis and respiration that turns inanimate substances into living organisms (see "15 — Water the Means, Medium and Message of Life" in this book, for further details).

It takes 12.56 eV to split water, an energetic photon of wavelength 98.7 nm in the soft X-ray region that would destroy life. But photosynthesis depends mainly on red light (~680 nm, 1.8 eV), and to some extent, blue light in the visible spectrum. So how is that possible? It is because of the excited water in the coherent domains.

More than 50 years ago, Nobel Laureate and father of biochemistry Albert Szent-Györgyi (1893–1986) had already suggested that water at interfaces is the key to life, and proposed that water at interfaces such as membranes is in the excited state, requiring considerably less energy to split than water in the ground state.[12] A sign of the excited water is that a voltage should appear at the boundary between interfacial water and bulk water, which was indeed observed not long after Szent-Györgyi made his prediction. Most, if not all, water in living organisms is interfacial water, as it is almost never more than a small fraction of a micron away from surfaces of membranes or macromolecules. Water at the interface is a coherent domain generated by the interaction of electromagnetic field and water, and stabilized by the interface, as Emilio and colleagues had suggested.[10]

Unexpected support for the theory came from a field study on aerosols generated at five waterfalls in the Austrian Alps carried out by Pierre Madl at Salzburg University in Austria with Del Giudice and other collaborators around the world. The aerosols showed a bimodal size distribution with small clusters a few nanometres in diameter consisting of a few hundred water molecules and larger aggregates about 100–200 nm in diameter with millions of water molecules.[13] Whereas the small clusters disappear very rapidly with distance from the falls, the larger aggregates are able to propagate for hundreds of metres. The aggregates detected, both large and small, are electrically charged with negative charge predominating for 85% of aggregates. The existence of surface electrical charge and unusual size distribution of the aggregates are both predictions of the quantum electrodynamics theory of water, although the precise details have yet to be worked out.

Understanding the Quantum Physics of EZ Water

Biophysicist Gerald Pollack at University of Washington Seattle in the United States has provided a most vivid demonstration of interfacial water in his laboratory, and the fascinating story is told in Pollack's excellent book, *The Fourth Phase of Water*.[14] A hydrophilic gel in a suspension of microspheres just visible to the eye created interfacial water that extends hundreds of microns from the surface of the gel, excluding the microspheres as well as other solutes such as proteins and dyes, and hence referred to as an "exclusion zone" (EZ).[15]

EZ water appears to be ten times more viscous than bulk water, and its refractive index and density are about 10% higher. It has many other distinctive properties indicative of structure, including a peak of light absorption at 270 nm. Emilio and colleagues suggested that EZ water is a gigantic coherent domain stabilized on the surface of the attractive gel, though the higher density of EZ water rules out the tetrahedral-directed ice-like hydrogen bonds proposed for water in the coherent domains in free solution. Pollack has indeed suggested an entirely different structure for EZ water.[14]

In line with the proposed formation of coherent domains in quantum electrodynamics theory, there *is* a negative electric potential associated with EZ water, which enables it to act as a battery, as demonstrated in the Pollack lab.[14,16] However, the surface charge of the EZ water depends on the surface charge of the hydrophilic gel on which it is formed. While a negatively charged gel resulted in a negative potential of the EZ with respect to the bulk water with a low pH zone immediately next to it, a positively charged hydrophilic gel gives a positive potential at the surface of the EZ with a high pH zone adjoining.[17] Further, the magnitude of the potential difference also seems to be determined by the hydrophilic gel, independently of the thickness of the EZ.

Also in line with quantum electrodynamics theory, EZ water appears to be formed by the interaction of light with the water, the most effective light being far infrared/microwave ~3 000 nm. The gap between the ionizing potential of 12.56 eV and the excited level of 12.06 eV predicted by Del Giudice and colleagues is 0.5 eV, equivalent to 2 479.7 nm, not too different from ~3 000 nm (equivalent to 0.413 eV); although the 3 000 nm absorption may have more to do with the rearrangement of CDs into EZ water rather than charge separation. This infrared absorption maximum in wave number, ~3 333 cm^{-1}, is close to the 3 350 cm^{-1} identified as the spectral signature of four-coordinated sites in water clusters,[18] as distinct from that of ice at 3 220 cm^{-1}.

The peak absorption of EZ water at 270 nm (4.59 eV) is much longer than the 102.8 nm equivalent of 12.06 eV predicted from quantum electrodynamics theory. It should be noted that the value of 12.06 or 12.07 eV is selected as the most likely excitation level that fits with phenomena (including that fact that photosynthesis requires energetic photons of only 1.8 eV), and apply to water in the gas phase.[9,14] Quantum chemical calculations of neutral and charged $(H_2O)_n$ water clusters modelling fragments of a real hydrogen-bond network of water and dynamic simulations of the clusters — either upon electron removal from a stable neutral cluster or upon the excitation of various cluster vibrations — enabled separate stages of structure reorganization to be distinguished.[19] The energy necessary for the ionization of superficial water layers at irradiation was estimated from the intermediate ionization potentials of water clusters to be 9.5 eV (130.5 nm). The experimental value obtained on a rotating quartz disc was 9.3 eV. That is still far higher than the 4.59 eV (270 nm) for the formation of EZ water. There are indeed lower levels of excitation for water. We shall return to this important issue later.

In summary, it is likely that EZ water is excited water, probably associated with the formation of CDs as described in quantum electrodynamics theory, but its properties and structure are distinct from those of the CDs. It is also likely that EZ water has multiple layers with different properties away from the immediate surface of the hydrophilic gel.

Understanding Structured Water

EZ water is structured water, according to spectroscopic and nmr studies as well as birefringence under polarized light microscopy.[14] That does not mean structure is absent in bulk water, as the extended hydrogen-bonded networks and coherent domains both predict structure in bulk water.

Indeed, stable water clusters tens of nanometres to millimetres in dimensions can be isolated from bulk water and imaged under the transmission electron microscope (TEM) and the atomic force microscope (AFM),[20–22] as demonstrated in the laboratory of Chinese-born American physicist Shui-Yin Lo at Institute of Quantum Medicine, Pasadena California (Fig. 16.1). The clusters consist of millions to billions of water molecules and come in a wide variety of shapes and sizes. They make up structures that are flexible, and can be deformed by the tips of the atomic force microscope probe if scanned in the contact mode. Otherwise, they remain stable for weeks, even months, at room temperature and pressure. They have all the characteristics of "soft matter" — liquids, liquid crystals, colloids, polymers, gels and foams — that form mesoscopic structures much larger than the molecules themselves, but small compared with the bulk material.

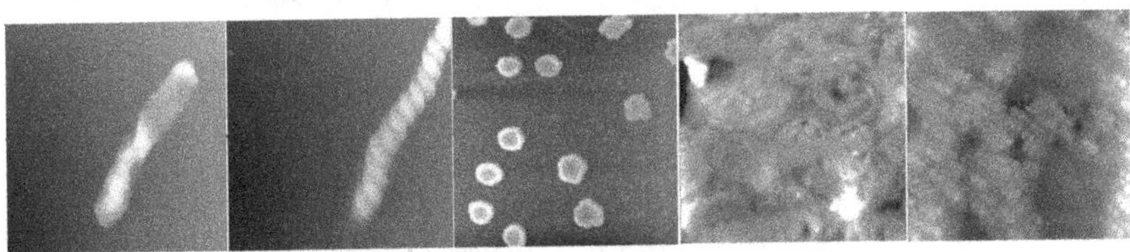

Figure 16.1 Different forms of supramolecular water clusters imaged with AFM; all fields are 5 microns square.

These structures were prepared by serially diluting a solution of pure sodium chloride with vigorous shaking (succussion) in ultrapure distilled de-ionized water in a low-dust room until it is about 10^{-6}M or 10^{-7} M, then placing drops to dry on a clean glass slide or some other substrate. Close-up, the diverse structures appear to be made of spherical "balls" 50–100 nm in diameter. Under the electron microscope, they are electron dense at the surface.

The effect of dilution, as explained by Lo and colleagues,[20–22] was to enable dipole interactions between water molecules to dominate over ionic interactions, and may be crucial in forming the supramolecular structures observed.

Similar structures have been produced since by several research groups, as reviewed elsewhere.[23] Significantly, they contain strong electric fields, with reported absorption maxima variously at 195 nm, 230.6 nm and 276.7 nm, perhaps not too dissimilar to the 270 nm of EZ water.

Italian chemist Vittorio Elia at University of Naples and his colleagues successfully imaged large water clusters in water repeatedly brought into contact with Nafion[24] or repeatedly filtered.[25] The "Nafionated water", which they equated with EZ water, showed an increase in electrical conductivity by up to two orders of magnitude. At the same time, there was a drop in pH from ~6 to 3, representing three orders of magnitude increase in proton concentration. The increase in conductivity was attributed to proton conduction,[26] suggesting that protons are present in the clusters. In other words, their clusters include not only EZ water but also water immediately next to it, which has been shown to be enriched in protons in the case of Nafion.[14,16] Lyophilization of 20 ml of the Nafionated water gave 1 to 2 mg residue, and AFM confirmed the presence of micron-size structures that look superficially similar to the large water clusters identified in the Lo laboratory.[20]

However, repeatedly filtered water, while showing the same increase in conductivity, was accompanied by an *increase* in pH compared with the starting distilled de-ionized water.[25] This suggests that the nature of the charged groups on the conditioning substrate — silicate, acetate and nitrate groups in the sintered glass, cellulose acetate and nitrocellulose filters versus sulphonate groups in Nafion — and the nature of physical contact will both affect the net charge of the conditioned water, the same as for the EZ water.[17] What the treated waters share in common is the separation or near-separation of charges due to coherent excitation of water as predicted by quantum electrodynamics theory.[6–11]

Ambient Electromagnetic Field Required for Nanostructures in Highly Dilute Solutions and Nanostructures Required for Biological Activity

Independently, work carried out in the laboratory of Alexander Konovalov at Russian Academy of Sciences, Kazan, Tatarstan, shows that highly serially diluted solutions, regardless of whether they involve organic salts, amphiphilic or hydrophobic compounds, spontaneously form clusters 100 to 300 nm in size with surface potential −2 to −20 mV.[27] The team characterized the nanostructures in solution in terms of conductivity, size, surface tension and surface potential at a wide range of concentrations. A significant finding is that the nanostructures fail to form when the diluted solutions are placed in a container shielded by permalloy to exclude electromagnetic fields.[28]

In further experiments, Konovalov and his team confirmed that stable structures containing millions or more water molecules in highly dilute solutions of a diverse range of chemical compounds are formed only in the presence of ambient electromagnetic fields; these structures exhibit physical properties distinct from bulk water and further, are essential for biological activity. I have reviewed their findings elsewhere.[29]

Over the past six years, Konovalov and his team have studied about 100 compounds at 10^{-2} to 10^{-20} M, diluted sequentially with succussion starting from the initial solution.[30,31] The list includes antioxidants, plant growth regulators, neuro-mediators, vitamins, tranquilizers, hormones, various drugs, as well as substances of unknown biological effects. The compounds range from simple molecules like glycine to complex macrocyclic compounds like porphyrins or calyxarenes.

They monitored electrical conductivity, surface tension, pH and in some cases, dielectric permeability and optical activity at different dilutions. To measure the size of nanostructures formed in solution, the team used dynamic light scattering (DLS). DLS is a physical technique generally used for determining the size distribution of small particles in suspension or polymers in solution.[32] Water chemists have discovered that it also enables the detection of nano-objects in highly dilute solutions that have very few solute molecules left, and this has greatly facilitated research on such solutions. At the same time, the technique determines the surface electrical (zeta) potential of the nano-objects.

Recently, the experiments were carried out both on the lab bench and within a three-layer permalloy (iron/nickel) container shielding out external electromagnetic fields. For example, the geomagnetic field was brought down to a thousandth of its normal level.

A quarter of solutions behaved "classically", i.e., highly dilute solutions become like bulk distilled de-ionized water in surface tension and electrical conductivity; but the majority, 75%, behaved non-classically, as for example, the antioxidants phenozane and α-tocopherol (vitamin E).

Phenozane, on being diluted in aqueous solution, exhibits a fall in surface tension by 10–20 mN/m at around 10^{-6} to 10^{-7} M, while electrical conductivity rises to 40 μS/cm (S, Siemen, is 1 Ampere/Volt) and keeps changing with subsequent dilution. These changes are accompanied by biological effects that vary significantly with dilution. At high dilutions of non-classical compounds, nanosize structures appear. Their dimensions change in subsequent dilutions — not in linear or monotonic fashion, but rather in jumps, and the formation of these nanostructures appears to be necessary for biological effects.

Samples kept in hypo-electromagnetic environments notably fail to form nanostructures below a certain dilution; the physical changes such as conductivity are absent; and the shielded solutions were also without biological effects at these high dilutions.

There appears to be a "boundary concentration", different for different compounds in the 10^{-5} to 10^{-8} range, beyond which nanostructures do not form inside permalloy containers, and where further biological effects normally show up.

A team at Emanuel Institute of Biochemical Physics, Russian Academy of Sciences in Moscow, carried out experiments to check on the hypothesis that biological effects are absent in the absence of nanostructures in highly diluted solutions. They looked at changes in the microviscosity of membranes exposed to potassium phenozane solutions. They found effects at 10^{-6} M corresponding to the usual maximum plus further peaks at 10^{-12} and 10^{-15} M for the dilute solutions kept on the lab bench; but the two additional peaks disappeared in solutions kept within the permalloy container (Fig. 16.2).

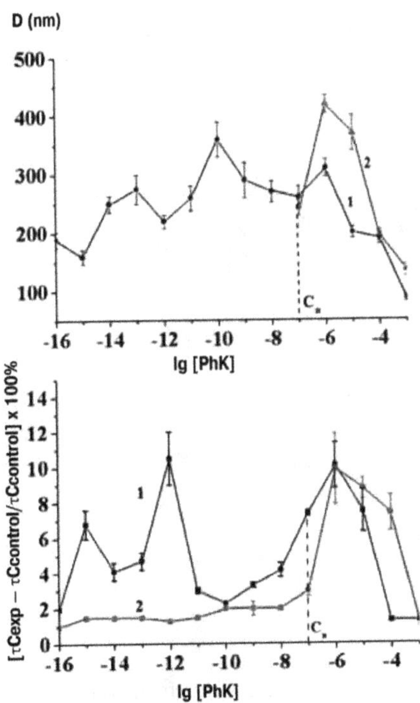

Figure 16.2 Formation of nanoassociates (top) and membrane microviscosity (bottom) in normal (black) and shielded (red) environments.

As a further proof of concept, MSc student Dmitry Konovalov experimented with solutions of cytyltrimethylammonium bromide. Under ordinary conditions on the lab bench, 240-nm nanostructures appear at 10^{-9}M; this does not take place within the permalloy container. However, if a 7-Hz field is generated within the container, the nanostructures form to about the same size as on the lab bench.[30] These experiments are extremely important, and should be replicated in other laboratories, as they could put homeopathic effects, for example, on a firm scientific footing.

Finally, using AFM, Konovalov and his team successfully imaged the nanostructures (Fig. 16.3). At the lowest dilution (10^{-6}M), spherical or semispherical particles are seen (most probably containing solutes).[31] At higher dilutions, however, nanostructures are detected in the semi-contact mode, which have the characteristics of soft matter, and quite similar to those first identified and imaged on AFM by Lo and his team (see above), including thread-like structures microns in length at the highest dilution of 10^{-10}.

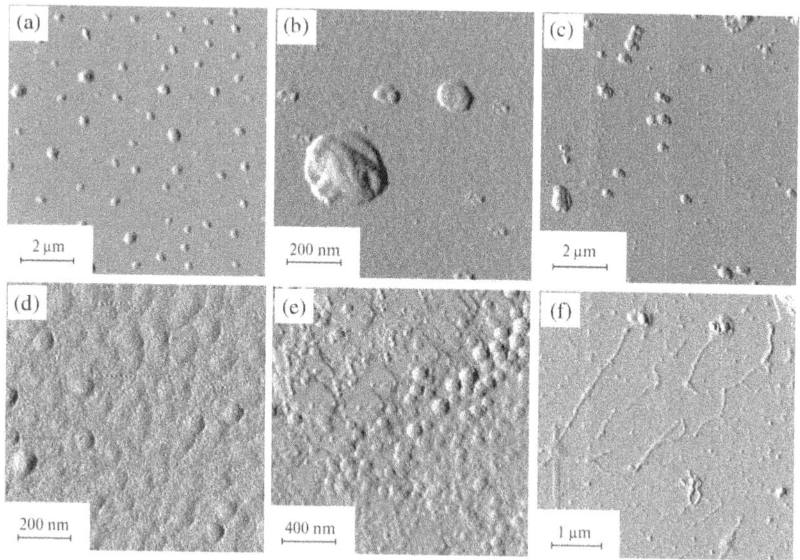

Figure 16.3 AFM images of nanostructures from solutions of amphiphilic calix[4] reorcinarene with tris(hydroxymethyl) methylamide at concentrations of 10^{-6} (a, b), 10^{-7} (c, d), 10^{-8} (e) and 10^{-10} (f).

Also as in Lo's laboratory, subunits ~100 nm in diameter are evident in the supramolecular structures (see Figs. 16.3c, 16.3d and 16.3e).

Symmetrical Spherical Dipoles from Coherent Domains

Lo and colleagues[16] suggested that the supramolecular structures are made of small unit "balls" ~ 100 nm that are "dipoles". However, these are not ordinary dipoles. Instead they may well be coherent domains predicted in quantum electrodynamics theory, as I have proposed.[23] They are about the right size, each having close to a million almost-free electrons at the periphery. At the same time, positively charged protons are most likely extruded outside the CD in common with what happens in EZ water[14–17] and Nafionated water,[24,26] which have an excess of protons. Consequently, these spherical CDs can mimic dipole interactions through negative charges on their periphery attracting positive charges just outside to form a three-dimensional potentially perfectly symmetrical giant electret (dipole). In other words, there will be an electric field measured in any direction as found by Lo[22] and also by Italian physicist Roberto Germano at Promete Srl San Giorgio a Cremano, working with Del Giudice and others on Nafionated water.[33] The formation of an electret also explains the enormous increase in conductivity in Nafionated[24,26] and repeatedly filtered water.[25] When isolated by drying on a substrate, symmetry is broken and the clusters adopt a variety of "snowflake"-like structures with the six-fold symmetry, which arise from close packing of spheres in my model (Fig. 16.4).

Another prediction from quantum electrodynamics field theory[8,10] is consistent with the structure of the clusters proposed here, and may account for their apparent stability. The coherent oscillations maintained by the electromagnetic field trapped within the CDs can occur not just between the coherent ground state and excited state of the water molecule, but also between two rotational levels, which produce correlations as large as several hundred microns, giving rise to a common dipole orientation, but a net zero polarization field (on account of its symmetry),

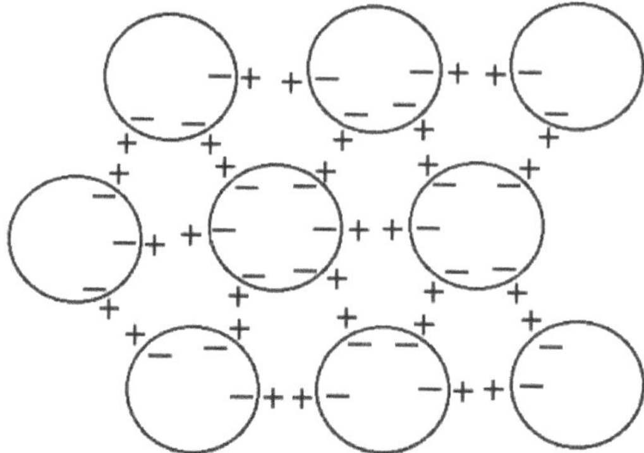

Figure 16.4 Spherical coherent domains forming a three-dimensional dipole structure; note the six-fold symmetry resulting from close packing of spheres.

unless and until the rotation symmetry is broken. The combination of the coherent oscillations and rotations, therefore, produces phase-locked coherent interactions among the CDs, resulting in stable supramolecular clusters with the electret structure depicted in Fig. 16.4.

Quantum Delocalization and Superconducting Protons

The separation of positive and negative charges in quantum electrodynamics theory of water is important for intercommunication, especially in the form of protons. I have considered proton conduction in living organisms since the first edition of *The Rainbow and the Worm*[3] published in 1993, and later hypothesized with David Knight that water aligned along collagen fibres in connective tissues could be the anatomical basis of the acupuncture meridians,[34] which enable the cells and tissues to intercommunicate via proton currents. I also proposed that water confined in nanospaces such as carbon nanotubes and collagen fibres could be proton superconducting.[35,36] There is now good evidence in support of that hypothesis,[37] which I shall briefly describe (see "15 — Water, the Means, Medium and Message of Life" in this book for more details).

It is generally accepted that the key to water's remarkable properties is the hydrogen bond interconnecting the water molecules. Linus Pauling was the first to suggest that the hydrogen bond is partly covalent in 1935.[38] In 1999, inelastic X-ray scattering on a carefully prepared slab of ice yielded results that supported Pauling's proposal; the data fit a quantum model rather than a classical electrostatic model for the hydrogen bond.[39]

In 2002, researchers at the FOM Institute of Atomic and Molecular Physics in the Netherlands used ultrafast femto-second pulses of infrared light to excite and probe the O–H covalent bond in liquid water at room temperature.[40] Again, the results support a quantum mechanical model in which the excited proton could be found simultaneously at a distance of the equilibrium O–H bond from both its neighbour oxygen on either side, and at a much reduced energy for the excited ($v = 2$) state than if the hydrogen bond did not exist.

The energy of excitation to the $v = 2$ delocalized state is estimated at 6 500 cm^{-1} (0.82 eV, or 1 538.5 nm), less than 20% of the O-H bond energy of 38 750 cm^{-1} (4.8 eV, 258.1 nm), estimated from O–H stretch vibration frequency. The energy required for excitation to the delocalized state is very much smaller than the 12.06 eV for excitation to coherent domains,[6] and the O-H bond energy at 4.8 eV is also much smaller than the value of 12.56 eV required for ionization.

I believe these results are telling us something very significant: it is much easier for H^+ to dissociate from the water molecule than for the water molecule to lose an electron. The energy of 4.8 eV is equivalent to light at 258.1 nm, which is close to the absorption peak of EZ water and consistent with the loss of H^+ from EZ, the H^+ accumulating at the interface of EZ with bulk water, suggesting that EZ water requires the dissociation of protons from water molecules. Consequently, the activity of protons derived from water may be much more important for living systems than electrons, and in the form of proton currents. Delocalization of protons — by absorption of ambient photons at 1 538.5 nm — increases the probability of proton transfer, i.e., proton conduction could take place much more readily and rapidly. It would be easy to check if the predicted

1 538.5 nm does promote proton conduction; if so, it could have important clinical applications.

Water confined in nanospaces is even more remarkable, and that applies to most biological water. It adopts new quantum states, resulting in proton conduction rates orders of magnitude higher than in bulk water.

New Quantum States of Water in Nanospaces and Superconducting Protons

Research led by George Reiter at University of Houston Texas uses deep neutron inelastic scattering to measure the momentum distribution of protons in water confined to nanospaces. The momentum of the proton is mainly determined by the wave-function of the proton's ground state (least energetic state).

The team investigated carbon nanotubes, glass sponges and Nafion membranes, and found similar results.[41–43] The confined water adopts a variety of new quantum states distinct from bulk water, and highly dependent on the precise dimensions of the nanospace. In other words, water adopts a single quantum state, wherever it is, which is quite remarkable in itself. (I invite you to think about the ocean, when it is clear and calm, and when raging with tsunami.[44]) Water confined in space dimensions of 2 nm or smaller has protons that are coherently delocalized in two momentum states.

In xerogel, a glass sponge, with Si-OH (silanol) groups lining the pores that can hydrogen-bond with water, the proton momentum distribution of water in 24 Å pores at room temperature is confined in a double-well potential.[41] For larger pores of 82 Å, the average momentum distribution was close to that of bulk water, though still quite distinct.

The perfluorosulphonic acid membranes Nafion 1120 and Dow 858 are polymers consisting of a hydrophobic poly(tetrafluoroethylene) PTFE backbone and randomly distributed side chains of perfluoroether terminating with sulphonic acids. When hydrated, nanophase separation occurs in which water is confined in domains a few nanometres in diameter surrounded by hydrophobic regions. The sulphonic acid group ($-SO_3H$) donates its proton to water when there is sufficient water in the pores, making them very good proton conductors. The momentum distributions at room temperature for the two membranes are dramatically different from water in the bulk: the kinetic energy for Nafion is higher by 107 meV/proton, and 124 meV/ proton higher for Dow 858 (Fig. 16. 5). They are qualitatively different quantum states from bulk water. At a concentration of 14 H_2O/SO_3H for both membranes, Dow 858 has a significantly higher conductivity by 70% than Nafion, as consistent with a higher degree of proton delocalization.

In a later publication, Aniruddha Deb at University of Michigan Ann Arbor joined up with George Reiter to investigate X-ray Compton scattering to probe the electronic ground state of nano-space confined water at the Japan Synchrotron Radiation Research Institute Sayo-Cho Hyogo.[45] They found that the difference in "bond disorder" between water confined in Nafion and bulk water is 17 times larger than that between bulk water just above the freezing point and just below the boiling point. That is not surprising given that the proton is coherent distributed in double wells separated by ~0.3 Å. The kinetic energy has also gone up because

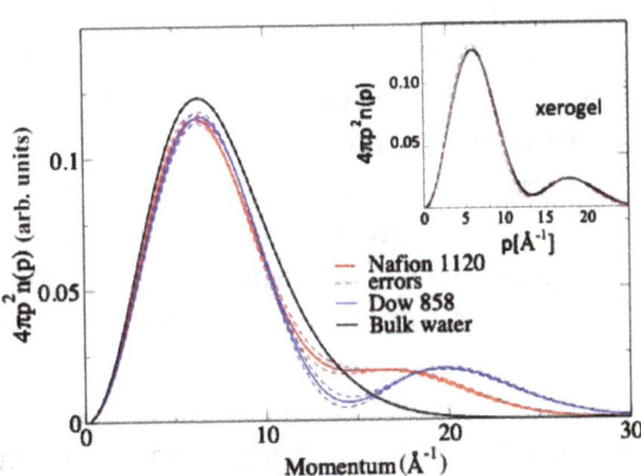

Figure 16.5 Proton momentum distribution of proton exchange membranes compared with bulk water; inset proton momentum distribution of xerogel (see text).

each of these wells is more tightly binding the proton than the covalent bond of the isolated water molecule. The kinetic energy (measured by deep inelastic neutron scattering) is 245 meV and 268 meV for Nafion and Dow, respectively, compared with 148 meV for bulk water at room temperature. The change in kinetic energy in going from 5°C to 90°C for bulk water is only 0.5 meV.

Further support for the new quantum states of nanospace confined water comes from excited-state proton transfer measurements of a fluorescent probe molecule, 8-hydroxypyrene-1,3,5-trisulphonate (HPTS). The molecule tends to stay in the middle of the water-filled regions in Nafion. When excited by a laser pulse, the proton in the OH group of HPTS is ionized and rapidly transferred to the surrounding solvent.[46] The recombination time depends on the transport processes affecting the free proton. A diffusion process leads to $t^{-1.5}$ dependence that was observed in bulk water. In Nafion, the observed rate is much slower at $t^{-0.8}$, because it requires the redistribution of electrons through the hydrogen bonded network. Similar results were obtained in reverse micelles. Both reverse micelles and the water-filled nanospaces in Nafion are similar to the nanospaces in the cell. And these findings will certainly have implications for biochemical reactions in the cell.

Hydrated Nafion consists of long parallel but otherwise randomly packed water channels surrounded by partially hydrophilic side branches, forming inverse-micelle cylinders. At 20% by volume of water, the water channels have diameters of between 1.8 and 3.5 nm, with an average of 2.4 nm.[47]

Nafion films have a proton conductivity of about 0.1 S/cm (S, Sieman = 1 Ampere/Volt), among the highest in proton exchange membranes (PEMs). For comparison, the electrical conductivity of copper, one of the highest, is 596 000 S/cm, and silicon, a semiconductor, 0.156 S/cm.[48] But the conductivity of a single high purity Nafion nanofibre 400 nm in diameter made by electrospinning reached 1.5 S/cm, an order of magnitude greater, as demonstrated by Yossef A Elabd and his colleagues at Drexel University Philadelphia.[49] This is due to alignment of interconnected ionic aggregates along the fibre axis direction as evidenced by X-ray scattering.

Proton Transport through Carbon Nanotubes

Very fast proton transport was indeed reported in carbon nanotubes, ~40 times the rate in bulk water in molecular dynamic simulations.[50] More recent *ab initio* (starting from first principles) path-integral (quantum approach) molecular dynamics simulations showed no energetic barrier to proton transfer in every case when quantum delocalization is taken into account. The main difference between bulk liquid water and water confined in a carbon nanotube is a favourable pre-alignment of water molecules in the latter case. Configurations where the excess proton is quantum delocalized over several adjacent water molecules along with continuous interconversion between different hydration states reveal that, as in bulk water, the hydrated proton under confinement is best described as a fluxional defect, rather than any individual hydration state of the excess proton propagated classically along the water chain.[51]

Wonjoon Choi at Massachusetts Institute of Technology Cambridge led an investigation into the diameter-dependence of proton transport through the interior of isolated, specially fabricated single-walled carbon nanotubes 1 mm long, and demonstrated a surprising five-fold enhancement of stochastic proton transport rates at a diameter of approximately 1.6 nm, dropping off sharply to either side.[52] The diameter 1.6 nm is near the transition between water behaving rather as it would in an ordinary capillary tube at diameters >1.6 nm and water behaving just the opposite way at diameters <1.4 nm in a temperature-diameter phase diagram.[53] At diameter ~1.5 nm, double- or triple-walled water nanotube structures have been proposed. Whether these structures are responsible for the sharp increase in proton conduction at 1.6 nm is not yet known.

Not only does nanospace confined water show high temperature superconductivity, it also exhibits superfluidity, with flow rate enhancement of 50- to 900-fold, depending on diameter, compared with that predicted from conventional fluid-flow theory.[54] This has large implications for water transport into cells and electrolyte balance, which is crucial for the health of cells and organisms.

These results of carbon nanotubes confined ware are also most relevant to water associated with collagen fibres.

Collagen Fibres and Acupuncture Meridians

Evidence dating back to the 1970s indicates that collagen does conduct protons. G.H. Bardelmeyer in The Netherlands found that electrical conductivity in bovine Achilles tendon is fully determined by the water of hydration, and the electric current is primarily carried by protons at water contents up to 45%, and by small ions at water contents beyond 65%.[55] Between water contents of 8.5% and 126%, conductivity went up by eight orders of magnitude. He estimated that pure water's dissociation constant is 10^{-5} that of absorbed water; i.e., adsorbed water is five orders of magnitude more likely to let go of protons. Later, Naoki Sasaki in Japan found that the conductivity of collagen increased markedly with water absorbed — at an exponent of 5.1–5.4 — between a water content of 0.1 to 0.3 g/g.[56] These results make sense in the light of the recent observations on proton superconduction in nanospace confined water.

The important discovery of superfast quantum tunnelling in chiral water polymers is dealt with in "15 — Water the Means, Medium and Message of Life" in this book. If similar quantum tunnelling occurs in the chiral water nanotubes along collagen fibres, it could mean almost *instantaneous* intercommunication between cells and tissues throughout the body.

Another important property of collagen discovered in the late 1980s is its capacity for *second harmonic generation* (SHG), i.e., combining photons interacting with it to form new photons with twice the energy and therefore double the frequency and half the wavelength.[56] Since then, *in vivo* SHG imaging has been widely developed for diagnostic purposes (Fig. 16.6). It should be noted that SHG was previously restricted to crystalline material such as

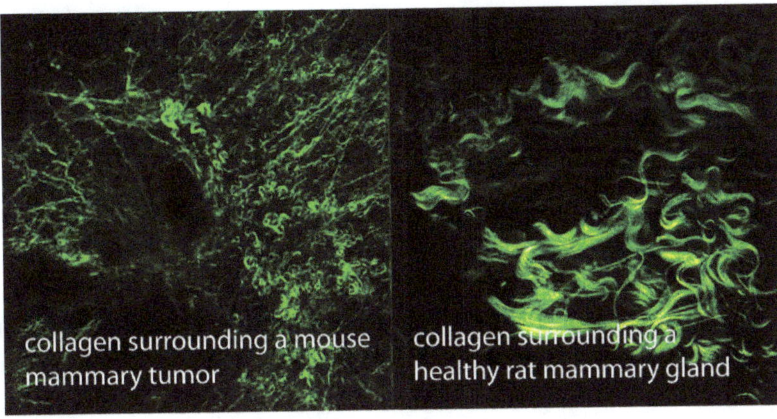

Figure 16.6 Second harmonic generation *in vivo* imaging of collagen, courtesy of Carolyn Pehlke, University of Wisconsin, Madison.[57]

quartz. And although it is clear that SHG in collagen depends on hydration (with liquid crystalline water[4]), the scientists have chosen to ignore that totally.

A paper[58] submitted to a conference in 2003 reported results of experiments in which Type 1 collagen bundles obtained from rat tails were structurally modified by increasing non-enzymatic cross-linking, or thermal denaturation, or by collagenase digestion, or dehydration. While they found that the hydration state significantly affected the polarization dependence of SHG, there was little or no change as a result of the extensive structural modifications short of complete disintegration. These results strongly suggest that the liquid crystalline water adsorbed in collagen is the source of the SHG.

A second paper[59] from a different research group published in 2005 showed that "SHG radiates from the shell of a collagen fibril rather than from its bulk." The effective thickness of the SHG shell was strongly dependent on the ionic strength of the surrounding solution, increasing as ionic strength decreases. However, the authors have not attributed the SHG shell to liquid crystalline water.

A lot remains to be done in this fertile area (see my review[61]). Metabolic/energetic regulation may well depend on the flow of protons via liquid crystalline water structured in nanospaces throughout the extracellular matrix (as *qi* along the acupuncture meridian system) into the interior of every single cell and its nanospaces.

Quantum Coherent Water Orchestrating Quantum Jazz

To me personally, Emilio's most significant contribution is in providing a concrete hypothesis on how the staggering molecular complexities of cells and organisms can be coordinated.

In my book *The Rainbow and the Worm*, I defined quantum coherence after quantum physicist Roy Glauber[62] (who later got the 2005 Nobel Prize for his work in quantum optics) in terms of *factorizability*. I later expressed factorizability as follows (thanks to Fritz-Albert Popp, who taught me almost everything I know about quantum physics, but that is another long story[4]): "A system is quantum coherent when its parts are so perfectly correlated that their cross-correlations factorize exactly as the product of the individual self-correlations, so that each appears paradoxically as though totally uncorrelated with the rest. It is a state of maximum local freedom *and* global cohesion; something that's impossible in a classical mechanical system."

Quantum coherence is a sublime state of being whole: a superposition of coherent activities over all space-times, a pure (ideal) dynamic state towards which the system tends to return asymptotically. I use the idea of "quantum jazz" to highlight the immense diversity and multiplicity of supramolecular, molecular and submolecular players, the complexity and the coherence of the performance, and above all, the freedom and spontaneity, with each and every player improvising from moment to moment yet keeping in step and in tune with the whole.

Quantum Jazz of Natural Genetic Modification

To appreciate the scope as well as the precision and finesse of quantum jazz, we need look no further than the "natural genetic engineering" or "natural genetic modification" — cut-and-splice operations on DNA and RNA — that cells and organisms need to do constantly in real time in order to survive.[63] Leading molecular geneticist James Shapiro at University of Chicago Illinois in the United States is so impressed with what he and others have been finding out over the past four decades, and especially since the human genome was sequenced, that he says evolution happens by natural genetic engineering and not by the natural selection of random mutations. In fact, there is almost nothing that's random inside the cell and organism. Organisms are constantly adjusting to the environment by turning on and off the right genes, creating new genes if need be, shaping the environment, and preparing for the future.

Just to produce a single protein — originally thought to be one continuous genetic message — requires elaborate cut-and-splice operations. The international research consortium project ENCODE (Encyclopedia of DNA Elements) data have revealed that vast areas of genomic DNA include many "non-coding" segments. The "gene" is actually scattered in bits across the genome, overlapping with bits of multiple other genes that have to be spliced together to make a messenger (m)RNA for translation into a protein.

When bacteria are starved and there is a substrate they cannot metabolize in the environment, they can mutate and/or cut and splice to make the right genes in order to enable them to use the substrate. This phenomenon of "directed mutation" has been studied by a number of geneticists including Shapiro (see "8 — Non-Random Directed Mutations and Quantum Electrodynamics" in this book). Many different proteins and DNA sequences have to come together in choreographed succession to form and rearrange the nucleoprotein complexes necessary for directing the precise cut-and-splice operations involved.

Geneticists are discovering more and more molecular nuts and bolts every day; the complexity of interaction networks is enough to give anyone except the most dedicated new breed of "systems biologists" a severe headache. Genes only occupy less than 2% of the genome. The rest, thought to be useless "junk" DNA not so long ago, is 85% transcribed, and thousands of non-coding RNAs belonging to several large families with important and specific regulatory functions have already been identified and more are emerging every day. Sorting out that morass of molecules is a primary preoccupation of battalions of dedicated geneticists. The question that's never asked is how these molecules with very specific functions can find one another and join up to do their job just at the right time and place. How does A know when and how to "recruit" B to join up with C, D, E and F to act on G at a specific site on the genome? And there are tens of thousands of such sites in a cell's genome.

One answer they have not considered is electromagnetic signaling and resonance, which I have suggested since the first edition of *The Rainbow and the Worm*,[3] following Colin McClare (1937–1977),[64] a pioneering physiologist whom few understood, but who was a major influence in my intellectual development. McClare not only pointed out the fact that resonating molecules attract one another, but also the precision with which interactions can occur, compared with the usual "lock and key" or "induced fit" hypothesis of how molecules come together due to random collisions in free diffusion. As it turns out, there is nothing like free diffusion possible in the living cell; it is jam-packed with molecules, membranes and organelles, and highly organized, due to self-assembly, which is probably nothing if not electromagnetic resonance at work. There is indeed independent evidence that macromolecules sharing the same function also share a common vibrational frequency.[65]

This makes even more sense in the context of quantum coherent water. Emilio and colleagues[6–11] propose that water CDs can be easily excited, and are able to capture surrounding electromagnetic fields to produce coherent excitation in the frequencies of the external fields. This in turn enables selective coherent energy transfer to take place. All molecules have their individual spectrum of vibrational frequencies. If the molecule's spectrum contained a frequency matching that of the water CD, it would get attracted to the CD, and become a guest participant in the CD's coherent oscillation, settling on the surface of the CD. Furthermore, the CD's excitation energy would become available to the guest molecules as activating energy for chemical reactions. Sequential reactions could occur because the new products would have a different vibrational spectrum. This could explain how entire pathways of reactions could be assembled. And it could explain how directed mutations can occur.[66] The substrate in the environment could send its electromagnetic signals to the enzyme breaking it down as well as to the gene encoding the enzyme, causing the gene to respond by attracting the transcription and requisite mutagenic machinery until the right mutation for making an active enzyme is achieved.

Del Giudice's CDs would also account for the so-called "memory of water" in homeopathic remedies (see Chapter 8 in *Living Rainbow H2O*[4]) and many other phenomena previously considered "occult", including that in one of the last articles he wrote for *Science in Society* with his wife Margherita[67] on the importance of quantum coherent phase information underlying the clinical practice of "butterfly touch".

Is it possible that cells or organisms as a whole also intercommunicate by means of electromagnetic and electric signals, as implied by the principle of minimal stimulus? This is completely uncharted territory as far as conventional cell biology is concerned, but evidence for intercellular communication has existed since the 1920s and has been rediscovered by many, including Fritz-Albert Popp and my friends and collaborators Franco Musumeci, Agata Scordino and Antonio Triglia at Catania University.[68]

To Conclude

It is the water in us trapping electromagnetic fields that gives us life and makes us sensitive to electromagnetic fields. That is what Del Giudice and colleagues are telling us. There is a distinct possibility that we are sensitive to the fields of other organisms, as we are sensitive to fields of the sun and the earth (reviewed elsewhere[69]) and possibly also from distant stars, all without our conscious knowledge.

Life appears quantum electrodynamical through and through, and water is at the heart of it all, much as Emilio has taught us.

It is a fitting conclusion to the gentle intellectual giant who will be immortalized in our collective memory and in the memory of generations to come.

Tribute

Emilio Del Giudice was the most brilliant scientist in the world to have inspired me among countless others. He was also the kindest, most generous, entertaining, wise and witty man I have known. I had been looking forward to discussing my latest ideas with him on mathematics, art, beauty, consciousness and the universe;[70] for such were his wide-ranging interests and passion. That was not to be. He was taken away suddenly and without warning.

This was what he said in a short video for our *Colours of Water* art/science/music festival in March 2013:[71]

> "I am totally at odds with the paradigm of conventional science that looks on matter as an inert entity pushed around by external forces. The paradigm of quantum field physics does not separate matter from movement as matter is intrinsically fluctuating.
>
> There is a possibility of tuning together the quantum fluctuations of a large number of bodies and creating coherence in matter through music. Human organisms could be a part of such coherence.
>
> Artistic experiences are resonances in the framework of our quantum field paradigm. Their relevance for the self-organization of matter has been recognized by artists and humanists long before the scientists.
>
> Conventional science is very far from the dreams, needs and wishes of people. It has no place for the spontaneous movement of organisms or the love between organisms. My fellow countryman, the philosopher G.B. Vico, used to say: 'Poetic truth is a wider truth than physical truth.' But he was referring to the mechanistic physics of his time.
>
> Our task today is to show that the modern quantum field paradigm is able to raise physical truth to the same level as poetic truth."

He was and is my soul mate; now he reaches my mind and heart directly through the ether.

Emilio has lit the fire for the new science of the organism, as Prometheus had for the classical science of mechanism. He actually saw himself (and fellow scientists) in that role in a beautiful essay[72] honouring German-born British theoretical physicist Herbert Fröhlich (1905–1991), whose idea of energy storage and coherent excitations for living systems provided an entry point for my own incursion into the physics of organisms in the late 1980s. I count myself among Emilio's many torch-bearers who will roam the earth and spread the flame to dispel the darkness with his dazzling light.

References and Notes

1. Ho, M.W. "Quantum Coherent Water and Life". *Sci Soc* 51 (2011): 26–29.
2. Ho, M.W. "Cooperative and Coherent Water". *Sci Soc* 48(2010): 6–9.
3. Ho, M.W. *The Rainbow and the Worm: The Physics of Organisms*, 3rd edition. World Scientific, Singapore, 2008a.
4. Ho, M.W. *Living Rainbow H2O*. World Scientific, Singapore and Imperial College Press, London, 2012a.
5. Ho, M.W. "Water Is the Means, Medium, and Message of Life. *Int J Des Nat Ecodyn* 9 (2014a): 1–12.
6. Arani, R., Bono, I., Del Guidice, E. and Preparata, G. "QED Coherence and the Thermodynamics of Water". *Int J Mod Phys B* 9 (1995): 1813–1841.
7. Del Giudice, E., Preparata, G. and Vitiello, G. "Water As a Free Electric Dipole Laser". *Phys Rev Lett* 61 (1988): 1085–1088.
8. Del Giudice, E. and Vitiello, G. "Role of the Electromagnetic Field in the Formation of Domains in the Process of Symmetry-Breaking Phase Transition". *Phys Rev A* 74 (2006): 022105.
9. Del Giudice, E. "Old and New Views on the Structure of Matter and the Special Case of Living Matter". *J Phys Conf Ser* 67 (2007): 012006.

10. Del Giudice, E., Spinetti, P.R. and Tedeschi, A. "Water Dynamics at the Root of Metamorphosis in Living Organisms". *Water* 2 (2010): 566–586.
11. Bono, I., Del Giudice, E., Gamberane, L. and Henry, M. "Emergence of Coherent Structure from Liquid Water". *Water* 4 (2012): 510–532.
12. Szent-Györgyi, A. *Introduction to a Supramolecular Biology*. Academic Press, New York, 1960.
13. Madl, P., Del Giudice, E., Voeikov, V.L., Tedeschi, A., Kolarž, P., Gaisberger, M. and Hartl, A. "Evidence of Coherent Dynamics in Water Droplets of Waterfalls". *Water* 5 (2013): 57–68.
14. Pollack, G.H. *The Fourth Phase of Water*, Ebner & Sons Publishers, Seattle, 2013.
15. Ho, M.W. "Water Forms Massive Exclusion Zones". *Sci Soc* 23 (2004): 50–51.
16. Ho, M.W. "Liquid Crystalline Water at the Interface: Just Add Sunlight for Energy and Life". *Sci Soc* 39 (2008b): 36–39.
17. Zheng, J.M., Wexler, A. and Pollack, G.H. "Effect of Buffers on Aqueous Solute-Exclusion Zones Around Ion-Exchange Resins". *J Colloid Interface Sci* 332 (2009): 511–514.
18. Hamashima, T., Mizuse, K. and Fujii, A. "Spectral Signatures of Four-Coordinated Sites in Water Clusters: Infrared Spectroscopy of Phenol-$(H_2O)_n$ (~50 > n > ~20)". *J Phys Chem* 115 (2010): 620–625.
19. Novakovskaya, Y.V. "Theoretical Estimation of the Ionization Potential of Water in Condensed Phase: II. Superificial Water Layers". *Protection of Metals* 43 (2007): 22–33.
20. Lo, A., Cardarella, J., Turner, J. and Lo, S.Y. "A Soft Matter State of Water and the Structures It Forms". *For Immunopathol Dis Therap* 3 (2012): 237–252.
21. Lo, S.Y. "Anomalous State of Ice". *Mod Physics Lett B* 10 (1996): 909–919.
22. Lo, S.Y. "Survey of I_E™ Clusters". In *Double-Helix Water* (D.L. Gann and S.Y. Lo, eds.), pp. 117–159, D and Y Publishing, Las Vegas, 2009.
23. Ho, M.W. "Large Supramolecular Clusters Caught on Camera: A Review". *Water* 6 (2014b): 1–12.

24. Elia, V., Ausanio, G., De Ninno, A., Gentile, F., Germano, R., Napoli, E. and Niccoli, M. "Experimental Evidence of Stable Aggregates of Water at Room Temperature and Normal Pressure After Iterative Contact with a Nafion Polymer Membrane". *Water* 5 (2013): 16–26.
25. Elia, V., Ausano, G., De Ninno, A., Germano, R., Napoli, E. andNiccoli, M. "Experimental Evidence of Stable Water Nanostructures at Standard Pressure and Temperature Obtained by Iterative Filtration". *Water* 5 (2014): 121–130.
26. Elia, V., Napoli, E. and Niccoli, M. "Physical-Chemical Study of Water in Contact with a Hydrophilic Polymer". *J Therm Anal Calorim* 112 (2013): 937–944.
27. Ryzhkina, I.S., Murtazina, L.I., Kiseleva, Y.V. and Konovalov, A.I. "Properties of Supramolecular Nanoassociates Formed in Aqueous Solutions of Biologically Active Compounds in Low or Ultra-Low Concentrations". *Dokl Phys Chem* 428 (2009): 196–200.
28. Ryzhkina, I.S., Murtazina, L.I. and Konovalov, A.I. "Action of the External Electromagnetic Field Is the Condition of Nanoassociate Formation in Highly Diluted Aqueous Solutions". *Dokl Phys Chem* 440 (2011): 201–204.
29. Ho, M.W. "Supramolecular Nanostructures in Highly Dilute Solutions Required for Biological Activity". *Sci Soc* 64 (2014c): 48–49.
30. Konovalov, A. "Nanoassociates: Terra Incognita". *Science in Russia* 1 (2014): 4–10.
31. Konovalov, A.I. and Rychkina, I.S. "Formation of Nanoassociates As a Key to Understanding of Physicochemical and Biological Properties of Highly Dilute Aqueous Solutions". *Russian Chemical Bulletin*, International Edition 63 (2014): 1–14.
32. "Dynamic Light Scattering", Wikipedia, accessed 13 August 2014, http://en.wikipedia.org/wiki/Dynamic_light_scattering.
33. Germano, R., Del Giudice, E., De Ninno, A., Elia, V., Hison, C., Napoli, E., Tontodonato, V., Tuccinardi, F.P. and Vitiello, G. "Oxyhydroelectric Effect in Bi-Distilled Water". *Key Eng Mater* 543 (2014): 455–459.
34. Ho, M.W. and Knight, D.P. "The Acupuncture System and the Liquid Crystalline Collagen Fibers of the Connective Tissues". *Am J Chin Med* 26 (1998): 251–263.

35. Ho, M.W. "First Sighting of Structured Water". *Sci Soc* 28 (2005): 47–48.
36. Ho, M.W. "Collagen Water Structure Revealed". *Sci Soc* 32 (2006): 15–16.
37. Ho, M.W. "Superconducting Quantum Coherent Water in Nanospace Confirmed". *Sci Soc* 55 (2012b): 48–51.
38. Pauling, L. "The Structure and Entropy of Ice and of Other Crystals with Some Randomness of Atomic Arrangement". *J Am Chem Soc* 57 (1935): 2680–2684.
39. Isaacs, E.D., Shukla, A., Platzman, P.M., Hamann, D.R., Barbiellini, B. and Tulk, C.A. "Covalency of the Hydrogen Bond in Ice: A Direct X-Ray Measurement". *Phys Rev Lett* 82 (1999): 600–603.
40. Bakker, H.J. and Nienhuys, H.K. "Delocalization of Protons in Liquid Water". *Science* 297 (2002): 587–590.
41. Reiter, G., Burnham, C., Homouz, D., Platzman, P.M., Mayers, J., Abdul-Redah, T., Moravsky, A.P., Li, J.C., Loong, C.K. and Kolesnikov, A.I. "Anomalous Behaviour of Proton Zero Point Motion in Water Confined in Carbon Nanotubes". *Phys Rev Lett* 97 (2006): 247801.
42. Kyakuno, H., Matsuda, K., Yahiro, H. *et al.* "Confined Water Inside Single-Walled Carbon Nanotubes: Global Phase Diagram and Effect of Finite Length". *J Chem Phys* 134 (2011): 244501.
43. Reiter, G.F., Kolesnikov, A.I., Paddison, S.J., Platzman, P.M., Moravsky, A.P., Adams, M.A. and Mayers, F. "Evidence for an Anomalous Quantum State of Protons in Nanoconfined Water". *Phys Rev B* 85 (2012): 045403.
44. Ho, M.W. "Science and Art of Water". *Sci Soc* 58 (2013a): 48–51.
45. Deb, A., Reiter, G.F., Sakurai, Y., Itou, M., Krishnan, V.G. and Paddison, S.J. "Anomalous Ground State of the Electrons in Nanoconfined Water". *Spring-8 Information* 19 (2014): 2–6.
46. Spry, D.B., Goun, A., Glusac, K., Moilanen, D.E. and Fayer, M.D. "Proton Transport and the Water Environment in Nafion Fuel Cell Membranes and AOT Reverse Micelles". *J Am Chem Soc* 129 (2007): 8122–8130.
47. Schmidt-Rohr, K. and Chen, Q. "Parallel Cylindrical Water Nanochannels in Nafion Fuel-Cell Membranes". *Nat Mater* 7 (2008): 75–83.

48. "Electrical Resistivity and Conductivity", Wikipedia, accessed 7 June 2014, http://en.wikipedia.org/wiki/Electrical_resistivity_and_conductivity.
49. Dong, B., Gwee, L., Salas de la Cruz, D., Winey, K.I. and Elabd, Y.A. "Super Proton Conductive High-Purity Nation Nanofibers". *Nano Lett* 10 (2010): 3785–3790.
50. Dellago, C., Naor, M.M. and Hummer, G. "Proton Transport through Water-Filled Carbon Nanotubes". *Phys Rev Lett* 90 (2003): 105901.
51. Chen, J., Li, X.Z., Zhang, Q., Michaelides, A. and Wang, E. "Nature of Proton Transport in a Water-Filled Carbon Nanotube and in Liquid Water". *Phy Chem Chem Phys* 15 (2013): 6344–6349.
52. Choi, W., Ulissi, Z.W., Shimizu, S.F.E., Bellisario, D.O., Ellison, M.D. and Strano, M.S. "Diameter-Dependent Ion Transport through the Interior of Isolated Single-Walled Carbon Nanotubes". *Nat Commun* 3 (2013): 2397.
53. Kyakuno, R., Matsuda, R., Yahiro, H., Inami, Y., Fukuoka, T., Miyata, Y., Yanagi, K., Maniwa, Y., Kataura, H., Saito, H., Yumura, M. and Iijima, S. "Confined Water Inside Single Walled Carbon Nanotubes: Global Phase Diagram and Effect of Finite Length". *J Chem Phys* 134 (2011): 244501.
54. Qin, X., Yuan, Q., Zhao, Y., Xie, S. and Liu, Z. "Measurement of the Rate of Water Translocation through Carbon Nanotubes". *Nano Lett* 11 (2011): 2173–2177.
55. Bardelmeyer, G.H. "Electrical Conduction in Hydrated Collalgen: I. Conductivity Mechanisms". *Biopolymers* 12 (1973): 2289–2302.
56. Sasaki, N. Dielectric Properties of Slightly Hydrated Collagen: Time-Water Content Superposition Analysis. *Biopolymers* 23 (1984): 1725.
57. "Second-Harmonic Imaging Microscopy", Wikipedia, accessed 15 December 2012, http://en.wikipedia.org/wiki/Second_harmonic_imaging_microscopy.
58. Pehlke, C. Microtechnology Medicine Laboratory, Biomedical Engineering @University of Wisconsin-Madison, http://mmb.bme.wisc.edu/personal%20research/CarolynPehlke/CarolynPehlke.htm?id=people.
59. Stoller, P.C., Reiser, K.M., Celliers, P.M. and Rubenchik, A.M. "Effects of Structural Modification on Second Harmonic Generation

in Collagen". Submitted to SPIE Conference on Visualization and Data Analysis, San Jose, California, 20–25 January 2003, https://e-reports-ext.llnl.gov/pdf/243951.pdf.

60. Williams, R.M., Zipfel, W.R. and Webb, W.W. "Interpreting Second-Harmonic Generation Images of Collagen 1 Fibrils". *Biophys J* 68 (2005): 1377–1386.
61. Ho, M.W. "Super-Conducting Liquid Crystalline Water Aligned with Collagen Fibres in the Fascia As Acupuncture Meridians of Traditional Chinese Medicine". *For Immunopathol Dis Therap* 2 (2012c): 221–236.
62. Glauber, R.J. "The Quantum Theory of Optical Coherence". *Phys Rev* 130(1963): 2529–2539.
63. Ho, M.W. "Evolution by Natural Genetic Engineering". *Sci Soc* 63 (2014d): 18–23.
64. McClare, C.W.F. "Chemical Machines: Maxwell's Demon and Living Organisms". *J Theor Biol* 30 (1971): 1–34.
65. Ho, M.W. "The Real Bioinformatics Revolution". *Sci Soc* 33 (2007): 42–45.
66. Ho, M.W. "Non-Random Directed Mutations Confirmed". *Sci Soc* 60 (2013b): 30–32.
67. Tosi, M. and Del Giudice, E. "The Principle of Minimal Stimulus in the Dynamics of the Living Organism". *Sci Soc* 60 (2013): 26–29.
68. Musumeci, F., Scordino, A., Triglia, A., Blandino, G. and Milazzo, I. "Intercellular Communication during Yeast Cell Growth". *Europhys Lett* 47 (1999): 736–742.
69. Ho, M.W. "Life Is Water Electric". *Sci Soc* 57 (2013c): 43–47.
70. Ho, M.W. "The Story of Phi (in Six Parts)". *Sci Soc* 62 (2014e): 24–44.
71. Colour of Water art/science/music festival, 12–28 March 2013, http://www.i-sis.org.uk/coloursofwater/.
72. Del Giudice, E. "Prometheus: The Passionate Soul of Scientific Reason". *Water* 6 (2014): 61–71.

17

Fractal Mathematics and the Fabric of Nature

17

Story of Phi

A non-mathematician's initiation into the secrets of the magic number in the heart of mathematics that is woven into the fabric of the living world and the universe.

The Basic Mathematics

The Perfect Cut

The golden ratio ϕ (a Greek letter pronounced *phi*) is an irrational number, one that is impossible to express as a simple fraction; and it is the most irrational of all irrational numbers because it is the most awkward to *approximate* as a simple fraction. Yet it turns up everywhere in nature, from chemical bonds in molecules to branching trees, spiral galaxies, fundamental quantum reality, the universe itself; it is embedded in our brain waves, in music, and of course, in the heart of mathematics.[1–3]

The numerical value of phi is 0.6180339 ...(continued on indefinitely), and is the solution to a problem that sounds so simple.

Suppose you have a line of any length, call it 1. You need to divide it so that the ratio of the larger segment to the whole, 1, is exactly the same as that of the small segment to the larger (Fig. 17.1).

From Fig. 17.1, we get the following:

$$\phi/1 = (1 - \phi)/\phi$$

$$\phi^2 = 1 - \phi$$

$$\phi^2 + \phi = 1$$

$$\phi^2 + \phi - 1 = 0 \qquad (1)$$

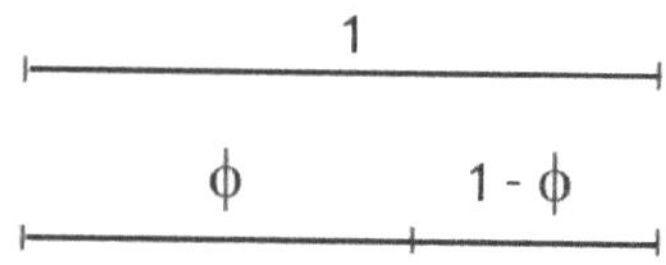

Figure 17.1 Golden ratio, the perfect cut.

Solving Eq. 1 for ϕ with the quadratic formula in basic algebra gives

$$\phi = (\sqrt{5} - 1)/2 = 0.6180339887\ldots$$

The reciprocal of *phi*, represented by the same letter in capital Φ (*Phi*) is also often referred to as the golden ratio, and is equal to

$$1/\phi = \Phi = (\sqrt{5} + 1)/2 = 1.6180339887\ldots = 1 + \phi$$

Amazingly, precisely the same numbers occur in *both* ratios after the decimal. Not only that, squaring Φ gives 2 plus exactly the same numbers after the decimal:

$$\Phi^2 = 2.6180339887\ldots = 2 + \phi$$

I leave you to do more sums and work out other marvels of *phi*.

The golden ratio is the perfect cut. It gives *self-similarity* or recursiveness at successive scales, and is considered the hallmark of beauty and balance in art and architecture. Most of all, *phi* has a host of magical properties enough to possess generations of mystics and scientists alike, beginning with the ancient Egyptian pyramid builders, who might have followed earlier forebears of this esoteric knowledge, now lost in the mists of time. Most of what is known in modern times came via the ancient Greek philosopher mathematician Pythagoras (~570BC–~495BC) and mathematician Euclid (323–283BC).

And I am among the latest to be smitten.

Being a scientist, I am far less interested in numerology than in what *phi* tells us about the very fabric of nature and natural processes, and ultimately the universe. But it will help to flesh out the mathematical picture before proceeding to nature and natural processes.

Fibonacci Sequence and Ideal Rabbits Breeding

The Fibonacci sequence is a sequence of numbers beginning with 1, 1, or 0, 1, in which successive numbers are the sum of the two previous:

1, 1, 2, 3, 5, 8, 13, 21, 34, 55, 89, 144, ...
0, 1, 1, 2, 3, 5, 8, 13, 21, 34, 55, 89, 144, ...

The sequence is named after Italian mathematician Leonardo Fibonacci (~1170–1250) who got it during his wide travels from Indian mathematics.[4] Fibonacci considered the growth of a biologically unrealistic rabbit population, assuming that a newly born pair of rabbits, one male and one female, are able to mate at the age of one month, so that at the end of the second month the female produces a fresh pair of rabbits (one male and one female), and the same for every month thereafter, as the rabbits never die. The question was: how many rabbits would there be at the end of the year?

Starting from a single pair, the following sequence of events takes place to build up the population:

At the end of the first month, they mate, but there is still only one pair.

At the end of the second month, the female gives birth to one pair, so there are two pairs.

At the end of the third month, the original female gives birth to a second pair, and this makes three pairs.

At the end of the fourth month, the original female produces a third pair, but the first pair of her offspring will also have produced *their* first offspring pair, making a total of five (Fig. 17.2). At the end of the nth month, the number of pairs of rabbits equals to the number of new pairs (the number of pairs in month n–2) plus the

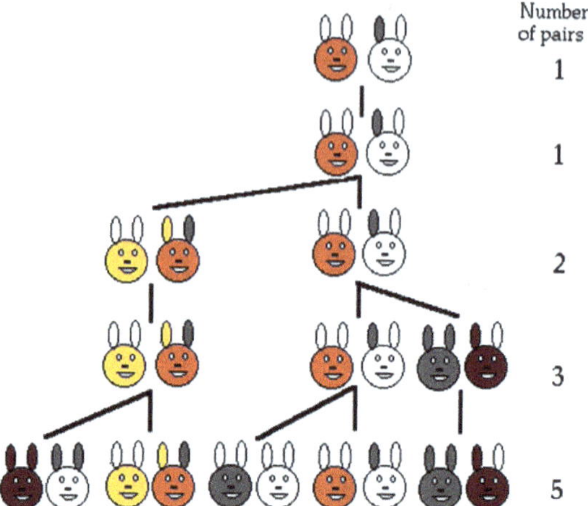

Figure 17.2 Growth in a population of imaginary rabbits follows Fibonacci sequence.[3]

number of pairs alive last month (n–1). This is the nth Fibonacci number:

$$F_n = F_{n-1} + F_{n-2}. \tag{2}$$

It was German mathematician and astronomer Johannes Kepler (1550–1517) who showed that the golden ratio is the limit of the ratio of consecutive Fibonacci numbers. In other words, the ratio of successive numbers get closer and closer to the value of the golden ratio as the sequence is continued *ad infinitum*. Now we know why *phi* is the most irrational of all irrational numbers.

The Fibonacci sequence and golden spiral

The relationship of the Fibonacci sequence and the golden ratio to growth with long-range correlation is seen in the spirals of Fig. 17.3.[5] As the spiral gets bigger and bigger, it becomes more and more similar to the golden spiral, one version of which is constructed from an initial golden rectangle divided into squares (Fig. 17.4).[5]

Both the Fibonacci spiral and the golden spiral *approximate* a logarithmic spiral, but is not actually a logarithmic spiral; and in

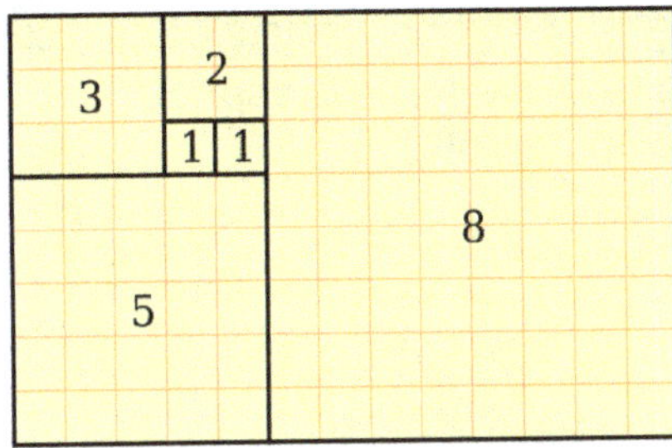

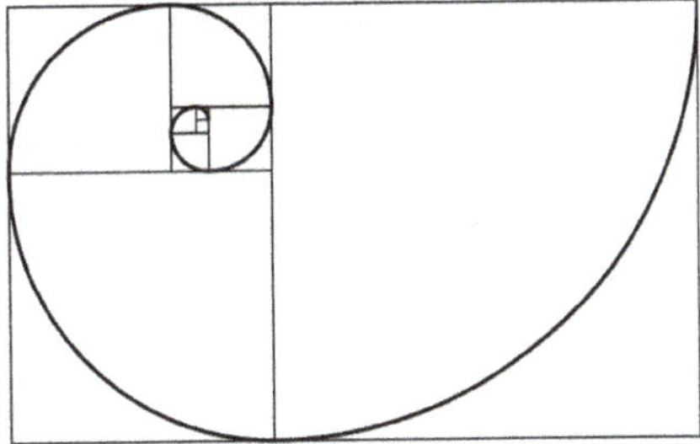

Figure 17.3 A spiral tiling with squares whose side lengths are successive Fibonacci numbers; the right diagram uses the squares of sizes 1, 1, 2, 3, 4, 8, 13, 21 and 34, and draws the quarter circle that fits into the squares; it approximates a golden spiral more and more as the Fibonacci numbers get bigger.

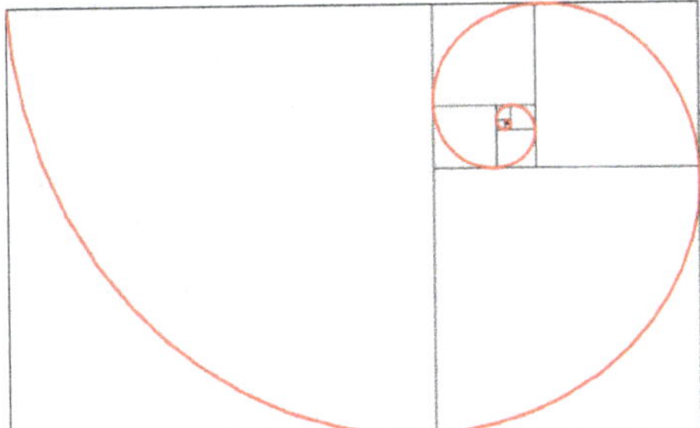

Figure 17.4 A golden spiral constructed by dividing an original golden rectangle into squares (left) and the Nautilus shell sectioned (right).

mathematics, even the golden spiral is not considered the real golden spiral. A *real* golden spiral is a logarithmic spiral whose growth factor is *Phi*. It gets bigger and further from its origin by a factor of Φ for every quarter turn or 90°.[6] The equation in polar (circular) coordinates for a logarithmic spiral is

$$r = ae^{b\theta} \tag{3}$$

where *r* is the radius, *e* is the base of natural logarithms, *a* an arbitrary positive constant, *b* the growth factor and θ the angle traversed. For the golden spiral, the golden ratio comes in when θ is a right angle: $e^{b\theta right} = \Phi$. So, the real golden spiral is not strictly a logarithmic spiral after all, as the growth between right angles is interpolated. The *Nautilus* shell (Fig. 17.4 right) consists of successive chambers and grows according to the logarithmic spiral, but by a factor of the golden ratio Φ for each complete turn of the spiral, rather than a quarter turn.

Penrose Tiling and Quasicrystals

The problem of tiling is to completely cover a surface area, as you would with tiles, and the golden ratio is a key to the problem, especially if you want interesting patterns that are not just based on squares. Tiling is easily done with squares (four sides), or triangles and hexagons (three and six sides, respectively). But it was long believed impossible to fill an area with five-sided tiles or pentagons.

In the early 1970s, British mathematical physicist Sir Roger Penrose at Oxford University discovered that a surface can be completely covered in an asymmetric, non-repeated, and aesthetically highly pleasing way, using just two different shaped tiles derived from the pentagon. Figure 17.5 shows two variations. You can flip the tiles and get many other patterns.

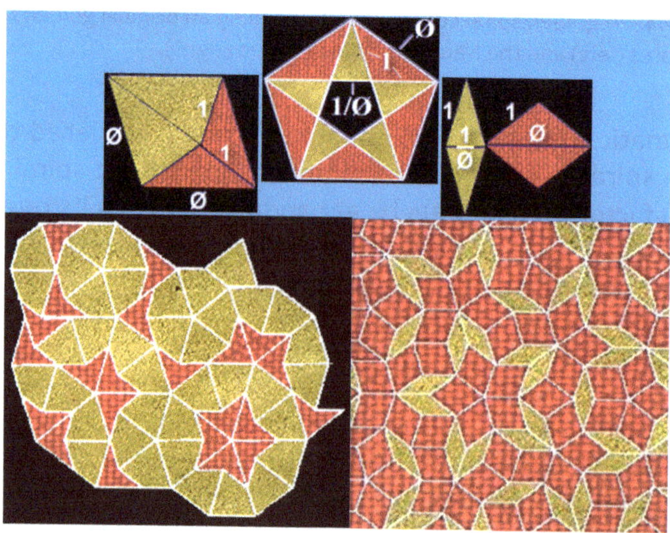

Figure 17.5 Penrose tiling based on the pentagon and the golden ratio.[6]

The pentagon not only embodies the golden ratio, but the resulting tiling also has a golden ratio of area covered by one kind of tile to the other. Five-fold symmetry is especially common in living organisms, and also exists in the enchanting world of quasicrystals, although it is absent from crystals.[7]

Watching the Daisies Grow

We start our exploration of ϕ in the natural world with the spiral arrangements of plant elements around the stem that is based on the golden ratio. Asking why that should be so leads us into a deeper realm of beauty.

Phyllotaxis the Eternal Fascination

In the study of plants, one of the most abiding mysteries is phyllotaxis, the arrangement of leaves, flowers, petals or florets around the stem. This problem fascinated British mathematician, computer pioneer and code-breaker Alan Turing (1912–1954) from an early age. A sketch made by his mother showed him distracted from his cricket game "watching the daisies grow". Tragically, Turing committed suicide before he could solve the problem,[8] and after having been convicted of homosexuality and forced to undergo "hormone therapy". It was not until 2009 that British Prime Minister Gordon Brown made an official public apology on 24 December 2013 on behalf of the British government that Turing received a posthumous pardon.[9]

Spiral Phyllotaxis and the Golden Angle

Spiral phyllotaxis is common among plants. The leaf-like elements come out one by one from the growing tip of the main stem, each at a constant divergence angle *d* from the previous, which is often close to the golden angle 137.5°, derived from the golden ratio $\Phi = 1.618$; the actual angle is $360/1.618 = 222.5°$, but it is usual to recognize the smaller reflex angle of $360 - 222.5 = 137.5°$.

Figure 17.6 The Aeonium plant has a 2, 3 parastichy.

When two or more elements come off a single node (or tier), referred to as multijugate phyllotaxis, each whorl is at a constant divergence angle from the previous. To classify the multijugate patterns, we count the number of spirals joining each element to its nearest neighbours. This gives a pair of numbers, one for the spirals going counterclockwise, and the other for the spirals going clockwise; the pair of numbers is the *parastichy*. Thus, the *Aeonium* plant has a 2, 3 parastichy, there being two spirals in the counterclockwise direction and three in the clockwise direction (Fig. 17.6).[10] The two numbers are successive members of the Fibonacci sequence (see above).

When the elements are tightly packed, there are many more spirals. Turing's daisy shows florets in the centre in 21 counterclockwise and 34 clockwise spirals, giving a 21, 34 parastichy (Fig. 17.7). These pairs are also successive numbers in the Fibonacci sequence.

Why Phi?

Scottish biologist, mathematician and classic scholar D'Arcy Thompson (1860–1948) had already established early in the last century that the angle between successive elements in the tightly wound spirals approximate the golden angle 137.5° based on *Phi*.[11]

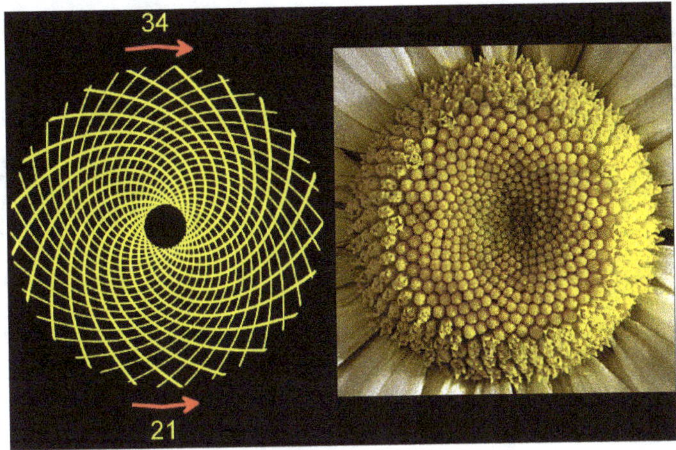

Figure 17.7 Florets in a daisy with 21, 34 parastichy (see text).

But no satisfactory explanation was forthcoming for a long time. Many trivial suggestions include adaptation and natural selection, or design by God; for the arrangement optimizes reception of light for photosynthesis.

Optimal Packing

From the snug way in which the florets fit together in the heart of the daisy, it would appear that the golden ratio and associated Fibonacci spirals do result in the optimum packing. This is easily demonstrated using a computer generative algorithm developed jointly by the research teams of Przemyslaw Prusinkiewicz at University of Regina, Saskatchewan, in Canada, and Astrid Lindenmayer at University of Utrecht in The Netherlands.[12] Remarkably, the optimum packing arrangement closest to the actual daisy is generated only when the divergence angle of 137.5° is used; slight deviations on either side result in very different patterns with large gaps between spirals (Fig. 17.8). This space-filling capacity of the golden ratio is very significant, as in the Penrose tiling above, and we shall come back to it later. It also gives a feeling of balance and beauty.

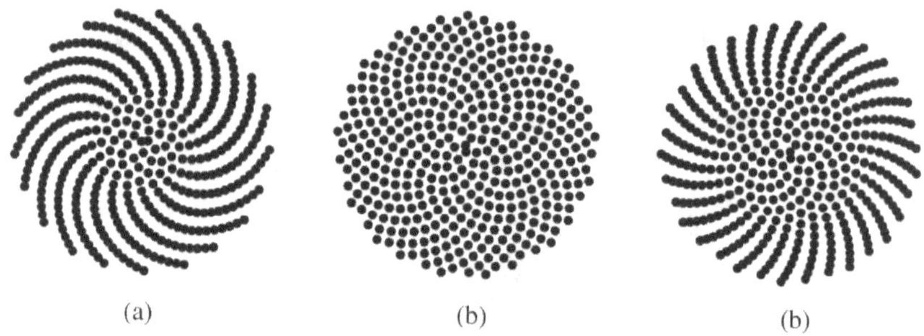

Figure 17.8 Phyllotaxis patterns generated on a disc with divergence angle 137.3° (a) 137.5° (b) and 137.6° (c).

Growth and Space-Filling

However, the importance of the golden angle still does not explain *how* it arises, without invoking divine design or its equivalent in neo-Darwinian "selective advantage"; for even if the particular arrangement of leaves and florets does benefit the plant, one needs to explain how the form was generated in the first place. We need to understand the mathematics, as well as the physics, chemistry and biology.

In 1992, French mathematicians Stéphane Douady and Yves Couder at French National Centre for Scientific Research (CNRS), Paris, made a breakthrough. They showed that the patterns arise spontaneously as the result of a simple growth process in which primordia appear periodically near the tip of the growing shoot and are pushed away (advected) from the centre as the tip elongates (or perhaps equivalently, increases in diameter). Identical elements are generated with a periodicity T at a given radius R_0 from the centre and advected at velocity V_0, and there is repulsion between the primordia, so new elements will appear as far as possible from the previous one. The results could be interpreted with a simple equation consisting of only one (dimensionless) parameter $G = V_0T/R_0$.[13]

To model the process physically, Douday and Couder set up a petri dish with a central cone filled with some viscous silicone oil,

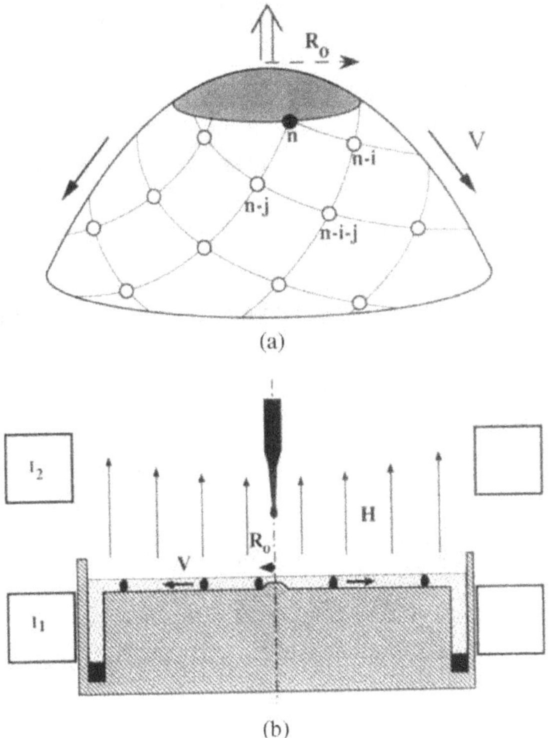

Figure 17.9 Diagram of the growth zone with successive primordial (a), and an experimental model of the process (b).

and released drops of ferroliquid onto the cone, which were advected towards the periphery by a magnetic field (Fig. 17.9).

When advection is strong, each drop is repelled by the one before and goes to the direct opposite (180°). But as G decreases, advection slows, or equivalently, the period between primordia shortens; each will be influenced by the two previous, and will go in the space between them at an angle of 150°. As G decreases further, each drop will be influenced by three or more previous, and the angle becomes 139°, quite close to the golden angle, when the spirals take on the Fibonacci values 3 and 5. The process approaches the golden ratio in the limit. It ends up filling the entire available space, creating a structure of minimum global interaction energy, a natural balance between opposing forces and tensions, which is where beauty lies.

The research also shows that the process gives reliable results regardless of the detailed biological or physical mechanisms; these characteristics of *reliability*, *robustness* and *genericity* are intimately associated with the balance and dynamic equilibrium embodied in the golden ratio, the kind of beauty and elegance that seduce scientists and mathematicians as well as artists through the ages.

The results were also successfully simulated by computer and can be used as the most natural system for classifying plants and animals sought by Swedish botanist and physician Carl Linnaeus (1707–1778) and generations of taxonomists after him, as demonstrated in a paper I wrote with Peter Saunders in 1994.[14]

Chaotic Dynamics and Golden Music of the Brain

Next, we explore how the chaotic dynamics of natural processes are related to the sublime music of information processing in the brain.

Self-Organized Criticality?

What could possibly connect cascading sand-piles, avalanches, music, brainwaves and more? It is "self-organized criticality", described in a paper published in 1987 co-authored by theoretical physicist Per Bak (1948–2002) who coined the term.[15] Self-organized criticality is a property of dynamical systems that look the same on different scales, and hence referred to as *self-similar*, or *scale-invariant*.[16] It is to be found at the critical point of a phase transition, as when water turns into ice, or a ferromagnet becomes fully magnetized.

Self-organized criticality is also a way of understanding complex systems in nature. It stems from the discovery that complexity could be generated as an emergent (new and unexpected) feature of extended systems from simple local interactions, for example, in work on cellular automata by Polish mathematician Stanislaw Ulam

(1909–1984) and Hungarian-American mathematician John von Neumann (1903–1957), plus a large body of work on fractal geometry — self-similar mathematical structures — by Polish-born French mathematician Benoît Mendelbrot (1924–2010). Simultaneously, extensive investigations on phase transitions in the 1960s and 1970s showed how scale-invariant fractals and power laws emerge at the critical point. A power law is a mathematical relationship often referred to as the $1/f$ law (after the $1/f$ electronic noise of transistors) in which the distribution of power density at different frequencies (power spectral density) is inversely proportional to the frequency, i.e.,

$$S(f) \propto \frac{1}{f^{\alpha}} \tag{4}$$

where f is the frequency and α an exponent generally between 0 and 2. Note that f can be any quantity, not just noise or frequency of the electronic signal. It could apply to size, or amplitude, or duration (of time).

Per Bak, Chao Tang and Kurt Wiesenfeld put all those ideas together in their paper.[15] They used the analogy of a pile of sand at the point of collapse, when the avalanches take on all sizes from small to large, but are distributed according to the $1/f$ power law, so that frequency of avalanches at different sizes scale as the inverse of size: small ones are more frequent than medium and medium more frequent than big ones. A log-log plot of frequency versus size results in a straight line, the slope of which gives the exponent.

The key result of the Bak-Tang-Wiesenfeld paper was to demonstrate how the complexity observed emerges *spontaneously* in a robust way that does not require any fine tuning or regulation, like the collapse of a sand pile under gravity at the point when it is piled high and big enough.

However, that is only part of the story of complex systems, and to this day, researchers are debating whether the $1/f$ power law is really applicable to natural phenomena, and what exactly it implies about the underlying mechanisms.

Before venturing further, let's see what the $1/f$ distribution can tell us about music.

Fractal Musical Time

Why do we love music? One reason we enjoy music lies in its balance of predictability and surprise, so researchers claim. They found musical pitch (frequency) spectrum following a $1/f$ power law, which achieves this balance of predictability and surprise; but what about musical rhythm? Musical rhythms, especially those of Western classical music, are highly regular and predictable; but are they? Daniel Levitin at McGill University Canada, Parag Chordia at Georgia Institute of Technology Atlanta and Vinod Menon at Stanford University California in the United States decided to put that to the test by analysing the rhythm spectra in 1 788 movements from 558 compositions of Western classical music.[17]

The rhythmic content of the compositions were systematically measured by noting the durations of the notes and of the rests, transforming the durations into Hz (cycle per second), plotting the data and finding the spectral exponent in the slope of the line obtained. They went through the works of 40 composers in 16 subgenres, and found an overwhelming majority of rhythms following the $1/f^{\alpha}$ power law with α ranging from ~0.5 to ~1. An exponent of 0 would be pure white noise, completely unpredictable, whereas an exponent of 2 and above would be highly predictable. Notably, classical composers whose compositions are known to exhibit nearly identical $1/f$ pitch spectra demonstrated distinctive $1/f$ rhythm spectra: the rhythms of German composers Ludwig van Beethoven (1770–1827) were among the most predictable, and Wolfgang Amadeus Mozart (1756–1791) the least behind African-American ragtime composer Scott Joplin (1867–1917), with Joseph Haydn (1732–1809) in between (Fig. 17.10). The difference in rhythmic predictability is such as to allow composers to identify their compositions uniquely and to distinguish them from works of their contemporaries.

Previous studies analysing classical compositions from the 18th to the 20th centuries reported nearly identical $1/f$ pitch structure

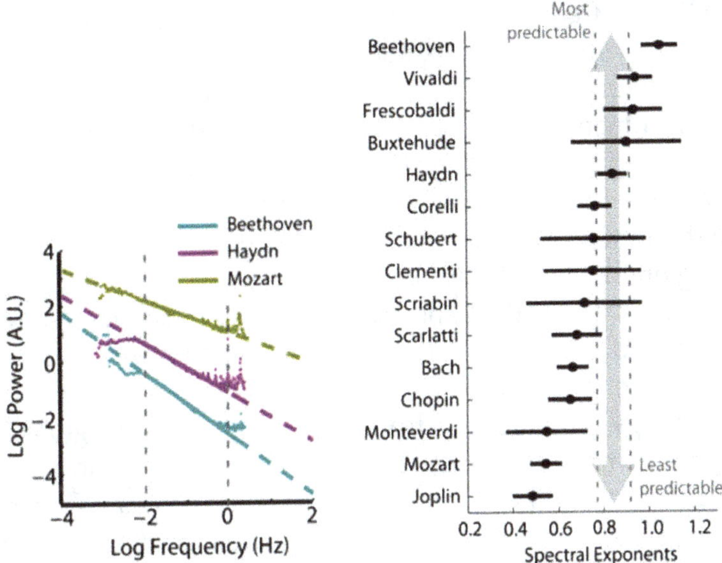

Figure 17.10 Power spectra and exponents of rhythms in Western classical compositions by composers.

among composers, with a very narrow range of spectral exponent 1.79–1.97. In contrast, the rhythm spectral exponent varied widely and systematically among composers, from 0.48 to 1.05. Even composers belonging to the same musical classical era — Beethoven, Haydn and Mozart — demonstrated distinctive rhythm spectra.

Organic Space-Time Fractal and Irrational Multiples for "Quantum Jazz"

Time structure in classical music turns out to be much more nuanced than previously suspected. This is of great interest to me, as this whole enquiry around the golden ratio began because I wanted to find out how it might be involved in the space-time structure of organisms, i.e., organic quantum space-time (as opposed to mechanical clock time of the Newtonian universe). Organic space-time has been a major theme of my book, *The Rainbow and the Worm: The Physics of Organisms,* since its first 1993 edition.[18] I surmised that organic space-time is fractal, as French astrophysicist Laurent Nottale has proposed that relativistic-quantum space-time is fractal.[19]

Organic space-time is organized as cycles of activity (or oscillations) over many scales,[18] generally recognized as the all-pervasive biological rhythms ranging from split seconds to days and years that have preoccupied physiologists and ecologists alike at least since the 18th century.[20] The cycles are coupled such that activities yielding energy transfer it directly to those requiring it with minimum loss or dissipation, *and the directions can be reversed whenever required.*[18] This scale-invariant space-time structure is the key to the efficiency and rapidity of energy transfer within living systems. Energy can be drawn from any local level to the global, and conversely, it can be concentrated to any local domain from the entire system. In the ideal, the system tends towards quantum coherence — an incredibly dynamic "quantum jazz" over all space-times — that maximizes both local freedom (autonomy) and global cohesion. How could that be achieved? How could we have local autonomy and be able to transfer energy rapidly from local to global and vice versa?

The only way that could be achieved, I suggested, is that the frequencies of biological rhythms are irrational multiples of one another[21] under steady-state default conditions (perhaps even multiples of the golden ratio). So the frequencies are close enough to rational multiples that they can slide towards rational multiples to achieve phase synchrony or phase-locking, thereby enabling energy transfer via resonance or positive interference.

There is no evidence as yet that biological rhythms in general are organized as irrational multiples. However, oscillations in local field potentials in the brain are found to approximate multiples of the most irrational number, the golden ratio, and this appears crucial in how the brain processes information or makes music.

Golden Frequencies of the Resting Brain

Electrical activities from the surface layers of the brain are readily recorded via an array of sensitive electrodes positioned on the scalp. This procedure, electroencephalography (EEG), records voltage fluctuations in the local fields, or local field potentials (LFPs) around groups of thousands of neurons close to each electrode.

Belinda Pletzer, Hubert Kerschbaum and Wolfgang Klimesch at the University of Salzburg in Germany have recognized the importance of brain frequencies that never synchronize in the resting brain. They reasoned that if the electrical oscillations in the brain are involved in communication between different brain structures and networks, it is important for groups of cells to keep their rhythms distinct without interference from other groups, and without interfering with the rhythms of other groups. And that is indeed achieved via frequencies in irrational multiples in the resting (default) brain.[22]

The brain functions in a massively parallel and distributed manner. Which processes are responsible for communication between the largely distributed brain structures? One possibility is oscillations. During the highly excitable phase, neurons are very likely to fire (spike in an action potential), whereas during the low excitability phase, activity is reduced or suppressed. Consequently, the oscillations facilitate simultaneous activation of common target cells and enhanced firing through an assembly of cells, and may be highly important in gating cognitive processes (especially in a quantum coherent brain).[23]

It is generally assumed that functional interplay between different cell assemblies is reflected by oscillatory coupling of several different kinds of phase synchrony and phase-locking. However, it is just as important to avoid spurious coupling that interferes with function. Therefore, frequencies in the default resting state should be arranged so that they never synchronize. The best way this could be achieved is to have frequencies in multiples of the most irrational number, the golden ratio. To prove their point, Pletzer and colleagues tabulated the typical EEG frequencies in recordings made mostly when subjects are in a "resting" state, not engaged in any mental processing tasks (Table 17.1).

The classical frequency bands of the EEG can indeed be described as a geometric series with a ratio between neighbouring frequencies approximating $\Phi = 1.618$. Not only does the ratio between neighbouring peak frequencies approximate the golden ratio, but

Table 17.1 Typical EEG frequency bands and subbands and corresponding periods

Frequency band		Frequency subband		Peak	Period
Name	[Hz]	Name	[Hz]	[Hz]	[ms]
delta	1.5–4	delta1	1–2	1.5	667
		delta2	2–3	2.5	400
theta	4–10	theta1	3–5	4	250
		theta2	5–8	6.5	154
alpha	8–12	alpha	8–12	10	100
beta	10–30	beta1	12–20	16	62.5
		beta2	20–30	25	40
gamma	30–80	gamma1	30–50	40	25
		gamma2	50–80	65	15
fast ripples	80–200	ripples1	80–120	100	10
		ripples2	120–200	160	6.25

also the successive frequencies are the sum of two previous, approximating the Fibonacci sequence. Synchronization of the excitatory phases of the two oscillations is impossible. Thus, the golden ratio gives a totally uncoupled (yet coherent) processing state, which most likely reflects the resting brain. However, the excitatory phases of the two oscillations occasionally come close enough to coincide. These coincidences are more frequent the higher the frequencies of f_1 and f_2. Thus, intuitively, at least, one can see that the golden ratio provides the highest physiologically possible desynchronized state, and at the same time, the potential for spontaneous diverse coupling and uncoupling between rhythms and a rapid transition from resting state to activity. This is borne out by findings from other laboratories.

Golden Music of the Brain

Dietmar Plenz at the US National Institute of Mental Health has identified "neuronal avalanches" — cascades of neuronal activities that follow precise $1/f$ power laws — in the excitatory neurons of the superficial layers in isolated neocortex preparations *in vitro* as well as in awake animals and humans *in vivo*.[24] The neocortex of mammals is a sheet of six layers folded inside the skull. Inputs to the cortex arrive at layer IV, whereas outputs to brain structures outside the cortex are provide by neurons in deep layers V and VI. The superficial layers II and III are where cortical neurons talk to one another only, with layer I mainly composed of fibre bundles supporting intra-cortical communication. A combination of experiments, theory and modelling showed that the neuronal avalanche of the default state with the $1/f$ signature of self-organized criticality gives the optimum response to inputs as well as maximum information capacity.

Most interestingly, the avalanche dynamics gives rise to *coherence potentials*, subsets of avalanches in which the precise waveform of the local field potential is replicated with high fidelity in distant network sites. The process is independent of spatial distance and includes near instantaneous neuronal activities as well as sequential activities over many timescales. Most coherence potentials are spatially disjunct, i.e., they do not obey nearest neighbour relationships. LFPs of successive coherence potentials are not similar, but they are practically identical within a coherence potential among all the participating sites, there being no growth or dissipation during propagation. This intriguingly suggests that the waveform of a coherence potential is a high-dimensional coding space in information processing of the brain. For decades, phase-locked neuronal activity has been reliably recorded using the LFPs or EEG and was found to correlate with the presentation of stimulus in animals and visual perception in humans.

Another team of researchers led by Miles Whittington at Newcastle University in the United Kingdom has begun to reveal the intricacies in the golden music of the brain by recording from multiple layers of the neocortex simultaneously. They found multiple local neuronal circuits supporting different discrete frequencies in the neocortex network, and the relationships between different frequencies appear designed to minimize interference and to allow the most diverse coupling of activities via stable phase interactions and the control of the amplitude of one frequency in relation to the phase of another.[25] There is even a transformation that combines the oscillations of two neighbouring frequencies sequentially to generate a third frequency whose period is the concatenated sum of the original two. With such an interaction, the intrinsic periodicity in each component local circuit is preserved: alternate, single periods of each original rhythm form one period of a new frequency, suggesting a robust mechanism for combining information processed on multiple concurrent spatiotemporal scales to generate what must be the most amazing golden quantum jazz.

The $1/f$ pattern of EEG is really a time-averaged smoothed collection of multiple, discrete frequencies, and does not represent all the frequencies and combinations of frequencies present in the brain. (It is like a recording of a Mozart symphony that averages out all the sounds made in discrete periods of time, so the music is completely buried.) Detailed observations made by the team have shown that at least three discrete frequencies δ (1–3 Hz), θ (6–9 Hz), and γ (30–50 Hz) are often expressed simultaneously, and can be associated with further much slower rhythms both *in vivo* and *in vitro*.

Discrete frequencies ranging from low δ to high γ can be produced from a single area of the isolated neocortex *in vitro*, with peak frequencies distributed according to the golden ratio. All attempts to generate single frequencies have failed, and the phenomenon has been referred to as spectral processing.

To keep simultaneously occurring frequencies apart and minimize interferences, the solution is to have ratios of frequencies that are irrational numbers. Coexistent $\gamma 1$ and $\beta 2$ rhythms in the cortex, for example, are generated in two different layers and survive physical separation of the layers. The ratio of peak frequencies is approximately Φ, resulting in a periodic pattern of change in low-level synchrony between the layers with a period equal to the sum of the two periods of oscillation present. This phenomenon can occur to some extent with any pair of co-expressed frequencies. But using Φ as a common ratio between adjacent frequencies in the

EEG spectrum enables the neocortex to pack the available frequency space (thereby maximizing the information processing capacity, or the capacity to produce the most music). If the cortex uses different frequency bands to process different aspects of incoming information, then it must also have the ability to combine information held in these bands to reconstruct the input; hence the importance of keeping them separate, as the golden ratio does.

Phase synchrony does happen, and has been seen in human recordings made with MEG (magnetoencephalography using an array of very sensitive superquantum interference device (SQUID) magnetometers) when frequency ratios are integer values. Stable phase relationships between frequencies with ratios of 2, 3 and 4 are seen during mental arithmetic tasks in localized regions of the neocortex; and the phenomenon has also been proposed to be involved in memory matching and attention.

The most readily observable form of cross-frequency interaction is that of "nesting". Here the amplitude (power) of a discrete frequency band is modified according to the phase of a lower frequency coexistent rhythm. This is seen when γ rhythms coexist with θ frequencies in the hippocampus. Hierarchies of nested rhythms are also seen. Nesting of δ, θ and γ rhythms exists both in the hippocampus and the neocortex. This arrangement ensures that successively high frequencies are maximally expressed in a manner dependent on the lower frequencies in the hierarchy and does not *per se* imply precise phase relationships, though stable phase relationships may be maintained.

It is also possible for a local circuit generating a single frequency rhythm to switch frequencies. Such changes are facilitated by a range of mechanisms including changes in neuronal intrinsic conductances and non-reciprocal interactions with other regions oscillating at a similar frequency. After stimulation, γ frequencies can transform to β frequencies (approximately halved) due to inhibitory postsynaptic potentials on the principal cells generating the action potentials.

An additional type of interaction, concatenation, involves a γ frequency generated by a superficial layer interacting with a $\beta2$

frequency in a deep layer combining into a new frequency, β1, but the intrinsic original frequencies are preserved in that alternate, single periods of each original rhythm form one period of the new frequency. Concatenation is possible for any given pair of rhythms.

Can interactions between multiple spatiotemporal scales of activity tell us anything about how the cortex processes sensory information?

In the temporal domain, the ability of a system to sort rapidly changing features of an input from more slowly changing features provides an efficient means of recognizing objects. A hierarchical arrangement of feature detection over a range of temporal scales can reproduce many properties of individual neurons in the visual cortex. Thus, from a computational perspective, it is an advantage for the cortex to process different temporal scales of information separately, using different frequencies. It has been shown that rhythms with larger temporal scales (slower frequencies) facilitate interactions over greater distances in cortical networks, i.e., they may synchronize over larger areas of the visual map in the retina of the eyes. Thus, different frequencies may have a role for processing sensory information on different spatial scales. In a visual task designed to test perceptual shifting from features of an object with low spatial frequency to those with high spatial frequency, a direct correlation was seen between spatial scale of the sensory object and the temporal scale (frequency) of associated cortical rhythms. Thus cross-frequency phase synchronization is a possible means of combining information from different frequency channels to fully represent a sensory object.

These results show that the golden ratio is woven into the fabric of our consciousness. But does that mean the golden ratio is somehow projected by our consciousness onto nature or real processes? No. We show from further mathematical analyses that the golden ratio arises from real processes that occur in cycles, and the golden ratio orchestrates all of nature's cycles to create organic space-time, which is fractal and quantum coherent. The story continues in the next essay.

References and Notes

1. Tanackov, I., Tepić, J. and Kostelac, M. (2011). "The Golden Ratio in Probabilistic and Artificial Intelligence". *Tehnički Vjesnik* 18, 641–647.
2. Merrick, R. *Interference: A Grand Scientific Musical Theory*, 3rd edition, 2011. Accessed 25 August 2015, http://www.Interference Theory.com.
3. Knott, R. (1998). *Fibonacci Numbers and the Golden Section.* Mathematics Department, University of Surrey, United Kingdom, 1998. Accessed 25 August 2015, http://www.maths.surrey.ac.uk/hosted-sites/R.Knott/Fibonacci/.
4. "Fibonacci Sequence", Wikipedia, accessed 27 December 2013, http://en.wikipedia.org/wiki/Fibonacci_number.
5. "Golden Spiral", Wikipedia, accessed 24 December 2013, http://en.wikipedia.org/wiki/Golden_spiral.
6. Meisner, G. "Penrose Tiling and Phi Φ = Phi $\approx$ 1.618, The Golden Number", 13 May 2012, http://www.goldennumber.net/penrose-tiling/.
7. See Ho, M.W. "Golden Mean Wins Chemistry Nobel Prize". *Sci Soc* 52 (2011): 10–11.
8. See Swinton, J. "Watching the Daisies Grow: Turing and Fibonacci Phyllotaxis". In *Alan Turing: Life and Legacy of a Great Thinker* (C. Teuscher, ed.), pp 477–498, 2004, http://www.dcc.ufrj.br/~luisms/turing/swinton.pdf; Saunders, P.T., ed. *Collected Works of A.M. Turing: Morphogenesis*. North Holland, Amsterdam, 1993.
9. "Alan Turing", Wikipedia, accessed 6 January 2014, http://en.wikipedia.org/wiki/Alan_Turing.
10. Gupta, R. and Saxena, K. Nature's Trademark — Phi: A Review Report on the Occurrence of the Golden Ratio in Nature", 2014, https://www.researchgate.net/publication/238086558_Nature's_Trademark_-_Phi_A_review_report_on_the_occurrence_of_the_golden_ratio_in_nature.
11. D'Arcy, T. *On Growth and Form*, Cambridge University Press, Cambridge, 1917.
12. Prusinkiewicz, P. and Lindenmayer, A. *The Algorithmic Beauty of Plants*. Springer Verlag, New York, 1990. Electronic version 2004, http://algorithmicbotany.org/papers/abop/abop.pdf.

13. Douady, S. and Couder, Y. "Phyllotaxis As a Physical Self-Organized Growth Process". *Phys Rev Lett* 68 (1992): 2098–2101.
14. Ho, M.W. and Saunders, P.T. "Rational Taxonomy and the Natural System As Exemplified by Segmentation and Phyllotaxis". In *Models in Phylogeny Reconstruction*, the Systematics Association Special Volume No. 52 (R.W. Scotland, D.J. Siebert and D.M. Williams, eds.), pp. 113–124, Clarendon Press, Oxford, 1994.
15. Bak, P., Tang, C. and Wiesenfeld, K. "Self-Organized Criticality: An Explanation of 1/f Noise". *Phys Rev Lett* 159 (1987): 381–384.
16. "Self-Organized Criticality", Wikipedia, accessed 5 November 2013, http://en.wikipedia.org/wiki/Self-organized_criticality.
17. Levitin, D.J., Chordia, P. and Menon, V. "Musical Rhythm Spectra from Bach to Joplin Obey a 1/f Power Law". *Proc Natl Acad Sci* 109 (2012): 3716–3720.
18. Ho, M.W. *The Rainbow and the Worm: The Physics of Organisms*. World Scientific, Singapore and London. 1st edition, 1993; 2nd edition, 1998; 3rd edition, 2008.
19. Nottale, L. *Scale Relativity and Fractal Space-Time: A New Approach to Unifying Relativity and Quantum Mechanics*, World Scientific, Singapore, 2011.
20. "Chronobiology", Wikipedia, accessed 12 December 2013, http://en.wikipedia.org/wiki/Chronobiology#History.
21. Ho, M.W. (2006). "Thermodynamics of Organisms and Sustainable Systems". Invited lecture in the conference on Environment, Agriculture, Food, Health and Economy, World Food Day, 17 October 2006, La Sapienza University, Rome, Italy, http://www.i-sis.org.uk/ThermodynamicsOfOrganisms.php; also Ho, M.W. "Circular Thermodynamics Of Organisms and Sustainable Systems". *Systems* 1 (2013): 30–49, http://www.mdpi.com/2079-8954/1/3/30.
22. Pletzer, B., Kerschbaum, H. and Klimesch, W. "When Frequencies Never Synchronize: The Golden Mean and the Resting EEG". *Brain Res* 1335 (2010): 91–102.
23. Ho, M.W. "Quantum Coherence and Conscious Experience". *Kybernetes* (1997): 26, 265–276, http://www.i-sis.org.uk/brainde.php.

24. Plenz, D. "Neuronal Avalanches and Coherence Potentials". *Eur Phys J Spec Top* 205 (2012): 259–301.
25. Roopun, R.K., Kramer, M.A., Carracedo, L.M., Kaiser, M., Davies, C.H., Traub, R.D., Kopell, N.J. and Whittington, M.A. (2008). "Temporal Interactions between Cortical Rhythms". *Front Neurosci* (2008), doi:10.3389/neuro.01.034.2008.

18

Space-Time is Fractal and Quantum Coherent in the Golden Mean

From philosophy and mathematics to quantum physics and far-from-equilibrium thermodynamics, biology and neurobiology, space-time appears fractal and quantum coherent in the golden mean: mathematically, our fractal universe is non-differentiable and discontinuous, yet dense in the infinite dimensional space-time. Physically, it is a quantum coherent universe consisting of an infinite diversity of autonomous agents all participating in co-creating organic, fractal space-time by their multitudinous coupled cycles of activities; biologically, this fractal coherent space-time is also the fabric of conscious awareness mirrored in the quantum coherent golden mean brain states.

Real Processes Do Not Happen at Points in a Space-Time Continuum

The English mathematician-philosopher Alfred North Whitehead (1861–1947) lived through an exciting era in Western science when the fabric of physical reality — Newton's flat, smooth and static universe — was being thoroughly ruffled by Albert Einstein's theories of special and general relativity and by quantum mechanics. The modern observer no longer views nature from the outside, being irreducibly quantum-entangled with the known, in theory with all entities in the entire universe.

These surprising lessons from nature became the basis of Whitehead's perennial philosophy,[1] ushering in a new age of the organism that inspired generations of scientists. He saw the universe as a super-organism encompassing organisms on every scale from galaxies to elementary particles, and argued it is only possible to know and understand nature both as an organism and with the sensitivity of an organism.

Most important and least understood was his rejection of the mechanical laws of classical physics and differential calculus for their failure to describe real processes. Not only do they leave out of account the essential knowing, experiencing organism, real processes occur in *intervals* of time associated with *volumes* of space. Absolutely nothing can happen at a point in an instant.

Instead of being a smooth, infinitely divisible continuum, space-time is discrete and discontinuous, and hence *non-differentiable*, as quantum physics has already discovered at the smallest scale. Unfortunately, mathematics had lagged behind physics. Both relativity theory and quantum theory inherited the mathematics of classical mechanics.

Roger Penrose's monumental tome — *The Road to Reality: A Complete Guide to the Laws of the Universe*[2] — is a "tour de force" as advertised. It charts the heroic and ingenious efforts of mathematical physicists to grasp hold of the post-Newtonian universe. They failed. The dream of uniting the two great theories of quantum physics and general relativity has remained unfulfilled, not least because these two modern theories are both based on the foundation of classical physics: a differentiable, continuous, space-time manifold.

Zeno's Paradox

The issue of continuity versus discontinuity of space-time did not originate in Newtonian mechanics. It can be traced back to ancient Greek philosophy, especially in the philosopher Zeno's paradoxes.

In one version, Achilles is running a race with the tortoise.[3] Achilles gives the tortoise a head start of 100 metres, say. If we suppose that both run at constant speed — Achilles very fast and the tortoise very slow — then after some time, Achilles will have run 100 metres, bringing him to the tortoise's starting point. But during that time, the tortoise has run a much shorter distance, say 10 metres. It will then take Achilles some further time to run that distance, by which time the tortoise will have gone ahead farther; and then he would need more time still to reach that third point, while the tortoise moves ahead, and so on. Thus, whenever Achilles reaches somewhere the tortoise has been, he still has farther to go. Therefore, because there are an infinite number of points where the tortoise has already been for Achilles to reach, he can never overtake the tortoise.

It is generally thought that Newton and Leibniz had both resolved Zeno's paradoxes with differential calculus, by inventing infinitesimal space and time intervals. Whitehead[1] and French philosopher Henri Bergson (1859–1941)[4] were among those who would not have accepted this "resolution".

Zeno's paradoxes were about the impossibility of motion as represented by an infinite sequence of static configurations of matter at points in time. Whitehead said in his *Concept of Nature* (p. 15):[5] "There is no holding nature still and looking at it." The absolute, infinitely divisible time and space of Newtonian physics are both abstractions from ever-flowing events of nature. He agreed with Bergson[4] in using the concept of "duration" for an interval of time experienced by the knower as (p. 53)[5] a *simultaneity* encompassing "a complex of partial events". I shall show that "duration" and "simultaneity" can be given very specific meanings in terms of characteristic time of processes and coherence time.

Mathematics of Discontinuity

Georg Cantor (1845–1918) discovered the Cantor set[6,7] in 1883, which is fundamental for discontinuous mathematics. The paradoxical nature of the Cantor set is that it is infinitely sub-divisible, but is completely discontinuous and nowhere dense. It is also a fractal with self-similar patterns on every scale. (See Box 18.1 for

Box 18.1

Some Informal Definitions of Mathematical Terms[11]

Set theory is the branch of mathematical logic about collections of mathematical objects. The modern study of set theory was initiated by German mathematicians Georg Cantor (1845–1918) and Richard Dedekind (1831–1916) in the 1870s.

A ***closed set*** contains its own boundary, its complement is an **open set** which does not contain its boundary.

A ***subset* A** of **B** is such that every element of **A** is also in **B**.

A subset **D** of a topological space (see below) **X** is said to be ***dense*** (or ***everywhere dense***) in **X** if the closure of **D** is equal to **X**. A ***nowhere dense*** set **N** in the topological space **X** is a set whose closure has empty interior. (Roughly speaking, it is a set whose elements are not tightly clustered together anywhere.) In the special case that **X** is a simple (metric) space, say the space of the real numbers, then the set **D** is dense in **X** if for any element of **X**, we can find a point in **D** arbitrarily close to it. ***Nowhere dense*** means that there is not even a tiny interval where the points are dense.

A ***Borel set*** is any set in a topological space (see below) that can be formed from open sets (or equivalently from closed sets) through the operations of countable union, countable intersection and relative complement. A countable set is one with the same number of elements as some subset of the set of natural numbers. The elements of a countable set can be counted one at a time, and although the counting may never finish, every element of the set will eventually be associated with a natural number. Union, denoted by $\cup$, of a collection of sets is the set of all distinct elements in the collection. Intersection of sets, denoted by $\cap$, is the set that contains only elements belonging to all the sets. The relative complement of set A in B is the set of elements in B but not in A.

A ***bijection*** is a mapping both one-to-one (an injection) and onto (a surjection); it is a function that relates each member of a set **S** (the domain) to a separate and distinct member of another set **T** (the range), where each member in **T** also has a corresponding member in **S**.

The classical triadic ***Cantor set*** is obtained by dividing the unit line into three equal parts, discarding the middle part *except for its end*

(*Continued*)

Box 18.1 (*Continued*)

points, and repeating the operation with the two remaining parts *ad infinitum*. In the ***random Cantor set*** it could be any of the three parts that is discarded at random after each division.

A ***metric space*** is a set for which distances between all elements of the set are defined. The most familiar metric space is the three-dimensional Euclidean space.

A ***Riemannian space*** is a topological space with metric properties that can be defined continuously from point to point (hence also called a ***Riemannian manifold***) including standard non-Euclidian spaces, i.e., spaces that are not flat.

A ***topological space*** is a set of points and a set of neighbourhoods for each point that satisfies a set of axioms relating to points and neighbourhoods. The definition of a topological space relies only on set theory[12] and is the most general notion of a mathematical space.

Topological dimension is the dimension of a topological space. For example, a point has topological dimension 0 whereas a line has topological dimension of 1; closing up the line into a circle makes no difference; it still has a topological dimension of one. Similarly, a flat sheet has a topological dimension of 2, the same for the surface of a cylinder, a sphere or a doughnut.

The ***Menger-Urysohn dimension*** is a generalized topological dimension of topological spaces, arrived at by mathematical induction. It is based on the observation that, in n-dimensional Euclidean space R^n, $(n - 1)$-dimensional spheres (that is, the boundaries of n-dimensional balls) have dimension $n - 1$. Therefore it should be possible to define the dimension of a space inductively in terms of the dimensions of the boundaries of suitable open sets.

The ***Hausdorff dimension*** generalizes the notion of dimension to irregular sets such as fractals. For example, a Cantor set has a Hausdorff dimension of ln2/ln3, the ratio of the logarithm to the base 2 of the parts remaining to the whole.

A ***fractal*** is a mathematical set that typically displays self-similar patterns, and has fractional dimensions instead of the usual integer, 1, 2, 3 or 4. Geometric examples are branching trees, blood vessels, frond leaves, etc.

(*Continued*)

Box 18.1 (Continued)

Deterministic Chaos describes dynamical systems with unpredictable behaviour that is highly sensitive to initial conditions, but is nevertheless globally determined, such that the trajectories are confined within a region of phase space called ***strange attractors***.

Transfinite is a term coined by Cantor; it means beyond finite, but not necessarily the absolute infinite.

informal definitions of mathematical terms; some, like "deterministic chaos", have no generally agreed definition.)

The mathematics of non-differentiable and discontinuous spaces is among the most significant discoveries/inventions, not only in mathematics, but especially for our understanding of physical reality. Cantor made a start in the late 19th century, but it did not really take off until well into the 20th century, reaching its peak in the science and mathematics of complexity associated especially with fractals[8] due to Polish-born French and American mathematician Benoit Mandelbrot (1924–2010), and deterministic chaos[9,10] due to American mathematician-meteorologist Edward Lorenz (1917–2008).

Continuous Non-Differentiable Fractal Space-Time

Canadian mathematician Garnet Ord was the first to propose a "fractal space-time" and to coin the term for it.[13] His starting point was the observation[14] of American theoretical quantum physicist Richard Feynman (1918–1988) that the paths of quantum are non-differentiable curves rather than straight lines when looked at on a fine scale. Moreover, relativistic interaction with particles at sufficiently high energies produces non-conserved particle number. These and other anomalies have encouraged quantum mechanics to abandon the concept of a point particle and its trajectory in favour of wave-packets or field excitations. Feynman's formulation in terms of path integral was an exception. In the same spirit, Ord set out to construct a continuous trajectory in space-time

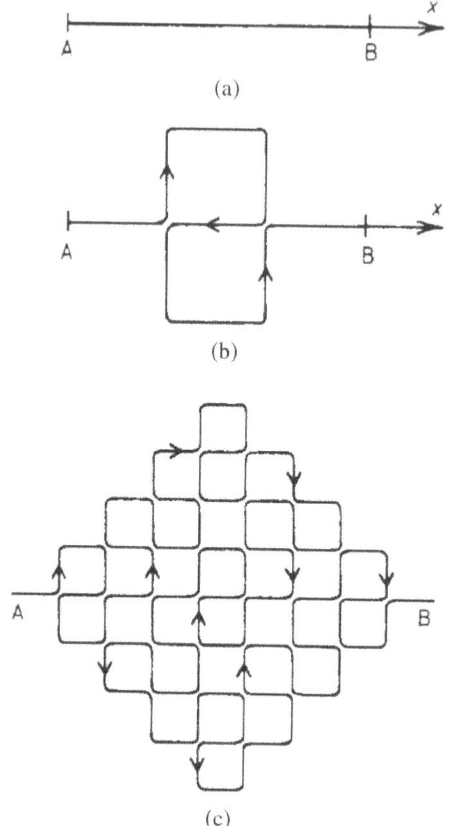

Figure 18.1 A particle's fractal trajectory at increasing resolution, at scale s = λ (a), s = λ/3 (b) and s = λ/9 (c).

that exhibits features analogous to those in relativistic quantum mechanics. He came up with a fractal trajectory exemplified by a Peano-Moore curve (Fig. 18.1). It is plane-filling with a fractal (Hausdorff) dimension of 2 instead of the classical linear path that has a dimension of 1.

Ord showed that, among other things, such a fractal trajectory exhibits both an uncertainty principle and a de Broglie relation (the wave length λ of a massive particle is related to its momentum p through Planck's constant h: $\lambda = h/p$). On a microscopic scale, the presence of fractal time is interpreted in terms of the appearance of particle-antiparticle pairs when observation energies are of the order of mc^2. On a macroscopic scale greater than the fractal wavelength, the free "fractalon" appears to move on a classical one-dimensional trajectory.

A more elaborate scale-relativity theory of fractal space-time was proposed by French theoretical physicist Laurent Nottale,[15] who was motivated by (pp. 4–5)[16] "the failure of a large number of attempts to understand the quantum behaviour in terms of standard differentiable geometry" to look for a possible "quantum geometry", i.e., fractals. His theory recovers quantum mechanics as mechanics on a non-differentiable space-time, and the Schrödinger equation is demonstrated as a geodesics equation.

Ord and Nottale have both proposed fractal space-times that are continuous and non-differentiable.

A more radical cosmology proposed by Egyptian-born Mohamed El Naschie is a fractal space-time based on the golden mean, which is *dis*continuous and non-differentiable (see below). It is closest to our intuitive notion of organic (as opposed to mechanical) space-time.[11]

E-Infinity Fractal Space-Time and Our Four-Dimensional Universe

El Naschie trained and practised as an engineer while indulging in his hobby of cosmology, and produced a startling new theory of space-time, which soon took over his life entirely. "E-infinity", as El Naschie calls it, is a fractal space-time with infinite dimensions, but has a Hausdorff dimension of 4.236067977 ... It means that at ordinary scales, it looks and feels four-dimensional (three of space and one of time), with the remaining of the infinite dimensions "compacted" in the 0.236067977 ... "fuzzy tail".[17]

One way to imagine such a universe is a four-dimensional hypercube with further four-dimensional hypercubes nested inside like Russian dolls[18] (see Fig. 18.2). The exact Hausdorff dimension of the infinite dimensional hypercube is $4 + \phi^3$, where $\phi = (\sqrt{5} - 1)/2$, the golden mean. The dimension $4 + \phi^3$ can be expressed as the

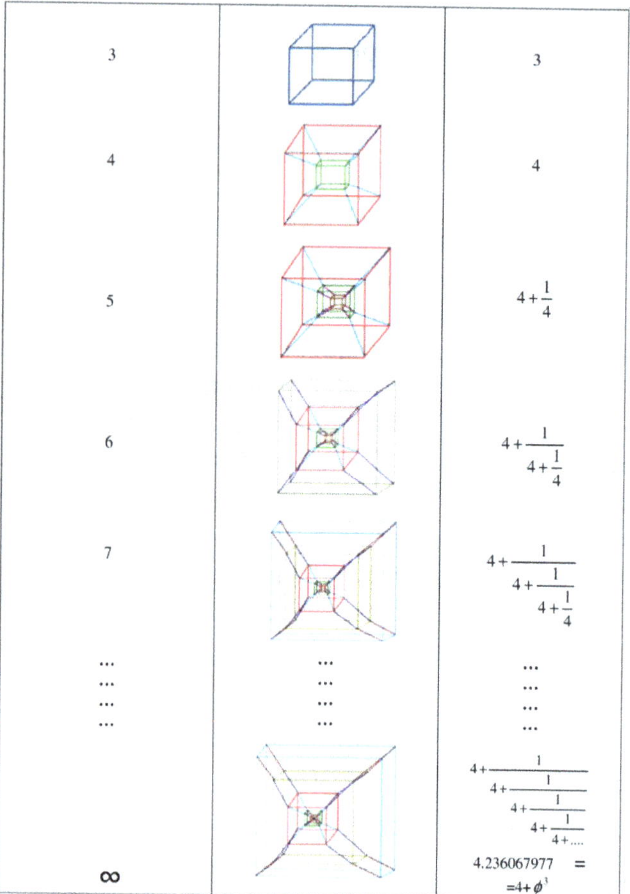

Figure 18.2 E-infinity fractal space-time in Euclidean representation as infinitely nested four-dimensional hypercubes.

following self-similar continued fraction (see Fig. 18.2) which sums to precisely $4+\phi^3$ at infinity:

$$4+\phi^3 = 4+\cfrac{1}{4+\cfrac{1}{4+\cfrac{1}{4+}}}$$

The four-dimensional hypercube is the Euclidean representation of the *E*-infinity universe. It is a challenge to represent *E*-infinity in its proper non-Euclidean form.

Type of fractal	Geometrical shape	Menger–Urysohn dimension	Hausdorff dimension	Corresponding random Hausdorff dimension	Embedding dimension	Corresponding Euclidean shape
Cantor Set		0	*ln 2/ln3 = 0.630929753*	$\phi = 0.61803398$	1	*Line*
Sierpinski gasket		2	*ln 3/ln2 = 1.584962501*	$\frac{1}{\phi} = 1.618033989$	2	*Square*
Menger sponge		3	D_{MS} = *ln20/ln3 = 2.7268*	$2 + \phi =$ 2.61803398	3	*Cube*
The 4 dimension random cantor set analogue of Menger sponge	*An artist impression of* $\varepsilon^{(\ )}$ *space-time*	4	$d_c^{(4)} =$ 4.236068	$4 + \phi^3 =$ 4.23606797	5	*Hyper cube*

Figure 18.3 Fractals and dimensions in the derivation of *E*-infinity space-time.

Mathematically, the *E*-infinity universe is a random Cantor set extended to infinite dimensions, and remarkably the Hausdorff dimension of the infinite extension is no larger than $4 + \phi^3$.

Fig. 18.3 illustrates the steps involved in deriving the *E*-infinity universe. Start from the one-dimensional Cantor set, the unit line, and divide it into three equal parts; take out the middle leaving the end points, and carry on the same operation on the two parts remaining, up to an infinite number of steps, eventually leaving nothing but isolated points, or "Cantor dust". The Menger-Urysohn dimension is 0, but its Hausdorff dimension is log2/log3 for the usual triadic Cantor set. However, for the *random* Cantor set, in which the section removed is not necessarily the middle one, but any of the three parts at random, the Hausdorff dimension turns out to be the golden mean $\phi = (\sqrt{5} - 1)/2 = 0.61803398...$ This significant result,[19] proven by American mathematicians Daniel Mauldin and S.C. Williams in 1986, is the key to the *E*-infinity universe.

A further property of the Cantor set is that its *cardinality* (number of points or elements) is exactly the same as the original continuous line. Thus, the Cantor set is a perfect compromise between the discrete and the continuum; it is a discrete structure that has the same number of elements as the continuum.

From the one-dimensional Cantor set, the higher dimensional random Cantor spaces can be constructed[17] (see Fig. 18.3). The two-dimensional version, the Sierpinski gasket with Hausdorff dimension $1/\phi = 1.61803398\ldots$, and the three-dimensional Menger sponge, with Hausdorff dimension $2 + \phi = 2.61803398\ldots$, are both well-known geometric shapes. The four-dimensional version with Hausdorff dimension $4 + \phi^3 = 4.23606797\ldots$, is given as an artist's representation. The four-dimensional version is the same as the *E*-infinity universe constructed from an infinite number of random Cantor sets, as will be made clear. Note that the diagram representing the four-dimensional Cantorian space-time is space-filling with smaller and smaller spheres. This space-filling property makes the Cantorian space-time non-differentiable and discontinuous, yet *dense* everywhere in space-time. It recalls the quasi-periodic Penrose tiling in two dimensions of Euclidean (flat) space where the golden mean is key.[20] Branching processes based on the golden mean are also space-filling,[21] as are spiral leave arrangement patterns with the golden angle between successive leaf primordia.[22]

El Naschie has presented several different formal derivations of *E*-infinity space-time; I give the simplest[17] based on the mathematical properties of Borel sets to which Cantor sets belong.

The expectation value of the Hausdorff dimension of the Cantor set extended to infinity is simply a sum over n, for $n = 0$ to $n = \infty$, of n multiplied by the Hausdorff dimension of the random Cantor set raised to the power n:

$$\langle \text{Dim } E - \infty \rangle_H = \sum_0^{\infty} n\left(d_c^{(0)}\right)^n \qquad (1)$$

where the superscript in $d_c^{(0)}$ refers to the Menger-Urysohn dimension of the random Cantor set, which is 0, while the corresponding Hausdorff dimension $d_c^{(0)}$ is ϕ. Summing up the infinite number of terms gives the answer $4 + \phi^3$ exactly as follows:

$$\begin{aligned}\langle \text{Dim } E-\infty \rangle_H &= (0)(\phi)^0 + (1)(\phi)^1 + (2)(\phi)^2 + (3)(\phi)^3 + \ldots \\ &= 4+\phi^3 \\ &= (1/\phi)^3 \\ &= 4.236067977\ldots \end{aligned} \quad (2)$$

The intersection rule of sets, the "bijection formula", relates the Menger-Urysohn dimension to the Hausdorff dimension. It shows that we can lift $d_c^{(0)}$ to any Menger-Urysohn dimension n to arrive at the correct Hausdorff dimension $d_c^{(n)}$ as follows:

$$d_c^{(n)} = (1/\, d_c^{(0)})^{n-1} \quad (3)$$

Taking $d_c^{(0)} = \phi$, and lifting to $n = 4$ dimensions gives

$$\begin{aligned} d_c^{(4)} = (1/\, d_c^{(0)})^{4-1} = 4 + \phi^3 = 1/\phi^3 = 4.236067977 \ldots \\ = \langle \text{Dim } E - \infty \rangle_H \end{aligned} \quad (4)$$

Thus, *the expectation value of the Hausdorff dimension of E-infinity universe is the same as that of a universe with a Menger-Urysohn dimension of 4*. That is why E-infinity is a hierarchical universe that looks and feels four-dimensional.

How *E*-infinity Relates to Penrose Tiling and Fibonacci Sequence

The E-infinity universe connects with Penrose tiling and the Fibonacci sequence[20] through E-infinity algebra[18,23] and the golden mean. The golden mean is mathematically an irrational number and like any other irrational number, it can be approximated by a fraction; for 22/7 is quite close to π. The usual way of obtaining such an approximation using continued fractions, converges more slowly for ϕ than for any other number, and in this sense, ϕ is the most irrational number.

In his *Noncommutative Geometry*[24] French mathematician Alain Connes identified Penrose's fractal tiling as a mathematical quotient space (a space of points "glued together" by an equivalence relationship), with the dimensional function:

$$D(a, b) = a + b\phi \tag{5}$$

where *a*, *b* are integers (whole numbers) and $\phi = (\sqrt{5} - 1)/2$. Writing $D_n(a_n, b_n)$ where both a_n and b_n satisfy the Fibonacci recurrence relation $x_{n+2} = x_n + x_{n+1}$, and starting with $D_0 = D(0, 1)$ and $D_1 = D(1, 0)$, the following dimensional hierarchy is obtained:

$$\begin{aligned}
&D_0 = D(0, 1) = 0 + \phi = \phi \\
&D_1 = D(1, 0) = 1 + (0)\phi = 1 \\
&D_2 = D(0 + 1, 1 + 0) = 1 + \phi = 1/\phi \\
&D_3 = D(1 + 1, 0 + 1) = 2 + \phi = (1/\phi)^2 \\
&D_4 = D(1 + 2, 1 + 1) = 3 + 2\phi = (1/\phi)^3 \\
&D_5 = D(2 + 3, 1 + 2) = 5 + 3\phi = (1/\phi)^4 \\
&D_n(a_n, b_n) = D\{(a_{n-1}, a_{n-2}) + (b_{n-1} + b_{n-2})\}\phi = (1/\phi)^{n-1}
\end{aligned} \tag{6}$$

It is notable that for D_4 (dimension 4), the Fibonacci number is $(1/\phi)^3 = 4 + \phi^3$, exactly the Hausdorff dimension of a Menger-Urysohn 4-dimensional space.

By induction,

$$D_n = (1/\phi)^{n-1} \tag{7}$$

and we get back the bijection formula from E-infinity algebra (see Eq. 3 above):

$$d_c^{(n)} = (1/\phi)^{n-1} \tag{8}$$

Summing random Cantor sets to infinity in the creation of E-infinity universe is evocative of space-time being *created* by actions over all scales, from sub-microscopic to macroscopic and beyond, as envisaged in *The Rainbow and the Worm: The Physics of Organisms*[25] following Whitehead[1] and German quantum physicist Wolfram Schommers.[26]

El Naschie has conjectured that E-infinity space-time can also resolve major paradoxes within quantum theory and produce new results as described elsewhere.[27–29] Here, we move on to the role of cycles in the organization of space-time and why the golden mean seems to be built into the fabric of life and the universe.

Cycles Everywhere for Stability, Autonomy and More

Nature abounds in cycles and oscillations, from sub-atomic vibrations to planetary motion, solar cycles and galactic rotations. Some, like Penrose and Vahe Gurzadyan, mathematician at Armenia's Yerevan Physics Institute, say even the universe cycles through deaths and rebirths, based on data collected by NASA's WMAP (Wilkinson Microwave Anisotropy Probe) and the BOOMERanG balloon experiment in Antarctica.[30] The importance of cycles and the golden mean for natural processes was considered in a previous article.[31] A brief recapitulation is given here.

Cycles are intimately tied to the study of dynamical systems, beginning with celestial mechanics. Isaac Newton (1642–1727) tried to describe the planetary cycles in terms of his laws of motion more than 300 years ago.[32]

Dynamical systems can be treated mathematically as oscillators. A harmonic oscillator has a certain natural frequency. When perturbed by an external force with the same frequency, resonance occurs and the motion of the oscillator becomes unbounded or unstable. For a typical nonlinear oscillator, resonance happens whenever the frequency of the perturbing force is a *rational* multiple of the natural frequency of the oscillator.

Russian mathematicians Andrey Kolmogorov (1903–1987) and Vladimir Arnold (1937–2010) and German mathematician Jürgen Moser (1928–1999), were responsible for the Kolmogorov Arnold and Moser (KAM) theorem, which is very important for understanding how cyclic activities (or oscillators) interact with one another.

The three mathematicians were investigating the behaviour of integrable Hamiltonian systems. The trajectories of Hamiltonian systems in phase space are confined to a doughnut-shaped surface, an invariant torus. Different initial conditions will trace different invariant concentric tori in phase space, separated by unstable chaotic regions, where the motion is irregular and unpredictable.

An important result of the KAM theorem is that for a large set of initial conditions, the motion remains quasi-periodic, and hence *stable*. KAM theory has been extended to non-Hamiltonian systems and to systems with fast and slow frequencies.

The KAM theorem becomes increasingly difficult to satisfy for complex systems; as the number of dimensions of the system increases, the volume occupied by the tori *decreases*. Those KAM tori not destroyed by perturbation become invariant Cantor sets, or *Cantori*; with frequencies approximating the golden mean.[33]

The golden mean effectively allows multiple oscillators within a complex system to coexist, and also leaves them free to interact globally (by resonance, which may have important application also in the study of the brain, see later). To get a better picture, we look at the circle map.

Cycles, Quasi-Periodicity, Golden Mean and Chaos

The circle map is a graph that maps the circle onto itself. The simplest form is[34,35]

$$\theta_{n+1} = \theta_n + \Omega - (K/2\pi)\sin 2\pi\theta_n \qquad (9)$$

Where the variable θ_{n+1} is computed mod 1, K is the coupling strength and Ω is the external driving frequency. This map is used to describe oscillatory systems from solid-state physics to heart rhythms. It tracks the universal behaviour of dynamical systems in transition from cycles to chaos via quasi-periodicity.

The most studied circle map involves a ratio of basic frequencies $w = \phi = (\sqrt{5} - 1)/2$, the "golden mean critical point" at $K = 1$ and $\Omega = \Omega_c = 0.60666106347011$ ($\approx \phi = 0.618033989 \ldots$), reported in many experiments in which universal numbers associated with the golden mean were observed.

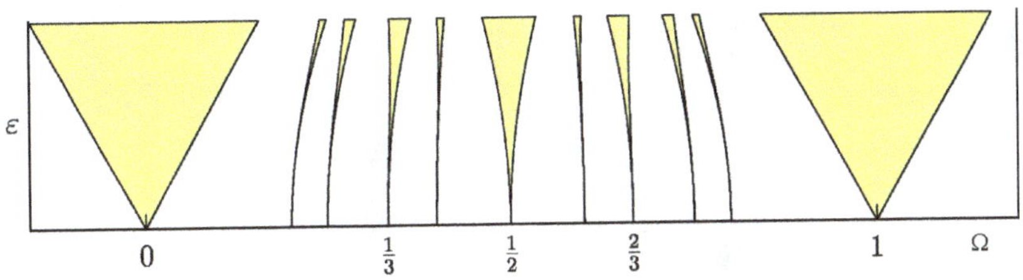

Figure 18.4 Some Arnold tongues in the standard circle map $\varepsilon = K/2\pi$ versus Ω.

Circle maps contain some key features. *Arnold tongues* of the KAM theorem are regions of circle maps with locally constant rational rotation (winding) numbers between the driver and the natural oscillator frequencies, p/q. Thus, the circle map of Eq. 9 contains regions when the oscillator is locked to the driving frequency (phase-locked, or synchronized). Here, θ is the polar angle with a value between 0 and 1; the two parameters are K, the coupling strength between the driver and the oscillator, and Ω, the driver frequency. A typical map with Arnold tongues is given in Fig. 18.4.

For $0 < K \leq 1$ and certain values of Ω, the map exhibits phase-locking; in the phase-locked regions, θ_n advances essentially in rational multiples of n; although it may do so chaotically on the small scale.

The phase-locked regions, called Arnold tongues, are shaded yellow in Fig. 18.4, while the quasi-periodic regions are white. Each yellow V region touches down to a rational value of $\Omega = p/q$ in the limit of $K \to 0$. This ability to lock on in the presence of noise is central to the utility of phase-locked loop electronic circuits.

The circle map in Fig. 18.4 is invertible or symmetrical around the mid line. For values of $K > 1$, the circle map is no longer invertible. The circle map in Fig. 18.5a[36] is extended to $K = 4$. Arnold tongues are in grey with winding numbers indicated inside the tongue. The white regions are quasi-periodic, and the stippled regions above the line $K = 1$ represent *chaos*. This map also depicts fractal self-similarity on different scales. Fractal self-similarity and chaos are closely related. A chaotic system has a fractal dimension and exhibits self-similarity over many scales.

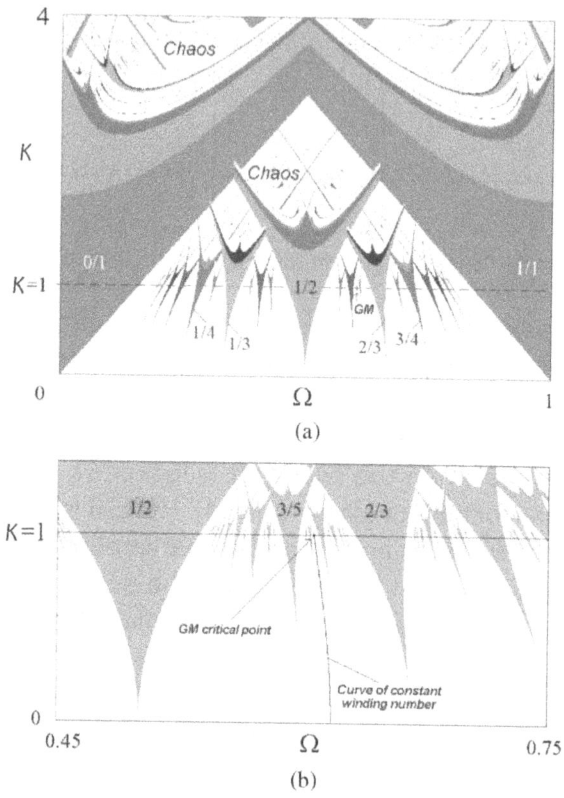

Figure 18.5 Extended circle map (see text for details).

Chaos does not mean random. Mathematically, *chaos is sensitive to initial conditions, but it is globally determined in the sense that it tends towards "strange attractors"* (see Box 18.1).

The golden mean critical point (GM) is where the curve of constant irrational winding number $\phi = (\sqrt{5} - 1)/2$ terminates on the line $K = 1$ (see Fig. 18.5b, a close-up of the main map above it). The curve goes through a region of quasi-periodic behaviour as the system undergoes transition to chaos. The critical point GM is marked by an infinite sequence of unstable orbits with periods given by the Fibonacci numbers.

The golden mean is literally located at "the edge of chaos", and its role is to keep the system of oscillators active without interfering with one another and away from chaos.

There are claims that the planetary orbits around the sun exhibit golden means according to the Fibonacci sequence numbers.[37,38] So is our solar system stable? According to some astrophysicists, the planetary orbits are chaotic and sensitive to initial conditions, hence unpredictable for longer than 100 million years into the future;[39] so there is no immediate cause for alarm even if the solar system is unstable.

Chaos and Strange Attractors

Edward Norton Lorenz is the generally acknowledged "father" of chaos theory.[40] According to one account, Lorenz was running a climate model on the computer described by 12 differential equations in the winter of 1961, when he decided to repeat a run with a small change. Instead of calculating to six decimal digits, he reduced that to three to save computing time, expecting to get the same results. But he didn't. That led to his discovery of the "sensitive dependence on initial conditions" of chaotic systems, which he described as the "butterfly effect". It makes long-term weather prediction impossible, for example. His "toy" equations produced the "Lorenz attractor" (Fig.18.6) (the prototype of "strange attractors" associated with chaotic systems), which

The Lorenz Fractal. Model: Clint Sprott

Figure 18.6 The Lorenz attractor by Clint Sprott.

serendipitously resembles butterfly wings, and became the emblem of the chaos era that followed. Numerous "strange attractors" have been created, mostly as computer artwork, beginning with the book[41] by American physicist Clint Sprott at University of Wisconsin, Madison. Chaos theory has found applications in describing natural processes in meteorology, physics, engineering, economics, biology and medicine.

The Lorenz attractor is a fractal, with self-similar structure on different scales. It has a fractal dimension of 2.06215 and lives in a space of at least three dimensions. It contains unstable periodic orbits that can be identified using various mathematical procedures,[42] and can also be regarded as twisted or "knotted" periodic orbits.[43]

Chaos theory has been taken up enthusiastically in every field including quantum physics, in the form of "quantum chaos", which finds chaos in the wavelike motion of quantum particles. Swiss-American quantum physicist Martin Gutzwiller (1925–2014) wrote:[44] "Phase space for a chaotic system can be organized, at least partially around periodic orbits, even though they are sometimes quite difficult to find."

Chaotic motion is characteristic of turbulent flows in fluids, gases and the atmosphere, traditionally regarded one of the most intractable problems in physics and mathematics. Indian meteorologist Mary Selvam at Institute of Tropical Meteorology, Pune, first proposed a theory of turbulent fluid flow based on fractal space-time fluctuations in 1990.[45]

Selvam described the fractal fluctuations on all space-time scales as a superposition of a continuum of vortices. Large-scale fluctuations result from integrating smaller-scale fluctuations within; and the growth trajectory traces an overall logarithmic spiral flow path with the quasi-periodic Penrose tiling pattern for internal structure.

The ratio of radii or circulation speeds corresponding to the successive growth steps of the large eddy generating the geometry of the quasi-periodic Penrose tiling pattern is the golden mean $\Phi = 1.618\ldots$[20]

Quantum Coherence and Circular Thermodynamics of Organisms

In *The Rainbow and the Worm*[25] I presented empirical evidence and theoretical arguments suggesting that organisms are quantum coherent, and derived a circular thermodynamics of organisms, which is also implied by quantum coherence. The circular thermodynamics depends on a coherent fractal organization of biological space-time. This argument has been applied to sustainable ecosystems and economic systems,[25] and most recently, sustainable cities as organisms.[46] The diagrams illustrating the concept of circular thermodynamics were first presented in "The Biology of Free Will"[47] published in 1996, a version of which appears under the same title in this book (see 9 — The Biology of Free Will). I shall be referring to the diagrams in that essay below.

The most important thermodynamic feature of organisms is that they are not heat engines, and do not make their living by heat transfer. They are isothermal systems far away from thermodynamic equilibrium, and depend on direct energy transfer by proteins and other macromolecules acting as "quantum molecular energy machines" at close to 100% efficiency.[25] In addition, there is also a prodigious kinetic advantage to the reactions involved.

It is well known that enzymes speed up chemical reactions enormously[48] by a factor of 10^{10} to 10^{23}, but they cannot do that without the water that constitutes 70 to 80% of cells and tissues. It is widely recognized that water gives flexibility to proteins, reduces the energy barrier between reactants and products and increases the probability of quantum tunnelling by a transient compression of the energy barriers. But there is more to how water actually *organizes* enzyme reactions in living organisms. Findings within the past decade suggests that interfacial water associated with macromolecules and membranes in cells and tissues is in a quantum coherent liquid crystalline state, and plays a lead role in creating and maintaining the quantum coherence of organisms, as elaborated in *Living Rainbow H2O*[49] and elsewhere.[50,51]

For isothermal processes, the change in Gibbs free energy ΔG, is

$$\Delta G = \Delta H - T\Delta S \tag{10}$$

where ΔH is the change in enthalpy (heat content), T is temperature in deg K, and ΔS is the change in entropy.

Thermodynamic efficiency requires that ΔS approaches 0 (least dissipation). That can also be achieved when $\Delta G = 0$ via entropy-enthalpy compensation, i.e., entropy and enthalpy changes cancelling each other out. We shall see how the organism manages to do that.

A Fractal Hierarchy of Coupled Cycles

To keep far away from thermodynamic equilibrium — death by another name — an organism must capture energy and material from the environment to develop, grow and recreate itself from moment to moment in a life-cycle, to reproduce and provide for future generations (see "9 — The Biology of Free Will", Fig. 9.1, in this book). The key to understanding the thermodynamics of the living system is energy capture and storage under energy flow to create a reproducing life-cycle. The dynamic closure implied by the life-cycle is the beginning of a circular thermodynamics that transforms and transfers energy and materials with maximum efficiency.[25,52]

The life-cycle is a fractal hierarchy of self-similar cycles organized by the characteristic space-times of the processes involved. All real processes have characteristic space-times. The heart (10^{-1} m) beats in a second, nerve cells (10^{-4} m) fire in a tenth of a second or faster, and protons (10^{-15} m) and electrons (10^{-17} m) move in 10^{-12} to 10^{-15} s. Cells divide in minutes, and physiological processes cycle in hours, a day, a month or a year. *This may well be the reason space-time tied to real processes is non-differentiable and discontinuous.* The coherent fractal hierarchy of living activities arises because processes with matching space-times interact most strongly through resonance, and also link up to the entire hierarchy. That is why biological activities come predominantly in cycles or biological rhythms. In the language of quantum physics, the organism is a superposition of coherent quantum activities over all space-times.[25] The possibility for cycles in the living world coupling and linking up to cycles in the physical universe is surely why life is possible, and indeed some would argue, as Whitehead did,[1] that the entire universe is alive.

The coupled cycles form a nested fractal self-similar structure. The life-cycle consists of smaller cycles each with still smaller cycles within, spanning characteristic space-times from sub-nanometre to metres and from 10^{-15}s to hours and years (see "9 — The Biology of Free Will", Fig. 9.2, in this book).

Minimum Entropy Production

Cycles enable certain activities to be coupled together, so that energy yielding processes can transfer energy directly to those requiring energy, and the direction can be reversed when necessary. This cooperativity and reciprocity resulting from the fractal hierarchy of coupled cycles is an extended form of Onsager's reciprocity relation that conventionally applies strictly only to near-equilibrium steady state. Lars Onsager (1903–1976) was a Norwegian-born American physical chemist. What it means in practice is that energy can be concentrated to any local point where it is needed, and conversely spread globally from any local point. In that way, the fractal hierarchy of coupled cycles *maximizes both local autonomy and global cohesion*, which is the hallmark of quantum coherence[25] according to the criterion of factorizability defined by American quantum-physicist Roy Glauber.[53]

To get an idea of such coupled cycles, one needs to look no further than charts of biochemical metabolic pathways.[54] Most, if not all, of the reactions go either way, depending on the local concentrations of reactants and products. In further accord with circular thermodynamics, biochemical recycling is ubiquitous; there are numerous scavenging or salvaging pathways for the recovery of building blocks of proteins, nucleic acids, glycolipids and even entire proteins.

The fractal hierarchy of coupled cycles confers dynamic stability and autonomy to the system on every scale. Thermodynamically, no net entropy is generated in the case of perfect cycles, and the system maintains its organization.

As explained in detail elsewhere,[25,52] the fractal structure effectively partitions the organism into a hierarchy of systems nested within systems *defined by the extent of equilibration of (dissipated) energies*. Thus, energies equilibrated within a smaller space-time will still be out of equilibrium in the larger encompassing system, where it is capable of doing work.

There are now two ways to mobilize energy efficiently with $\Delta S \rightarrow 0$: very slowly compared to the characteristic time so it is reversible at every point, or very rapidly compared to the characteristic time, such that in both cases the energy remains stored (in a coherent non-degraded form) as it is mobilized. Consequently, the organism *simultaneously* achieves the most efficient equilibrium *and* far-from-equilibrium energy transfer.

The nested fractal dynamical structure also optimizes the kinetics of energy mobilization. Thus, biochemical reactions depend strictly on local concentrations of reactants, which are extremely high, as their extent of equilibration is typically nanometres (in nanospaces).

In the "zero-entropy" ideal — approached most closely by the healthy mature organism and a healthy mature ecosystem — an overall internal energy conservation and entropy compensation ($\Sigma\Delta S = 0$) is achieved. The system organization is maintained and dissipation minimized; i.e., the entropy exported to the environment also approaches zero, $\Sigma\Delta S \geq 0$ (Fig. 18.7).

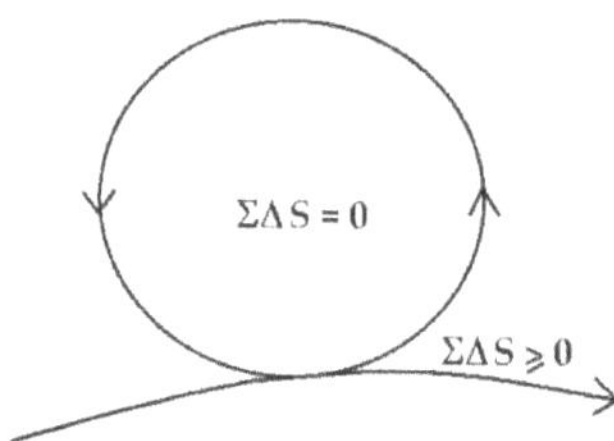

Figure 18.7 The zero-entropy ideal of circular thermodynamics.

Internal entropy compensation (and energy conservation) necessitates a free variation in microscopic states within the macroscopic system; i.e., *a violation of the internal microscopic detailed balance at every point of classical steady-state theory*. (This is also the basis for the extension of the Onsager reciprocity relationship valid near-to-equilibrium to far-from-equilibrium condition.)

For an organism, it implies that detailed energy balance is not necessary at every point. Parts of it indeed can be in deficit, and very much so, as when running away from a tiger; the energy can be replenished after a successful escape from distributed energy stores in other parts of the organism. The same applies to ecosystems: all species are storing energy and resources (nutrients) for every other species via complex food webs and other symbiotic relationships.

These considerations result in the prediction that a sustainable system maximizes cyclic, non-dissipative flows while minimizing dissipative flows, i.e., it tends towards *minimum entropy production* even under far-from-equilibrium conditions, as conjectured by Belgian theoretical chemist and Nobel laureate Ilya Prigogine (1917–2003) many years ago.[55]

Such a system has a hierarchy of coherent energy storage space-times so that within the coherence volume, there is no time separation, and within the coherence time, there is no space separation. Thus, organic space-time exists as a hierarchy of simultaneities, or durations, precisely as envisaged by Bergson[4] and Whitehead.[1,5]

The golden mean most likely enters into the fractal structure in the form of the golden fractal. As the most irrational of all numbers (see previous section "How E-infinity relates to Penrose tiling and Fibonacci sequence"), it allows the maximum number of non-resonating activities to coexist (representing maximum coherent energy storage). On the other hand, it is also arbitrarily close to a maximum number of rational numbers so specific resonances can easily be established for energy transfer. We shall see more clearly how the golden fractal of neuronal activities is the key to optimum intercommunication and information processing in the final section.

Fractals, Quantum Coherence, Dissipative Systems and Quantum Electrodynamics

The coherent fractal structure maximizes global connectivity and local autonomy, the hallmark of quantum coherence, as mentioned earlier. A very significant discovery was made by Italian quantum field theorist Giuseppe Vitiello at University of Salerno. He showed that a functional representation of self-similarity — the most distinguishing feature of fractals — is mathematically isomorphic with squeezed quantum coherent states.[56] Thus, quantum coherence appears to underlie the ubiquitous fractal self-similarity of natural processes. And indeed, some quantum physicists regard the world as quantum, which implies quantum coherence, as I argued elsewhere.[57]

Heisenberg's uncertainty principle states that there is a fundamental limit to the precision with which certain pairs of complementary variables such as position x and momentum p can be known simultaneously, i.e., $\Delta x \Delta p > h/4\pi$, where h is Planck's constant. It was first introduced by German quantum physicist Werner Heisenberg (1901–1976) in 1927. A squeezed coherent state is where this relationship is a minimum,[58] i.e., $\Delta x \Delta p = h/4\pi$. The simplest such state is the ground state of the *quantum harmonic oscillator*, the analogue of the classical harmonic oscillator and one of the most important model systems in quantum mechanics for which an exact analytical solution is known. In another publication, Vitiello extended the isomorphism between fractals and quantum coherence to quantum electrodynamics and dissipative systems.[59]

Vitiello points out it is well known and well established by empirical observations that systems characterized by ordered symmetry such as crystals, ferromagnets and superconductors may be described in quantum field theory by coherent boson condensation in the ground state. The simplest coherent ground state is indeed the squeezed coherent state. The squeezed coherent state also provides a representation of the system as coupled damped/amplified oscillators, a prototype of dissipative systems and the environment or bath in which it is embedded. In other words, the dissipative system is dynamically closed, by a bath that stores the energy dissipated and sends it back, keeping the system stable. In this regard, it is formally equivalent to the dynamic closure required in the living system that gives rise to circular thermodynamics and quantum coherence (see above). Another well-known result in classical and quantum electrodynamics is the conservation of the energy-momentum tensor (a linear relation between the energy and momentum vectors), which again forms a closed system between the matter and the electromagnetic field that exchanges energy and momentum. Vitiello then proved that isomorphism exists also between electrodynamics and the system of damped/amplied oscillators, and hence the squeezed coherent states.[59] He has expanded on this argument in a paper written jointly with myself and El Naschie.[60]

Topology of Quantum Space-Time

The isomorphism among quantum coherence, fractals, dissipative systems and electrodynamics suggests a considerably deeper relationship. In a series of earlier papers[61–64] El Naschie showed that E-infinity theory gives predictions on the mass spectrum of elementary particles and quarks in high energy physics based on the golden mean that are in remarkable agreement with those generally accepted experimentally and theoretically in the literature. Further, the same predictions are provided by a physical interpretation of four-dimensional fusion algebra, of Connes' non-commutative geometry and related theories such as Freedmann topological theory of four manifolds (wild topology), and Penrose space X (of which E-infinity is a higher dimensional space). The results strongly suggest that the mass spectrum based on the golden mean is a reflection of the quantum topology of space-time.[63] This conjecture is reinforced, as El Naschie pointed out, in the simplest mechanical model for E-infinity theory: two golden mean coupled-oscillators[64] with frequencies of vibrations $\omega_1 = \phi$, $\omega_2 = 1/\phi$. (This model to some extent anticipated Vitiello's discovery described above.) In the case of quarks, what we consider to be a particle can be thought of as a highly localized vibration, a standing wave simulating a particle (as in string theory). Generalizing *n* such nested oscillators, Marek-Crnjac was able to use Southwell and Dunkerley summation theorems for structural stability to obtain the masses of the elementary particles of the standard model, as well as the current quarks and constituent quarks.[65] The results are again very close to the theoretical and experimental masses found previously.

Marek-Crnjac also noted that the golden mean plays a central role in dynamic stability in KAM theory (see earlier), pointing out that string theory eliminates the wave-particle duality by using the highly localized vibration of a Planck length string that gives rise to a particle when perceived on lower energy levels.[65] In this way, the golden mean enters into the mass spectrum via the KAM theory. As the golden mean is "the most irrational number" (in the sense described in the previous section "How E-infinity relates to Penrose tiling and Fibonacci sequence"), it plays the key role in the stability of periodic orbits and the onset of global chaos. The ratio of Ω/ω (driver to oscillator frequency) is decisive as to whether the motion is localized (regular) or dissipative (stochastic). If the ratio is rational, the torus is destroyed. However, if it is irrational, the torus persists. A well-known application of this theory is the Kirkwood gaps — dips in the distribution of the main belt asteroids, which coincide with orbital resonances with Jupiter. What this means is that a particle is observed only when the energy is localized in a highly coherent vibration in an irrational frequency, and the most irrational frequency is ϕ.

Marek-Crnjac's argument applies also to the circular thermodynamics of organisms and the fractal hierarchy of living activities developed earlier. The mass spectrum of elementary particle and living activities are both a reflection of the universal quantum topology of space-time.

Coherent Brain Waves, Scale-Free Laws and Fractals

Laboratory observations on the brain have consistently detected spatially extended regions of coherent neuronal activities, most thoroughly described in recent years by American neurophysiologist Walter Freeman and colleagues at University of California Berkeley.[66–68] Observations clearly show that the coherent cortical activities cannot be fully accounted for by the electric field of the extracellular dendritic current or the extracellular magnetic field from the high-density electric current inside the dendritic shafts, nor by chemical diffusion. Spatially extended patterns of phase-locked oscillations are intermittently present in resting, awake subjects, and in the same subjects actively engaged in cognitive tasks. They are best described as properties of the background activity of the brain modulated upon engagement with the environment. These packets of waves extend over spatial domains covering much of the hemisphere in rabbits and cats[69–72] and over regions of 19 cm linear dimension in the human cortex with near-zero phase dispersion.[66,73,74]

Although Freeman himself does not say so, the observations strongly suggest quantum coherence in the activities of the brain. As Vitiello points out, this is best explained in quantum field theoretical terms,[59,60] which can describe transition between physically distinct phases, from a disordered state to an ordered state with long-range correlations. Specifically, Vitiello invokes the electric dipole vibration modes of the water molecules that constitute the matrix of neurons and glia cells.[75,76] The spontaneous breakdown of the rotational symmetry of electrical dipoles of water and other molecules gives rise to coherent condensation of dipole wave quanta, which are the agents responsible for neuronal coordination.[75–77] In particular, it is found that the memory state is a squeezed coherent state expressed mesoscopically as amplitude and phase modulation of the carrier signal observed in electroencephalograms (EEGs) and electrocorticograms (ECoGs). Laboratory observations show that self-similarity characterizes the brain ground state. Measurements of the durations, recurrence intervals and diameters of neocortical EEG phase patterns show power-law distribution with no detectable minima. Spectral densities in both time and space that conform to power law distribution[69,70] is a hallmark of fractals, which is therefore consistent with the squeezed coherent state for memory found earlier.

A confirmation of the brain scale-free behaviour comes also from the group led by Dietmar Plenz at the US National Institute of Mental Health. They have identified "neuronal avalanches" — cascades of neuronal activities following precise $1/f$ power laws — in the excitatory neurons of the superficial layers in isolated neocortex preparations *in vitro* as well as in awake animals and humans *in vivo*.[78] The neuronal avalanche of the default (resting) state with the $1/f$ signature of self-organized criticality also gives the optimum response to inputs as well as maximum information capacity.[79]

Most significantly, the avalanche dynamics give rise to coherence potentials consisting of subsets of avalanches in which the precise waveform of the local field potential is replicated with high fidelity in distant brain sites. The process is independent of spatial distance and includes near-instantaneous neuronal activities as well as sequential activities over many time scales. Most coherence potentials are spatially disjunct. Local field potentials (LFPs) of successive coherence potentials are not similar, but they are practically identical within a coherence potential among all the participating sites, there being no growth or dissipation during propagation. This suggests that the waveform of a coherence potential is a high-dimensional coding space in information processing of the brain. For decades, phase-locked neuronal activity has been reliably recorded using the LFPs or EEG, and found to correlate with the presentation of stimulus in animals and in human visual perception.

Also in living systems fractals must satisfy the state of quantum coherence which maximizes global cohesion as well as local autonomy, and enables energy from any local level to spread to the global and conversely concentrate energy to any domain from the entire system. Earlier discussion suggests that fractals with fractal dimension in the golden ratio (golden fractals) might be the most effective in giving autonomy to the greatest number of cycles. On the other hand, global cohesion is also ensured, because cycles in a fractal hierarchy are quantum coherent; energy can be shared between global and local. The golden ratio is further close to an infinite number of rational ratios, so special resonances or correlations can be easily established. Thus, one would expect biological rhythms in general to conform to the golden mean, although no such survey has yet been carried out. The golden mean figures prominently in the EEG frequencies of the resting brain[79] (see "17 — Story of Phi", Table 17.1, in this book).

It has been proposed that brain frequencies that never synchronize in the resting brain may play an important role in the organization of groups of cells, by keeping their rhythms distinct, and free from interference with one another, and this can be achieved via frequencies in irrational multiples in the resting (default) brain.[80]

A team of researchers led by Miles Whittington at Newcastle University in the United Kingdom have been studying the intricacies in the golden music of the brain by recording from multiple layers of the neocortex simultaneously. They found multiple local neuronal assemblies supporting different discrete frequencies in the neocortex, and the relationships between different frequencies appear designed to minimize interference and to allow very diverse coupling of activities via stable phase interactions.[81] (See "17 — Story of Phi" in this book for further details.)

Most fascinating is the suggestion that interaction between multiple spatiotemporal scales of activity is involved in processing sensory information. In the time domain, the ability to distinguish rapidly changing features of a sensory input from more slowly changing features provides an efficient means of recognizing objects. It has been shown that rhythms with larger time scales (slower frequencies) facilitate interactions over greater distances in the cortex, i.e., they may synchronize over larger areas of the visual map in the retina of the eyes. Thus, different frequencies may have a role for processing sensory information on different spatial scales. In a visual task designed to test perception shifting from features of an object with low spatial frequency to those with high spatial frequency, a direct correlation was found between the spatial scale of the sensory object and the temporal scale (frequency) of associated cortical rhythms. Thus cross-frequency phase synchronization is a possible means of combining information from different frequency channels to fully represent a sensory object.

To Conclude

We have considered the fabric of space-time from the widest perspectives, drawing on findings from mathematics, quantum physics, far-from-equilibrium thermodynamics, biology and neurobiology. The totality of findings converges to the startling conclusion that space-time is indeed fractal and quantum coherent in the golden mean. Mathematically, the fractal universe is non-differentiable and discontinuous, yet dense in infinite dimensional space-time. Physically, it is a quantum coherent universe consisting of an infinite diversity of autonomous agents all participating in co-creating organic, fractal space-time by their multitudinous coupled activity cycles. Biologically, this fractal coherent space-time could also provide the fabric of conscious awareness mirrored in the quantum coherence of our brain states. This view depicts a new organic cosmogony consonant with that of Whitehead,[1] resolving major paradoxes associated with classical mechanics, and paving the way to reconciling or transcending quantum theory and general relativity.

References and Notes

1. Whitehead, A.N. *Science and the Modern World*, Lowell Lectures 1925, Collins Fontana Books, Glasglow, 1975.
2. Penrose, R. *The Road to Reality: A Complete Guide to the Laws of the Universe*. Vintage Books, London, 2005.
3. "Zeno's Paradoxes", Wikipedia, accessed 20 October 2014, http://en.wikipedia.org/wiki/Zeno's_paradoxes.
4. Bergson, H., translated by Pogson, F.L. *Time and Free Will: An Essay on the Immediate Data of Consciousness*. George Allen & Unwin, Ltd., New York, 1916.
5. Whitehead, A.N. *Concept of Nature*. Tarner Lectures delievered in Trinity College, Cambridge, November 1919, published as Gutenberg eBook, 16 July 2006.

6. Cantor, G. Über unendliche, lineare Puntkmannigfaltigkeiten V. *Mathematische Annelen* 21 (1869): 545–591.
7. Rosser, J.B. (2011). *Complex Evolutionary Dynamics in Urban-Regional and Ecologi-Economic Systems*. Springer Science+Business Media, LLC., Berlin, DOI 10.1007/978-1-4419-8828-7.
8. Mandelbrot, B.B. *The Fractal Geometry of Nature*, W.H. Freeman, San Francisco, 1983.
9. Lorenz, E.N. "Deterministic Non-Periodic Flow". *J Atmos Sci* 20 (1963): 130–141.
10. Lorenz, E.N. *The Essence of Chaos*. University of Washington Press, Seattle, 1993.
11. Ho, M.W. "Golden Geometry of E-Infinity Fractal Spacetime: Story of Phi Part 5". *Sci Soc* 62 (2014a): 36–39.
12. "Topological Space", Wikipedia, accessed 15 October 2014, http://en.wikipedia.org/wiki/Topological_space.
13. Ord, G.N. "Fractal Space-Time: A Geometric Analogue of Relativistic Quantum Mechanics". *J Phys A Math Gen* 16 (1983): 1869–1884.
14. Feynman, R.P. and Hibbs, A.R. *Quantum Mechanics and Path Integrals*, McGraw-Hill, New York, 1965.
15. Nottale, L. *Fractal Space-Time and Microphysics: Towards a Theory of Scale Relativity*. World Scientific, Singapore, 1993.
16. Nottale, L. *Scale Relativity and Fractal Space-Time: A New Approach in Unifying Relativity and Quantum Mechanics*, Imperial College Press, London, 2011.
17. El Naschie, M.S. "A Review of E Infinity Theory and the Mass Spectrum of High Energy Particle Physics". *Chaos Solitons Fractals* 19 (2004): 209–236.
18. El Naschie, M.S. "The Theory of Cantorian Spacetime and High Energy Particle Physics (An Informal Review)". *Chaos Solitons Fractals* 41 (2009): 2635–2646.
19. Mauldin, R.D. and Williams, S.C. "Random Recursive Constructions: Asymptotic Geometric and Topological Properties". *Trans Am Math Soc* 295 (1986): 325–346.
20. Ho, M.W. "The Story of Phi Part 1". *Sci Soc* 62 (2014b): 24–26, http://www.academia.edu/7718556/The_Story_of_Phi_Part_1_The_Mathematics.

21. Romanach, J. "There Is Something about Phi, Chapter 9: Fractals and the Golden Ratio". YouTube, accessed 1 March 2014, http://www.youtube.com/watch?v=BURjKRfOA9g.
22. Ho, M.W. "Watching the Daisies Grow". *Sci Soc* 62 (2014c): 27–29, http://www.academia.edu/7718580/Story_of_Phi_Part_2_Watching_the_Daisies_Grow.
23. Marek-Crnjac, L. "The Hausdorff Dimension of the Penrose Universe". *Phys Res Int* 2011 (2011): 874302.
24. Connes, A. (1994). *Noncommutative Geometry*, Paris, http://www.alainconnes.org/docs/book94bigpdf.pdf.
25. Ho, M.W. (2008). *The Rainbow and the Worm: The Physics of Organisms*. World Scientific, 1993, 2nd edition, 1996, 3rd edition, 2008, Singapore and London. http://www.i-sis.org.uk/rnbwwrm.php.
26. Schommers, W. "Space-Time and Quantum Phenomena". In *Quantum Theory and Pictures of Reality* (W. Schommers, ed.), pp. 217–277, Springer-Verlag, Berlin, 1989.
27. El Naschie, M.S. "Quantum Collapse of Wave Interference Pattern in the Two-Slit Experiment: A Set Theoretical Resolution". *Nonlinear Sci Lett* A 2 (2011): 1–8.
28. El Naschie, M.S. "Topological-Geometrical and Physical Interpretation of the Dark Energy of the Cosmos As a 'Halo' Energy of the Schrödinger Quantum Wave". *J Mod Phys* 4 (2013): 591–596.
29. Ho, M.W. "E Infinity Spacetime: Quantum Paradoxes and Quantum Gravity". *Sci Soc* 62 (2014d): 40–43.
30. "Our Universe Continually Cycles Through a Series of 'Aeons'", *The Daily Galaxy*, 26 September 2011, http://www.dailygalaxy.com/my_weblog/2011/09/we-can-see-through-the-big-bang-to-the-universe-that-existed-in-the-aeon-before-.html.
31. Ho, M.W. "Golden Cycles and Organic Spacetime". *Sci Soc* 62 (2014e): 32–34.
32. Eugene, W.C. (2008). "An Introduction to KAM Theory". Preprint January 2008, http://math.bu.edu/people/cew/preprints/introkam.pdf.
33. "Kolmogorov-Arnold-Moser Theorem", Wikipedia, accessed 24 January 2014, http://en.wikipedia.org/wiki/Kolmogorov%E2%80%93Arnold%E2%80%93Moser_theorem.

34. "Effect of Noise on the Golden-Mean Quasiperiodicity at the Chaos Threshold", http://www.sgtnd.narod.ru/science/noise/2noise/eng/2noise.htm
35. "Arnold Tongue", Wikipedia, accessed 4 February 2014, http://en.wikipedia.org/wiki/Arnold_tongue.
36. Ivankov, N.Y. and Kuznetsov, S.P. "Complex Periodic Orbits, Renormalization, and Scaling for Quasiperiodic Golden-Mean Transition to Chaos". *Phys Rev E* 63 (2001): 046210.
37. Lombardi, O.W. and Lombardi, M.A. The Golden Mean in the Solar System". *The Fibonacci Quarterly* 22. (1984): 70–75, http://www.fq.math.ca/22-1.html.
38. "Phi and the Solar System", 13 May 2013, http://www.goldennumber.net/solar-system/
39. Tremaine, S. "Is the Solar System Stable?" Scott Institute for Advanced Study, Summer 2011, http://www.ias.edu/about/publications/ias-letter/articles/2011-summer/solar-system-tremaine.
40. Sprott, J.C. "Honors: A tribute to Dr Edward Norten Lorenz". *EC Journal* (Winter 2008): 55–61, http://sprott.physics.wisc.edu/lorenz.pdf.
41. Sprott, J.C. *Strange Attractors: Creating Patterns in Chaos*. M&T Books, New York, 1993.
42. Viswanath, D. "Symbolic Dynamics and Periodic Orbits of the Lorenz Attractor". *Nonlinearity* 16 (2003): 1035–1056.
43. Birman, J.S. and Williams, R.F. "Knotted Periodic Orbits in Dynamical Systems: 1. Lorenz's Equations". *Topology* 22 (1983): 47–82.
44. Gutzwiller, M. (2008). "Quantum Chaos". *Scientific American*, January 1992, republished 27 October 2008, http://www.scientificamerican.com/article/quantum-chaos-subatomic-worlds/.
45. Selvam, A.M. "Cantorian Fractal Space-Time Fluctuations in Turbulent Fluid Flows and the Kinetic Theory of Gases". *Apeiron* 9 (2002): 1–20.
46. Ho, M.W. "Sustainable Cities As Organisms: A Circular Thermodynamics Perspective". *Int J Des Nat Ecodyn* 10 (2015): 127–139.
47. Ho, M.W. "The Biology of Free Will". *J Conscious Stud* 5 (1996): 231–244.

48. Kraut, D.A., Carroll, K.S. and Herschlag, D. "Challenges in Enzyme Mechanisms and Energetics". *Ann Rev Biochem* 72 (2003): 517–571.
49. Ho, M.W. *Living Rainbow H2O*, World Scientific, Singapore, 2012.
50. Ho, M.W. "Water Is the Means, Medium, and Message of Life". *Int J Des Nat Ecodyn* 9 (2014f): 1–12.
51. Ho, M.W. "Illuminating Water and Life". *Entropy* 16 (2014g): 4874–4891.
52. Ho, M.W. "Circular Thermodynamics of Organisms and Sustainable Systems". *Systems* 1 (2013): 30–49.
53. Glauber, R.J. "Coherence and Quantum Detection". In *Quantum Optics* (R.J. Glauber, ed.), Academic Press, New York, 1969.
54. "Metabolic Pathways". Sigma Aldrich, accessed 27 October 2014, http://www.sigmaaldrich.com/content/dam/sigma-aldrich/docs/Sigma/General_Information/metabolic_pathways_poster.pdf.
55. Prigogine, I. (1977). "Time, Structure and Fluctuations". Nobel lecture, 8 December 1977, http://www.nobelprize.org/nobel_prizes/chemistry/laureates/1977/prigogine-lecture.pdf.
56. Vitiello, G. "Fractals As Macroscopic Manifestation of Squeezed Coherent States and Brain Dynamics". *J Phys Conf Ser* 380 (2012): 012021 (13 pp).
57. Ho, M.W. "Quantum World Coming Series". *Sci Soc* 22 (2004): 4–15.
58. "Squeezed Coherent State", Wikipedia, accessed 1 November 2015, https://en.wikipedia.org/wiki/Squeezed_coherent_state.
59. Vitiello, G. "On the Isomorphism between Dissipative Systems: Fractal Self-Similarity and Electrodynamics". Toward An Integrated Vision Of Nature. *Systems* 2 (2014): 203–216.
60. Ho, M.W., El Naschie, M.S. and Vitiello, G. "Is Spacetime Fractal and Quantum Coherent in the Golden Mean?" *Global Journal of Science Frontier Research: A. Physics and Space Science* 15 (2015): 61–80.
61. El Naschie, M.S. "Quantum Loops, Wild Topology and Fat Cantor Sets in Transfinite High-Energy Physics". *Chaos Solitons Fractals* 13 (2002a): 1167–1174.
62. El Naschie, M.S. "Wild Topology, Hyperbolic Geometry and Fusion Algebra in High-Energy Physics". *Chaos Solitons Fractals* 13 (2002b): 1935–1945.

63. El Naschie, M.S. "On the Exact Mass Spectrum of Quarks". *Chaos Solitons Fractals* 14 (2002c): 369–376.
64. El Naschie, M.S. "On a Class of General Theories for High Energy Particle Physics". *Chaos Solitons Fractals* 14 (2002d): 649–668.
65. Marek-Crnjac, L. "The Mass Spectrum of High Energy Elementary Particles via El Naschie's $E^{(\infty)}$ Golden Mean Nested Oscillators, the Dunkerly-Southwell Eigenvalue Theorems and KAM. *Chaos Solitons Fractals* 18 (2003): 125–133.
66. Freeman, W.J. *Mass Action in the Nervous System*. Academic Press, New York, 1975.
67. Freeman, W.J. *Neurodynamics: An Exploration of Mesoscopic Brain Dynamics*. Springer-Verlag, Berlin, 2000.
68. Freeman, W.J. *How Brains Make Up Their Minds*. Columbia University Press, New York, 2001.
69. Freeman, W.J. "Origin, Structure, and Role of Background EEG Activity: Part 1. Analytic Phase". *Clin Neurol* 115 (2004a): 2077–2088.
70. Freeman, W.J. "Origin, Structure, and Role of Background EEG Activity: Part 2. Analytic Amplitude". *Clin Neurol* 115 (2004b): 2089–2107.
71. Freeman, W.J. "Origin, Structure, and Role of Background EEG Activity: Part 3. Neural Frame Classification". *Clin Neurol* 116 (2005a): 1118–1129.
72. Freeman, W.J. "Phase Transitions in the Neuropercolation Model of Neural Populations with Mixed Local and Non-Local Interactions". *Biol Cybern* 92 (2005b): 367–379.
73. Freeman, W.J. and Burke, B.C. "A Neurobiological Theory of Meaning in Perception: Part 4. Multicortical Patterns of Amplitude Modulation in Gamma EEG". *Int J Bifurcat Chaos* 13 (2003): 2857–2866.
74. Freeman, W.J. and Rogers, L.J. "A Neurobiological Theory of Meaningi Perception: Part 5. Multicortical Patterns of Phase Modulation in Gamma EEG". *Int J Bifurcat Chaos* 13 (2003): 2867–2887.
75. Vitiello, G. "Dissipation and Memory Capacity in the Quantum Brain Model". *Int J Mod Phys B* 9 (1995): 973.

76. Jibu, M. and Yasue, K. *Quantum Brain Dynamics and Consciousness*. John Benjamin, Amsterdam, 1995.
77. Del Giudice, E., Preparata, G. and Vitiello, G. "Water As a Free Electric Dipole Laser". *Phys Rev Lett* 61 (1988): 1085–1088.
78. Plenz, D. "Neuronal Avalanches and Coherence Potentials". *Eur Phys J Spec Top* 205 (2012): 259–301.
79. Ho, M.W. "Golden Music of the Brain". *Sci Soc* 62 (2014h): 30–31+44.
80. Pletzer, B., Kerschbaum, H. and Kliesch, W. "When Frequencies Never Synchronize: The Golden Mean and the Resting EEG". *Brain Res* 1335 (2010): 91–102.
81. Roopun, R.K., Kramer, M.A., Carracedo, L.M., Kaiser, M., Davies, C.H., Traub, R.D., Kopell, N.J. and Whittington, M.A. "Temporal Interactions between Cortical Rhythms". *Front Neurosci* (2008), doi:10.3389/neuro.01.034.2008.

Index

www.ingramcontent.com/pod-product-compliance
Lightning Source LLC
Chambersburg PA
CBHW070741220326
41598CB00026B/3724

* 9 7 8 9 8 1 3 1 0 8 8 6 8 *